THE
SHIPWRECK
HUNTER

THE SHIPWRECK HUNTER

A Lifetime of Extraordinary
Discoveries on the Ocean Floor

DAVID L. MEARNS

PEGASUS BOOKS
NEW YORK LONDON

The Shipwreck Hunter

Pegasus Books Ltd
148 West 37th Street, 13th Floor
New York, NY 10018

ISBN: 978-1-68177-760-3

10 9 8 7 6 5 4 3 2 1

Printed in the United States of America
Distributed by W. W. Norton & Company, Inc.

For Sarah, Samuel, Alexandra and Isabella

Dedicated to the survivors and the families
who preserve the memories of those lost at sea

Contents

Prologue

For seven days the shipwreck, lost for fourteen years and lying broken somewhere on the abyssal plain over four kilometres beneath us, had eluded our new-fangled sonar: so brand new it was still on its first full operational dive. This was to be the last of nine search lines covering the 430 square nautical miles where we thought the best chance of the wreck being located was. If this line, like the eight we had already searched, was negative, we would be left with one simple question: was the shipwreck we were trying to find hidden within the mountainous terrain that occasionally cropped up throughout our search box, or were we simply looking in the wrong location?

I was doing my best to hide my uncertainty and inexperience from both my team and the clients, who were on board the small support vessel with us, but I could feel the pressure rising. It would make no difference to anyone that my company had performed amazingly well to design, build and mobilize in the ridiculously short period of five months all the specialist equipment we were using, or that the actual search operation had gone remarkably smoothly, without a second of lost time. Unless we found the wreck, our work would be deemed a failure. We had won this important and potentially lucrative search contract in the face of fierce competition from two far more experienced companies. When they vigorously protested the award and predicted that we would fail, in part because I personally was too inexperienced to lead such a challenging project, it put even more pressure on us to succeed.

Yet even the huge gamble my bosses had taken with the company's future and reputation in tackling this complicated project paled in comparison with what was at stake for our clients. For them it was quite literally a matter of life and death. The life in question belonged to the man who was being criminally prosecuted for sinking the ship, while the deaths were those of the crew he was accused of callously causing. There are a number of reasons why someone might be compelled to spend several million dollars to find a shipwreck lost in the deep ocean. To solve a multiple murder case is arguably the most sensational.

I wasn't sure what made me more nervous: that we still hadn't found the wreck despite having nearly completed our search box; that having been appointed the expert witness for the search I would be held personally responsible should we fail; or that the trial judge was actually at sea with us monitoring every move and decision I was making. To make matters worse, I was getting on badly with the judge. We spoke different languages, and a dispute about how my company was being paid meant our relationship had been fraught with tension and distrust for the past several days. I wouldn't have been surprised if he lost all faith in the search, and us, and terminated it when the current search line was finished.

At the extreme depth at which we were working, it took four seconds for the sonar's acoustic wave, travelling at about 1,530 metres per second, to emanate sideways and scatter off mud, rocks and any man-made objects before returning to the sonar to be electronically converted into brightly coloured images displayed on the computer screens in front of us. At the end of each four-second cycle I would tilt my head closer to the screen to get a clearer look at the latest strip of sea floor revealed, hoping for the distinctive signs of wreckage to appear. After seven days, the constant tilting of my head forwards and backwards like the slow beat of a metronome was so ingrained in me it happened without conscious control. But while my body was stuck on autopilot, trapped in the tempo of the ongoing search, my mind was racing ahead with the possible repercussions of failure. As much as I hated admitting defeat – and I was careful to keep such negative thoughts to myself – I felt I had to be prepared in case the worst happened. And then the next four-second cycle changed everything.

Without warning, a bright yellowish-white rectangle appeared on my screen. The target was unlike any others we had seen during the search, mainly because it was sitting alone in the expanse of orangey-red that depicted the soft muddy sediment cover of the flat abyssal plain. In the time it took my heartbeat to accelerate, another 'ping' from the sonar illuminated what was undoubtedly a very hard object. Was it the start of a geological rock field, or just one of the countless pieces of marine trash that littered the seabed? Or would this target justify my quickening pulse?

A minute or two passed before I could sensibly assess what we were seeing. The images scrolled down the screen like a multicoloured waterfall, revealing what lay below as if the ocean had been magically drained of all its water. This god-like power was what had first attracted me to working with sonars. In my mind nothing could compare with the excitement of deciphering previously unexplored expanses of the sea floor like this. Radiologists could use X-rays and MRIs to peer into the body, but I could do pretty much the same with no less a subject than the earth itself.

Not long after the hard target appeared, it petered out and with it my hopes that we had found something significant. There were still a handful of faint yellow pixels scattered to either side, indicating that the target wasn't a solitary object, but soon enough the seabed became as flat and featureless as before. By now, everyone in the twenty-foot container that served as our operations control centre was crowding my screen and weighing in with their opinions of what we had detected. My rough measurement of the target's size showed it wasn't even close to the 140-metre length and 18-metre breadth of the ship we were after. This wasn't our shipwreck, but could it be a piece of it?

Our wreck had supposedly been sunk by a large bomb hidden in one of the two main cargo holds and timed to explode far away from any land mass, so we were expecting at least some of the hull to be blown to pieces. While it was disturbing to think of the shipwreck as a crime scene and the grave site of six innocent people, that was exactly what it was. All the available research indicated that no other modern ships had sunk in this area, but I also needed to recognize that our search box was situated along a major east–west shipping route, so the potential for unreported shipping losses to have occurred here was high. I believed I had designed a conservative search plan to give us the best possible chance of distinguishing our shipwreck from other targets, but after seven days of no results my initial confidence was seriously waning.

I waited in anticipation for more targets to follow the initial scattering, but none appeared. Slowly my head stopped its forward-and-back motion and began to drop. Was that it? A brief flutter of excitement and promise, followed by the empty feeling of failure? Was my career as a deep-ocean shipwreck hunter destined to start in disappointment, from which a second chance might never materialize?

Suddenly the screen began to glow brightly again with not one but a string of hard targets stretching across the starboard side of the line we were searching. Whatever the identity of these targets, their number and spread was about right for the dimensions of our shipwreck. This appeared to be the debris field of the exploded cargo hold section, but without a single large target representing the aft superstructure of the ship, there was no way to be positive. By now my face was just inches from the screen and I was praying for more targets. Finally my prayers were answered and one large beauty appeared.

I jumped up and felt a surge of exhilaration course through my body. 'We've found it!' I shouted. 'That's got to be our shipwreck! Look how it's been blown to smithereens by the bomb!' I threw both arms in the air and began to physically grab everyone in the container, although it was clear they weren't as convinced by the images as I was. Quickly I started measuring the main targets to prove to them what I intuitively sensed: that the combined dimensions matched the ship very well. Whether they were persuaded or not, they all began to feed off my excitement.

Word began to filter through the ship that we had found something, and pretty soon our small container was heaving with people, including the judge and his three advisers. I repeated my explanations to the new arrivals, using zoomed-in views of the targets to reinforce my interpretations. Unlike photographs or video, sonar images are not visual, so they rely on experts to explain exactly what they represent. As I was one of only two people on board with any real experience interpreting such images, everyone had to more or less accept what I was telling them.

In truth, an additional higher-resolution sonar image of the wreckage would have to be produced to show that we had found the wreck, and even then only a photograph or video taken by an ROV would be acceptable in court as proof of the ship's identity. I knew that the judge and his advisers would want a systematic investigation of the wreckage, and that several more weeks of hard work lay ahead, but now we would have all the time we needed to produce this evidence.

The search was not destined to end in failure as I had once feared. Instead it was to be a technical triumph and an enormous success for everyone involved. We had found one of the world's deepest and most notorious shipwrecks, a wreck that was at the heart of an extremely high-profile murder trial with national importance in the country in which it was being held. We would no longer be viewed as novices amongst the small band of companies that operated in the deep ocean. We had proved that our equipment was superior and that our team was one of the best in the business. With this success under our belts, a string of clients would line up to hire our services and keep us in constant work for the next two years, thus transforming the value of our small company.

After all the excitement died down, I walked out to the fantail of the ship to have a moment to myself and appreciate what we had just achieved. I was elated by our success but also extremely relieved. Many months of hard, stressful work had paid off in the few minutes it took our sonar to reveal the shipwreck, leaving me with a feeling of satisfaction unlike any I had ever experienced before. If this was what it felt like to find a deep-ocean shipwreck, I was determined that it would not be the last time for me. In fact I was dead set on doing it again and again.

Introduction

'How do I get a job like yours?' Of all the questions I'm asked when I speak publicly about my work as a professional shipwreck hunter, this is the one I know will occur every single time. Invariably it comes from a young man, mid twenties at most, and by his tone of voice I can tell that he seriously thinks I might have advice that will change his life forever.

It's a tough question to answer. For one, I never plotted out a path to this most unusual of professions myself, so I have no sure-fire strategy to offer. Secondly, I know of no school or university anywhere in the world that teaches all the skills needed to be a successful shipwreck hunter. Finally, and most importantly, there isn't a ready-made industry out there looking to hire prospective shipwreck hunters.

My own story starts in Union City, New Jersey. Lost in the shadows of Manhattan's skyscrapers across the Hudson River, Union City seems an unlikely location to inspire a future marine scientist. In fact its only claim to fame is that it is the most densely populated city in America, with nearly 60,000 people crammed into an area of just over one and a quarter square miles. To make matters worse, the nearest body of water is the Hackensack Reservoir Number Two, a man-made lake of 69 million gallons ringed by a two-metre-high fence to stop kids like me scaling it for a sneaky dip on a hot summer's afternoon.

1

Not that New Jersey doesn't have water; it is a coastal state with over 200 miles of seaboard fronting the Atlantic Ocean. But in the 1960s, when I was growing up, you hardly ventured beyond the three or four blocks that made up your immediate neighbourhood. When you weren't in school, you were playing with your friends in the street between two long rows of parked cars. It might have been touch football, stickball, or skully – a game in which players shoot soda bottle caps along the road surface in between the passing cars – but whatever the game, it was generally within a few yards of your own front porch.

For lower-middle-class families like mine, holidays were a week or two each summer in a rented cottage down on the Jersey shore. These were great times with my brother and two sisters, when nearly every hour of the day was spent on the beach or in the water. We would body-surf on cheap Styrofoam boards until the skin on our chests was rubbed raw, or until our mother had to drag us home for dinner.

Occasionally a dorsal fin would be sighted just off the beach and the lifeguards would hurry everyone out of the water, but more often than not the offending beast was just a harmless sunfish. Sometimes, on a very clear day, a passing ship might be spotted steaming into Port Elizabeth to offload its cargo. Other than these moments, I would hardly ever think about what lay beyond the horizon or beneath the pounding surf. Despite my love of the ocean, these holidays weren't the inspiration that drew me to a life working at sea.

Ironically, it was trips inland to my grandmother's house in Honesdale in north-eastern Pennsylvania that were the real catalyst for my decision to study marine biology at university. As the youngest of four children, I would be packed off by myself on the bus to spend a week each summer visiting my cousins and uncles. Although Honesdale was barely a hundred miles away from Union City and could be reached by car in about two hours, it was the complete opposite of the city and like a whole new world for me. It was spacious, green, uncrowded and filled with lakes and ponds where I learned to fish and appreciate the natural beauty of the countryside.

My mother's parents had emigrated to America from southern Italy around the turn of the twentieth century and eventually settled in Honesdale, where they had ten children, although two sadly died as infants during a diphtheria outbreak. They were a family of shopkeepers. First there was a general store and candy shop; later on, when my uncles were older and could help with the business, they ran a bar and restaurant that was widely known for making the best Neapolitan pizza in that part of Pennsylvania. The family was well respected within the tight-knit community, and when the call came to serve in World War II, three of my uncles fought on various fronts with the army and navy.

After the war ended, my two eldest uncles, Pete and Vince, together with some friends, bought 350 acres of private woodland to form the Bucks Cove Rod and Gun Club, where they could fish on several lakes and hunt for deer. Over the years, they bought additional land, nearly doubling the club in size, and added another lake, which they constructed themselves. These lakes were where my uncles taught me to fish for bass, catfish and the occasional toothy pike. We'd take a small rowing boat and stay out all day, releasing most of the fish we caught but keeping a few of the biggest catfish for supper. I loved those days with Pete and Vince. I never wanted them to end and would stay at the water's edge deep into the night, casting my line to land one more fish, leaving only when the bats and the swarming mosquitoes forced me inside.

It never occurred to me then that something I loved doing so much could form the basis of a career, or that working in a natural outdoors environment was an option for me. My father was an antiques dealer who travelled in to Lower Manhattan every day to sell furniture to shops in Greenwich Village, while my mother was a registered school nurse looking after 1,200 students in a large city high school. Watching them go to work each day, I naturally assumed that I too would wind up working in a big city; probably Manhattan, where most people from across the river in New Jersey gravitated.

This mindset was forever changed, however, when I was sixteen and Vince took me to see a large freshwater fishery. I loved looking at the fish, of course, but mostly I was struck by the scientific approach that was used to produce the largest number of healthy fish possible. I remember leaving the fishery that day with the distinct idea that this was something I would enjoy and was capable of doing. When I came to think about what I would do after graduating from high school, I was torn between becoming an antiques dealer like my father and going on to university to study science. While I had become quite adept at the antique trade, having helped my father for many years, the people who inspired me the most were the astronomer Carl Sagan and the heart transplant surgeon Dr Christiaan Barnard. I found Sagan to be amazingly eloquent and passionate when speaking about the cosmos, while in my eyes Barnard was a bona-fide hero for being the first surgeon to successfully transplant a human heart.

My teachers had always marked me out as someone quite bright, but I wasn't the best student because – in their words – I needed to apply myself more in class. Basically they all said I wasn't working to my full potential. Not long after visiting the freshwater fishery, I went to the local library and began investigating careers that would allow me to work outdoors and ideally in conjunction with water and fish. I found information on becoming a fisheries biologist, which was immediately appealing, but the job that really fired my imagination was marine biology.

Marine biology ticked all the right boxes for me. It was a serious professional career in science that also offered the prospect of conducting research at sea on marine organisms including fish. Compared with being a freshwater fisheries biologist, which I thought might restrict me to jobs within the continental United States, the idea of being free to work on all the world's oceans seemed limitless and exciting. There was also a very important practical matter.

One of the universities listed as offering a specialized BSc degree in marine biology was Fairleigh Dickinson University, whose Teaneck campus was a mere twenty-minute drive from my house. This meant I could live at home whilst attending university. I certainly wasn't looking forward to negotiating the notoriously unattractive highways of northern New Jersey every day for four years, but I had to take into consideration the practical benefits of going to a local university virtually on my doorstep. I was also realistic enough to know that getting accepted to an out-of-state university was probably going to be difficult for someone like me with average grades and without the money to pay the high cost of tuition and living expenses.

The other attraction of FDU's marine biology programme was that it included a half-year semester at the university's own private laboratory on the Caribbean Island of St Croix, in the US Virgin Islands. The university's prospectus promised students the opportunity to study in a setting in which the classroom was a pristine coral reef, with lessons taken underwater whilst scuba diving. Although there were more prestigious marine biology and oceanography programmes dotted around the country, none of them offered such a marvellous facility in this kind of exotic location. For the first time in my life I was seriously motivated to work harder in school and improve my grades, so that I could get accepted into FDU's marine biology programme.

My only sense of what university would be like came from a short visit to my older brother during his freshman year at the University of Wisconsin in Madison. UW-Madison in the early 1970s was a hotbed of student activism, and it was also ranked by *Playboy* magazine as one of the top ten party schools in America. Consequently I saw very little academic work of any type the week I spent on campus. Most people I came across in the dorms were, it seemed, preoccupied with either the latest developments in the Watergate scandal or getting high, or both.

When I did finally get into university, it was nothing like I had experienced in Wisconsin. FDU was primarily a commuter school, with only a small percentage of students living on campus. It also had large international and graduate student bodies, so in general everyone's attitude was to get your work done and get out of school rather than sticking around to party. My time at FDU in Teaneck was all about attending lectures and studying, and when I wasn't doing that, I was working part-time several days a week in the biology department or in other odd jobs to help pay for my tuition. I saw university very much as a means to an

5

end, and I wasn't about to jeopardize everything I was working towards for a bit of fun. It was time to take my education seriously.

Unfortunately, my poor preparation from high school caught up with me during my freshman year. I struggled badly to keep up with the pace of lectures, and at one point my grades were so poor – my grade point average (GPA) had dropped below 2.0 – that I was put on academic probation, which shocked and frightened me. This meant I had just one more semester to turn things around, otherwise I would be placed on suspension, which I knew would spell the end of my university career right then and there. I was incredibly frustrated, because my problems weren't for lack of trying; I spent every waking hour studying, but the results were simply not coming for me. That summer holiday I had to face the prospect that I wasn't good enough to get the degree I wanted so badly.

I didn't find the answer to my problem until the start of my second year, when I was placed in a microbiology lab class with a handful of bright students who all became very good friends. Whereas in my first year I'd studied by myself, now I began working in a group with my new friends, who showed me how to learn effectively and efficiently. The difference in my work, and most importantly my exam results, was immediate. In a single semester I went from nearly flunking out to getting some of the highest grades in many of my classes, especially the core science subjects that dominated my course. I still worked very hard, but armed with some new techniques and the benefit of studying in a group, I was able to put my disastrous freshman year behind me and improve my GPA to a respectable figure.

Improving my GPA became particularly important, because by the middle of my junior year I had decided to continue my education and apply for graduate school, targeting about a dozen top universities on both coasts that offered PhD and MSc degrees in marine biology. To do this I would need a 3.5 GPA – over and above the 3.0 minimally required for acceptance into graduate school – in order to be seriously competitive with the other PhD and MSc candidates vying for scholarships and the best research opportunities. I had dug a huge hole for myself with my disastrous freshman year, and the only way out of it was to get virtually perfect grades from now on.

∞∞∞∞∞∞∞∞∞∞∞∞∞∞∞∞∞

I left for my semester in St Croix in early January, just after celebrating the new year with my family. My father had died from a heart attack the previous September, at the start of my senior year, so this was the first Christmas we'd had without him. Although he had been ill, his death was still a major shock that left a huge hole in my life. I'd been extremely close to him, having spent a large portion of my childhood helping him with his antiques business. It wasn't an easy way to make a living. Every day he was faced with having to find and buy antiques in New Jersey and sell them later that same day in Greenwich Village, where most of his customers' shops were based. The constant need to create these profitable market opportunities was wearing on him, and I think this was one of the reasons he was happy to have me along for company. It also helped that, with his tutelage, I had developed an eye for spotting valuable items.

Some of the best times for my father's business were the periodic 'clean-up' weeks, when the nearby cities and townships in New Jersey would allow their residents to put any objects, no matter the size, out on the streets to be collected the following morning as rubbish. In the process of emptying their attics and basements of trash, people would also unknowingly throw away treasures that only we would recognize as valuable antiques. I would happily run along the sidewalk spotting and collecting the worthwhile objects whilst my father followed me slowly in his van. Working together this way we were able to cover far more ground than he could have done alone.

This might not have been an ideal activity for a young boy to be engaged in on a school night, but I couldn't have been happier when I found something of real value, like an oriental rug. I was making life easier for my father and contributing to the family's finances by literally turning people's trash into cash. Of course my mother wasn't enamoured with our late-night excursions, as she saw it as yet another excuse for me to avoid my homework. The way I looked at it, I was just helping with the family business like a lot of children were expected to do. If we had lived on a farm, no doubt I would have been spending early mornings tending to animals or taking whole days off from school during the harvest.

I learned so much from my father and I missed him dreadfully when he died. It pained me to think he wouldn't be around to hear about my experiences in St Croix or watch me graduate from university. Despite my grief at his death, however, I was able to meet the target I had set of getting A grades in every class, and I left for St Croix with confidence that I would be able to handle whatever workload they threw at us. Despite the fact that I was the youngest in the family, I was on track to be the first to graduate university, as my brother ultimately dropped out and my two sisters worked for years before going back to college. I had studied extremely hard to recover from my nightmare first year, and now, with all my goals back within touching distance, I was determined to make the most of this next opportunity.

The moment you step off a plane into a hot tropical climate, you know instantly that you've been transported to a foreign land, very different from where you've come from. The heat doesn't just hit you in the face; it envelops your entire body and permeates all your senses. This was how I felt landing in St Croix on a blazing sunny day after flying in from New Jersey, which was still firmly gripped in the dead of winter.

If there were any lingering doubts about how different St Croix was going to be, they all faded away when I caught my first sight of the shallow lagoons just off the eastern tip of the island where FDU's West Indies Lab (WIL) was based. The water was a colour I had literally never seen before; a brilliant aquamarine, flashing as the sunlight reflected off ripples dancing across its surface. Despite the long flight, several of us immediately ditched our bags and hiked down to a remote lagoon to go snorkelling. I wasn't prepared for how beautiful the corals, fish and anemones would be. Everything was so spectacularly bright and colourful. As soon as I entered the water I knew one thing instantly: that I would never go swimming in the polluted and visibly impenetrable waters of the Hudson River ever again.

The locals call St Croix a paradise, and that is never more evident than when you begin exploring the different types of coral reefs that populate the waters surrounding this small Caribbean island. Fortunately, the structure of the classes taught at FDU's lab allowed us plenty of time for scuba diving. After morning lectures, nearly every afternoon was spent in the field – generally underwater – observing and learning about the local flora and fauna. Our first assignment was to recognize by sight all the fish, coral and invertebrate species, which meant spending countless hours in the water using waterproof guidebooks and checklists written on slate to keep tabs of everything we had to identify during our dives.

Having learned to dive in a cold flooded quarry in New Jersey, where the most interesting thing to see was a stolen car dumped there after a botched crime spree, diving in St Croix was a revelation. To begin with, the visibility was spectacular and the water so warm it made wetsuits unnecessary. The main attraction, however, had to be the communities of fish that lived within the coral reefs and made them a constant hive of activity. One of the most amazing spectacles was the daily migration of juvenile grunts, a mostly yellow and blue striped fish, which leave the reef en masse each day at twilight to feed in the adjacent beds of seagrass. When we were told that these large schools of fish would swim off in single file at a specific time every evening, I remember feeling a little sceptical, but as I lay in wait for the grunts along with the rest of my class, sure enough, at exactly the predicted time, a single grunt led a procession of his mates off the reef in a perfect line that any drill sergeant would be proud of.

As fascinated as I was with the ecology and behaviour of reef fishes, as the semester wore on I grew slowly disillusioned with the idea of becoming a marine biologist. Coral reef ecology is an enormously complex subject that relies on small advances in a wide range of different fields before even a simple understanding of the whole ecosystem is possible. I felt the pace of discovery was too slow and didn't want to devote a large chunk of my life to a field of study in which I would only make a small contribution.

For my own research project I decided to choose a topic that I felt was of greater significance to the overall health and fate of the island's reefs, which had begun to show the first signs of the deadly coral disease that was to decimate many species the following decade. By measuring the quantity and productivity of calcium-carbonate-producing plants and animals, including corals, spread across one of the island's shallow bays and adjacent fringing reefs, I was able to put together the first bio-geochemical carbonate production budget ever established for St Croix.

I liked this project for several reasons. Firstly, it had a mixture of marine biology and geology that I found stimulating. I had taken no courses in geology until arriving at the WIL and was surprised to find how much I enjoyed this new field. Secondly, the fieldwork was very physically demanding as I had a lot of underwater acreage to cover while counting and measuring all the carbonate producers along the half-mile transect I'd chosen to survey. I was spending up to four hours under-water every day, with one long dive in the morning and a second in the afternoon. Because the water was so shallow, decompression sickness wasn't an issue and I was free to extend my dives by means of carefully controlled breathing until I literally ran my tank dry. All the hard work meant I was probably as physically fit as I've ever been in my life.

When I submitted my final paper, I received the A grade I so desper-ately wanted and needed, but even that was less important to me than the comment made by our marine geology professor, who wrote that the quality of the paper was so good, it actually deserved to be published in a peer-review journal, something that was extremely rare for a student paper. My work resulted in the first quantitative snapshot of the state and health of this important environment, which would also serve as a comparative baseline for similar studies in the future.

When my final GPA was calculated, however, I still fell short of the 3.5 score many of the graduate schools were telling me I needed to qual-ify for places and scholarships. Despite getting straight A's in every class in my senior year, the highest I could pull my GPA up to was 3.34.

In truth, I never really had a chance after my disastrous freshman year. However, I was proud that my GPA in the core science subjects that made up 60 per cent of my overall curriculum was 3.71. I prayed this was enough for at least one graduate school to take a chance on me.

When I heard that the University of South Florida (USF) were upgrading their small St Petersburg-based Institute of Marine Science to department status, with the addition of about a dozen new faculty and thus many more places for incoming graduate students, I made sure they were on the list of schools I applied to. I had been advised to apply primarily for PhD programmes on the basis that prospective PhD students had a much better chance of getting the research funding and/or financial scholarship I would need to further my education. Unfortunately, this advice proved to be misguided, for the simple reason that I wasn't a strong enough candidate to compete for these highly coveted places. This became distressingly clear to me as the rejection letters started arriving in our mailbox.

I was getting to be quite an expert at sniffing out rejections with just the briefest of glimpses at the first couple of sentences. After reading the full contents of the first few letters, which left me with an awful feeling of having been kicked in the stomach, I would thereafter only open the top fold to glance at the opening paragraph. If the phrases 'After careful consideration' or 'We regret to inform you' showed up, there was really no point in reading the rest of the letter. This routine continued with sickening regularity until the letter from USF arrived.

At first I couldn't tell whether I was in or out, because the first paragraph did include some of the phrases I had learned to dread. But as I began to make sense of the letter, it was clear that USF, while not accepting me outright as a PhD student, had thrown me a lifeline. In short, the letter explained that they would be happy to take me on for an MSc degree, following which I could pursue a PhD if I wanted to continue at USF. To say I was overjoyed would be a huge understatement. After so many rejections, I was relieved that *any* school was giving me´ a chance, and the fact that marine science was an up-and-coming department at USF – having recently been made a state-wide Center of Excellence and with ambitions for becoming nationally prominent – made my acceptance of their offer one of the easiest and happiest decisions I ever had to make.

Although I entered the MSc programme at USF as a marine biologist, it wasn't long before I decided to change to marine geology. I had spent part of my first year at USF improving the research paper I had written at WIL, and my interest in marine geology was increasing at the direct expense of marine biology. I could also see that the marine geology students had more seagoing research opportunities and that they were a more tightly knit group than the biologists, who, because they pursued such different research projects, didn't collaborate as much. In addition to what my heart was telling me, my head also liked the fact that marine geologists were far more employable than marine biologists, so making the switch was unlikely to hurt my job prospects when I finished with graduate school.

In order to change courses, I had to have one of the marine geology faculty take me on as their student, and I was very fortunate that Al Hine agreed to do just that. Al was an assistant professor who had a growing reputation as a talented sedimentologist specializing in shallow carbonate environments. Most importantly for me, he was a proper seagoing scientist, meaning that I would have plenty of opportunities to go to sea myself and get hands-on experience with the high-resolution geophysical instruments that were the tools of his trade. Al believed in taking all his graduate students to sea and giving them full responsibility for operating the expensive and scientifically powerful equipment. On my first training cruise with him, he put me in charge of a high-frequency side-scan sonar system that we used to image the sea-floor geology, and it was without question an experience that transformed my life forever.

The sonar was a decrepit EG&G 259-4 unit that was in such bad shape I had to spend a week breaking it down and repairing it just to get it working to about half its rated capability. The 259-4 was one of the earliest commercially available side-scan sonars and was notoriously difficult to operate. However, when I finally got it going and it started to produce sonar images of the West Florida continental shelf as we slowly steamed away from the coast, I knew immediately that working with this type of equipment was what I wanted to do for the rest of my life.

The images of sand waves and rocky outcrops, even taking into account their relative poor quality, were an absolute revelation to me. I felt this was probably the most perfect mix of science and underwater exploration that I could ever hope to find. What I found particularly powerful was the ability of the sonar to scan wide strips of the sea floor to reveal both geology that was millions of years old and sedimentary patterns created during the most recent change of tide. Before our short cruise was finished, I had already decided to learn everything I could about this amazing technology and become an expert in its use.

In order to fulfil my ambition to become an expert seagoing geophysicist, I joined every single scientific cruise I possibly could during my years at graduate school. One way or another I managed to go to sea on ten different cruises, totalling about three months in duration, which enabled me to become a proficient equipment operator. The high-resolution geophysical gear that we used included acoustic profilers, sparkers and air guns, but my main interest was always the side-scan sonar. I quickly became the department's resident expert, and by the end of my tenure at USF, Al Hine was having me give the lecture on the theory and use of side-scan sonar to the incoming class of students.

As Al's field research was primarily focused on the continental shelf, we were generally working in less than 200 metres water depth. I had no deep-water experience to speak of, so when I got the opportunity to join one last cruise that happened to be leaving from our dock to study a newly discovered community of cold-seep clams and tube worms found at the base of the West Florida escarpment, I jumped at it. This cruise appealed to me for a number of reasons. First, the co-principal investigator was the legendary Dr Fred Spiess of the Scripps Institute of Oceanography, who famously pioneered the use of deep-tow instruments for scientific mapping of the deep seabed. Second, the research vessel was going to be the RV *Knorr*, which had hit the headlines just the previous year when she was used to find the wreck of the *Titanic*. And finally, the study area was at a depth of 3,600 metres, which meant I would be exposed to a whole new set of instruments and methodologies.

I couldn't dream of a better way to cap off my graduate school career, which had just finished with the successful defence of my MSc thesis. This was an incredible opportunity to take part in cutting-edge deep-water research and to work with some of the best scientists in the world on the biggest and best ship I had ever boarded. Probably even more importantly, I had just run out of money for food and rent and was living on a friend's sofa while I waited to hear if I was going to be offered the job at Eastport International for which I had recently interviewed. So I could easily add self-preservation to the list of benefits in joining a three-week cruise that offered a warm bunk and all the food I could eat.

As I had anticipated, the cruise was an unforgettable experience and I learned an enormous amount. I came away with a whole new understanding of how to deploy and operate side-scan sonars in the deep ocean and with a confidence that I was ready for my career to move on to the next level. Partly out of necessity, and partly as a tactic, I gave Eastport a deadline as to when I would be making my decision about which job offer to accept, which meant they were going to have to call me at sea on an expensive shore-to-ship connection to let me know their intentions. While I knew that the radio office of the *Knorr* had been the scene of some extraordinary calls over the years, especially when Dr Robert Ballard announced they had found the *Titanic*, I can't imagine that many people have used it, either before or since, as the place to receive and negotiate their first serious job offer. Eastport wanted me to grow their geophysical survey and search business and I was delighted to accept. I was also greatly relieved to know that when the *Knorr* docked back in St Petersburg, I would be moving on to a new and exciting life and not back to my friend's sofa.

From my first day at Eastport International, I knew that I had been given a marvellous opportunity to excel at a seriously cool and exciting company, but also that I would need to work extremely hard to get up to speed on a technical level. Eastport's main business at the time was as a contractor for the US Navy Supervisor of Salvage (SUPSALV), maintaining and operating their deep-water remotely operated vehicles (ROVs), which were often called on in emergencies to recover US Government assets lost at sea. The company was full of engineers, mechanics, electronics technicians and pilots to 'fly' the ROVs – all highly skilled technical experts. My background as a scientist certainly impressed the management and helped me get my foot in the door, but among the guys I would be going to sea with it was only a curiosity. To earn their respect and become a valued member of the team I needed to know my stuff, backwards and forwards, and prove myself where it counted: offshore.

When I joined Eastport in September 1986, their forte was the recovery of downed aircraft from deep water. The company operated a SUPSALV ROV called *Deep Drone 6000* that could dive to a depth of 1,800 metres and either pick up small bits of aircraft wreckage with its two manipulator arms or attach lifting lines to recover an entire Chinook helicopter if necessary. Together the navy and Eastport had recently completed the recovery of the wreckage of the ill-fated Space Shuttle *Challenger*, which exploded in spectacular fashion shortly after take-off, killing all seven of its crew members, including Christa McAuliffe, who was poised to become the first schoolteacher in space. The accident, caused by a failed O-ring seal on the right-hand solid rocket booster (SRB), was a horrific tragedy seen live by millions of Americans – including me, watching from the opposite coast of Florida in St Petersburg.

The recovery of *Challenger* was an enormously difficult project because the wreckage was spread over a huge area and pieces of the problematic SRB had sunk in the deep, swift-moving waters of the Gulf Stream. The salvage teams coped well with the difficult environmental conditions, but afterwards the navy decided it needed a more powerful ROV that could handle the worst-case combination of high currents and great depth down to a maximum of 6,000 metres – more than three times deeper than the existing *Deep Drone* could go. No ROV had ever dived so deep, and the development of this new ROV, called *CURV III*, turned into a race against another US Navy R&D centre to see who could reach this important milestone first. Because of my background in marine science, the first assignment of my new job was to compile a worldwide database of surface and subsurface currents and worst-case current profiles that would establish the key performance criteria for the new *CURV*. It would never have occurred to me as I watched the contrails of *Challenger*'s debris fall to earth on that sad day in late January that in a way I would owe my first job to that horrible tragedy.

Dealing with deadly accident scenes was obviously something I was going to have to get used to. Objects that need to be found and recovered from the bottom of the ocean generally only get there because some disaster has occurred. Sometimes we were tasked to recover inanimate objects like test torpedoes that failed to float to the surface at the end of their run, or live Tomahawk cruise missiles that never reached their intended target. But mostly it was helicopters or jet aircraft, and in those cases the chances that someone had lost their life were usually quite high. I guess that everyone has their own coping mechanism, although it wasn't a big topic of discussion offshore. Personally I found it helpful to remember that the job we were doing to assist the accident investigation teams was vital in the prevention of future accidents.

It wasn't long before I started to develop my own commercial clients outside Eastport's normal base of US Navy and Defense customers. In the main these were industrial projects like the survey of proposed pipeline routes for the oil and gas industry. They were technically demanding, high-pressure tasks that were satisfying to complete but not as interesting as searching for lost objects. I had nothing against 'big oil', but working as a geologist in the industry, either as a geophysicist in the exploration for oil deposits or to determine the best locations for the placement of platforms and pipelines, simply didn't appeal to me. So while these industrial projects were good, profitable work, what really got my blood flowing was when my small team was called on to search for something lost underwater.

Often the key to a successful search is getting out on the water as soon as possible after the object is lost, when the memory of eyewitnesses is still good and physical evidence is fresh. The tendency for critical information to degrade with the passage of time is well known, so speed of response is of the utmost importance. We had to be ready, with our equipment fully operational and prepared to go, at all times, which meant being on virtually constant call twenty-four hours a day. This was especially true in the search for sunken boats on Chesapeake Bay, the nearest large body of water, just ten miles away from Eastport's offices. Our proximity to the bay was certainly bad luck for one boat owner, who, when putting his brand-new speedboat through its paces, stupidly decided to try jumping the wakes of larger ships for fun. Unfortunately for him, he mistimed one jump so badly the boat landed bow first, driving him and the entire vessel violently into the water.

From his hospital bed the next morning, the boat owner claimed the sinking was caused by delamination of the speedboat's fibreglass hull: basically a manufacturing defect. This claim placed the blame squarely on the boatbuilders, who reacted immediately by hiring us to locate and recover the vessel to determine the true cause of the sinking. It wasn't just the $25,000 replacement cost of this single boat that concerned them; the delamination claim, if proven, would jeopardize future sales. Not wanting the owner's claim to gain any momentum, they asked us to move as fast as possible to start the search.

Several hours later, I was mobilized with my side-scan sonar on a small boat south of the Francis Scott Key Bridge in Baltimore, trying to decide the best place to start the search. A couple of eyewitnesses had come forward to point out the general area they had seen the speedboat jumping wakes the day before. However, because their information was quite vague – no one had actually observed the boat crash and sink – I had no idea how long the search would take. Thirteen minutes later, I had the answer. A perfect boat-shaped target appeared on my sonar trace, ending the quickest search I have ever conducted, before or since. Less than an hour later, the boat had been raised by diver's airbags and was under tow back to the dock so the builders could examine the damage carefully. When they saw that the bow had been smashed in, with no signs of delamination, they drove straight to the hospital to confront the owner. By that time my job had been done and I was on my way back to the office, but I would have liked to have seen the owner's face when he was told his claim was being rejected, and that if he had any doubts he was free to inspect his boat himself as it was currently waiting in the hospital car park.

After the first few years at Eastport I began to take on larger and more complex projects. The company was growing steadily and new milestones were being achieved all the time, including winning the race for *CURV* to be the first ROV to dive below 6,000 metres, and setting a new depth record for aircraft salvage (4,500 metres) by an ROV with recovery of a South African 747 jumbo jet that crashed while on fire into the Indian Ocean. My bosses were very happy with my work but it was impossible to be truly satisfied as we still hadn't achieved our objective of improving the company's side-scan sonar search capabilities to the same deep-water standards as our ROVs. Major advances like this often relied on the strength of a single project, and until then such a project hadn't presented itself to us. That situation was about to change when in the summer of 1990 we were controversially awarded the contract by an Austrian criminal court to find the wreck of the MV *Lucona*.

I

MV *Lucona*

MURDER AND FRAUD ON THE HIGH SEAS

MV *Lucona*

SUNK 23 JANUARY 1977

6 died
6 survived

For a land-locked country like Austria, it might seem odd that the longest and most expensive criminal trial in its legal history revolved around the sinking of an ordinary merchant ship on the high seas. However, while the Motor Vessel (MV) *Lucona* was indeed ordinary, Udo Proksch, the man who chartered her for his own fraudulent purposes, was anything but.

Although it required years of technical preparation, Proksch's plan was brazenly simple. It went like this: charter a merchant vessel to carry your costly cargo to a non-specified Far East location; insure the cargo for 31 million Swiss francs (about $18.5 million) with an insurance company carefully selected so as not to question your claim; instead of the expensive uranium processing plant documented on the manifest, load the vessel with 288 tons of antiquated coal-mining and repainted wheat-mill equipment along with other worthless cargo; while supervising the loading of this fraudulent cargo secretly take on board a time bomb packed with enough explosives to completely obliterate *Lucona*'s

21

steel hull; carefully set the bomb's timer so that it explodes weeks into the vessel's voyage when the ship is far from land and in very deep water, ensuring the crew has no chance of survival; and finally, after the bomb explodes and the ship's shattered hull has plunged to the bottom of the Arabian Sea, entombing everyone on board in a steel coffin, submit your claim to the cargo insurers for reimbursement of your $18.5 million financial loss.

The *Lucona* affair kept the chattering classes of Vienna entertained for years with tabloid stories of espionage, corruption and murder, culminating in the most sensational criminal trial Austria has ever seen, which claimed the careers and freedom of numerous conspirators and accomplices. The political fallout from the scandal was just as damaging, leading to the 1986 downfall of Udo Proksch's beloved socialist party and the subsequent resignation of key government ministers connected to Proksch. In a country accustomed to scandal, the *Lucona* affair is considered to be the scandal of the century.

Through the story's various twists and turns, the one constant was that Proksch was strongly suspected of foul play right from the start, and his accusers and the authorities were determined not to give up on bringing him to justice. It may have taken thirteen years, but finally, in January of 1990, Proksch found himself on trial for six counts of murder, six counts of attempted murder, aggravated insurance fraud and wilful endangerment with explosives. His political connections, which he had counted on for intervention in the past, were now out of power and thus unable to help. For once in his life, he was on his own.

⁕⁕⁕⁕⁕⁕⁕⁕⁕⁕⁕⁕⁕⁕⁕

The case against Udo Proksch in the Vienna Regional Criminal Court ultimately came down to two central questions: was his cargo what he said it was, and did he have the *Lucona* blown up? The only sure way to get the answers to these questions was to search for and find the wreck. Proksch himself realized this and said as much during a dramatic monologue in reply to one of the state prosecutor's questions:

'Where is this ship? If it's sunk, it must have hit a mine or been sent to the bottom by a submarine. I don't think the ship is lying on the bottom of the ocean. Find the *Lucona*!'

Without this physical proof, the chance that Proksch could slip off the hook was very real. His defence was that the *Lucona* could have been lost in a number of different ways, all plausible and completely unconnected to him. It was even possible that the ship had been taken by pirates and was still in use, sailing around the world with a new crew and a different name. All these scenarios were designed to complicate the trial and introduce doubts that made it impossible for the state to prove its case without first finding the wreck and examining it. This probably explains why Proksch was so confident about standing up in court and virtually challenging the judge to find the *Lucona*. He could never have imagined it was a challenge that would be taken seriously.

What Proksch and his defence team hadn't counted on was the independent and slightly eccentric nature of the presiding judge, Hans Christian Leiningen-Westerburg, who decided that actually finding the *Lucona* was the only way to guarantee a fair trial. Many people in the Ministry of Justice opposed a search for the wreck, but Leiningen-Westerburg stuck to his guns and initiated an international request for tenders for the search. Although only a handful of companies around the world had experience of working in the extreme depths (over 4,000 metres) in which the *Lucona* had supposedly sunk, a remarkable twenty-two bids were received by the court.

At the time, although Eastport was internationally known for its record-setting aircraft recovery work, mainly on behalf of the US Navy SUPSALV office, we owned none of the deep-water equipment needed to find and film the *Lucona*. Despite this, my bosses submitted a bold and highly speculative bid based on building or buying every single piece of equipment required and hoping to have it ready to start the search barely five months later, in early January 1991, as per the court's timeline for the trial.

Whilst we certainly had a better chance than most of the other bidders, Eastport were by no means favourite for the contract. Even when the list of bidders was reduced to thirteen and finally to a shortlist of three, we still figured we were only third best on the grounds of having no real track record of finding shipwrecks in such deep waters. In the deep ocean, experience is all-important, and there is no easy way to fill that void if it is missing from your team. Our guess was that our fiercest rival and competitor, Oceaneering International, a NASDAQ-listed company that could have bought us ten times over (and ultimately did three years later), was in pole position. So when a fax from Judge Leiningen-Westerburg arrived in late July 1990 naming Eastport as the court's chosen search contractor and inviting our president Craig Mullen and vice president of operations Don Dean to Vienna for technical discussions, we were delighted and surprised in equal measure.

Even if the names Udo Proksch and *Lucona* were largely unknown outside Austria, there was no doubt that this case, and the search, was going to generate a great deal of news coverage and public interest around the world. Solving the mystery of a high-seas mass murder by finding the wreckage of a ship destroyed by a time bomb could transform the fortunes of a company like ours virtually overnight, but only if we could deliver a successful outcome. With so much at stake, failing to find the *Lucona* would be disastrous for both the court and Eastport, with potential lifelong damage for the reputation of everyone involved.

As Craig and Don set off for Vienna, it was still too early to know what my exact role in the project would be, but as the designated side-scan sonar and search expert in Eastport's bid, I knew I would be a key figure whatever the case. Strangely, that question was answered by Judge Leiningen-Westerburg himself.

∞∞∞∞∞∞∞∞∞∞∞∞∞∞∞

While it was exciting to be involved in such a high-profile criminal trial, I mainly saw the *Lucona* project as the opportunity we had been waiting for to develop Eastport's deep-water search-and-recovery business. The financial structure of our agreement with the court allowed us to purchase a state-of-the-art 6,000-metre-rated side-scan sonar system, including all the ancillary deck and cable handling equipment, and build from scratch a 7,000-metre-rated ROV. Taking the project on was still a huge technical and financial gamble, and we were under pressure straight away from our two closest competitors, who in truth were better placed than us to win the contract based on their superior track records. When their letters of protest to the court were forwarded to us for comment, I couldn't help but feel personally slighted, as by then I had been named as the project's manager. Nevertheless, we fought these protests as fiercely as we fought for the initial award, and soon there was no turning back for either the court or us.

We had barely five months' lead time to specify, acquire, design, build, assemble, integrate and test approximately sixty tons of equipment. It was an insanely ambitious schedule driven by the court's procedural requirements, which meant that the first time all this new equipment would be together in the same location was when it arrived on the loading dock in Singapore, where it was to be mobilized on board an offshore vessel hired to support the search. Integration and testing would therefore have to occur in the field, which, though completely contrary to best engineering practice, was the only way we could meet the schedule dictated by Leiningen-Westerburg. That it placed an inordinate amount of responsibility on me and my small team to deal with any technical problems that arose was something we simply had to accept.

During the five-month build period, my main responsibility was to assemble the integrated navigation system we would be using and look after the purchase of the side-scan sonar search system, which was being built to my specifications by a small Seattle-based company. Judge Leiningen-Westerburg also visited our offices in Maryland, along with one of his technical experts, the naval architect Dr Gerhard Strasser, to check on our progress. Because of his technical training and better command of English, Dr Strasser did most of the talking, although he would often confer with Leiningen-Westerburg in German. This led me to assume that it would be Strasser who would be joining us during the search, so I was quite taken aback when he told me that the judge would also be coming to sea with us, in addition to two other explosives experts. I could already imagine just how cramped the small vessel we had hired was going to be.

The final surprise came when Leiningen-Westerburg swore me in as an official expert witness to the court. I had been warned that this might be required, but it still caught me a little off guard. Other than being responsible for directing the search operation and collecting evidence for the trial, I had no idea what else might be expected of me. Would I have to give evidence in court? At this point nobody, including the judge, seemed to know the answer to that.

Prior to this I had given little thought to Udo Proksch and the crimes he had allegedly committed. My sole focus was what it would take to find and film *Lucona*'s wreck. Now, however, I thought I had better start reading up on Herr Proksch and his controversial life.

∞∞∞∞∞∞∞∞∞∞∞∞∞∞∞

Short, balding and grubby but undeniably charming, especially to women and politicians, Udo Proksch was the owner of the famous pastry shop and chocolatier Der Demel, once purveyors to the Imperial and Royal Court of Austria-Hungary. He was a colourful character, full of contradictions, who was perhaps best described as 'a chaotic mixture of Salvador Dali and Orson Welles'. Over a period of some fifteen years, he held court in an exclusive club he established above Der Demel, cultivating and corrupting an elite cadre of friends who, wittingly or not, readily assisted his fraudulent business ventures.

Born in Rostock, East Germany, in 1934, to a family of Nazi loyalists, Udo Proksch was sent by his father to study at one of the National Political Institutes of Education – otherwise known as NAPOLA – secondary boarding schools for the production of future SS officers and members of the Nazi armed forces. In spite of his small size, he ultimately became one of the elite boys in his class, helped in part by a visit from Heinrich Himmler himself, bearing greetings to Udo from his father. It is believed that NAPOLA is where the young Udo acquired his love of the military and his lifelong taste for the use of explosives. His view of the intrinsic human penchant for war was encapsulated in his declaration that 'War is the father of all things. We will continue to fight and kill, this is what is inside us.'

Proksch's first jobs in East and West Germany were menial positions – he was a swineherd, a coal miner and a corpse washer – that gave no indication of the ruthless flair for making money that became apparent when he moved to Austria. He always worked with partners or close friends in his business dealings, which were generally characterized as being secretive and dubious, if not outright illegal. Unusually for an avowed warrior, he went on to study industrial design at the Academy of Applied Arts in Vienna and became an award-winning designer and art director for an eyewear company, where through the 1960s, under the alias Serge Kirchhofer, he designed the famous Carrera and Porsche lines of sunglasses.

In 1972, with the fortune he had amassed from his design work and other business ventures, Proksch acquired Der Demel directly from the Demel family, who had owned and run the business since 1857. Despite its grand history (it was first established in 1786) and status, his real interest appeared to be the third-floor suite of mirrored and chandeliered rooms above the pastry shop. There, with the urging of his close friend Leopold Gratz, who was mayor of Vienna and later Austria's foreign minister, he formed the infamous Club 45, whose elite members from business, the arts, law, science, media and politics would meet in private, engage in debauched parties and hatch illicit money-making schemes.

Everyone who was anyone in the Social Democratic Party (SDP) or the nation's socialist elite was a member of Club 45. In addition to the founders, Proksch, Gratz and Austria's finance minister Hannes Androsch, this included such leading government figures as interior minister Karl Blecha, defence minister Karl Lütgendorf, justice minister Harald Ofner and science minister Heinz Fischer. Proksch could count on all of them for favours and political protection when needed. He viewed Club 45 as his personal tool to gain control over its members by having intimate knowledge of their weaknesses and vices: '*Now the proles have taken over the rudder, I'll give them what they don't have: a place where they can dance, gorge themselves and booze it up – but they'll dance to my tune.*'

His connections were not confined to Austria, however. He claimed to be a friend of Russia's Nikita Khrushchev, King Hussein of Jordan and Ferdinand Marcos of the Philippines. He was alleged to be a spy, either for the KGB or the East German secret service, and was also suspected of being the middleman in various illegal high-tech exports and weapon sales to the Soviet bloc. But it was within the smoke-filled rooms of Club 45 that he hatched what he thought would be his biggest and surest money-making scheme. It depended on a ship called the *Lucona* not reaching its intended destination.

Like the thousands of other unremarkable ships that transport trade goods from port to port around the world, the MV *Lucona* was a run-of-the-mill cargo vessel. Dutch-owned, the ship had been exclusively chartered by Udo Proksch via his Swiss-based front company, Zapata AG. Having loaded the allegedly precious cargo – the expensively insured uranium processing plant – in Chiogga, Italy, it was instructed to steam for Hong Kong, where it was to receive further instructions about its final destination. Its voyage was being carefully monitored by Proksch, who, unusually, demanded to be updated with its precise position at least every three days.

The *Lucona* had safely navigated the Adriatic, Mediterranean and Red seas and was now crossing the Arabian Sea, heading for the eight-degree channel marked by Minicoy, an Indian atoll north of the Maldive Islands. It was 23 January 1977, the seventeenth day of what had so far been an uneventful voyage for Jacob Puister, *Lucona*'s Dutch master. Although he had never sailed this route before, Puister had little to be concerned about, as his ship was making good progress in near-perfect weather conditions. On board with him were his wife, ten ship's crew and the fiancée of the chief engineer.

His watch having ended at noon, Puister handed over to the chief mate, Jacobus van Beckum, and then attended to some paperwork in his office before heading off to bed for a rest. The Arabian Sea in January is still very warm, so Puister, like most of the crew, had taken to sleeping without any clothes on. Suddenly, without any warning, he was knocked violently out of his bunk and thrown to the floor, where he stood naked trying to comprehend what had just happened. He knew rough weather wasn't to blame, as the sea had been flat calm just an hour or so before, and he was sure he could still hear the engine running normally. Something was seriously wrong, though, as he could feel the ship shuddering to a halt. Whatever it was, he needed to get topside. Without stopping to get dressed, he rushed to the stairway leading to the bridge, where he found his wife looking shocked and panicked.

As soon as Puister reached the bridge, he realized that something calamitous must have happened, because all the wheelhouse windows were broken and glass was strewn everywhere. The thick toughened glass that protects a ship's bridge crew from the elements is designed to withstand the most horrendous storms and powerful wave impacts, so whatever had caused *Lucona*'s windows to smash must have involved enormous force. As he struggled to make sense of the scene around him, Puister's training kicked in and he pulled the engine speed lever back from full speed to dead slow. Looking ahead, he saw a greyish-yellow pall of smoke about ten metres distant. From the same direction there was a groaning, hissing sound.

He ran first to the port door opening and then over to the starboard side to look at how the surface of the sea around his ship was reacting. By now the dense cloud of smoke had drifted aft, enveloping the bridge, so he could see no further than two metres in any direction. When his calls to van Beckum went unanswered, he walked out to the bridge wing and was shocked to see the water level just below his feet and at an extreme angle to the railings. Their dire situation was now clear to him: *Lucona* was going down rapidly by the bow. One of the mates called for him to release the ship's lifeboats. Puister saw that there was no time for that and shouted to everyone to jump as fast as possible. With that, the *Lucona* lurched to starboard and he found himself being dragged underwater.

Although he kept his eyes open, he was unable to make out what had caught his leg and was pulling him down with the ship. All he could see was the light from the sun straight above, and he pulled his body upwards with great convulsive strokes that ultimately tore him free. A few more strokes propelled him back to the surface, where he began looking desperately for his wife, praying that she and the others had been able to jump clear. His vision was blocked, however, by the unbelievable sight of *Lucona*'s upended stern descending into the bubbling froth that marked where the bow had already been swallowed by the sea. Her single propeller, the last part of her structure still visible, was turning slowly as she sank.

Scarcely forty seconds had elapsed since the moment the slumbering Puister was knocked to the floor of his cabin. He hadn't had time to fully understand or appreciate his predicament, but now, as he paddled amongst the floating remains of the 1,200-ton ship that had sunk inexplicably beneath his feet, he feared for the lives of everyone on board, especially his wife.

In the end, half of those on board the *Lucona* lost their lives that day. The ship sank so fast that those who were unlucky enough to be working below deck or lying off-shift in their cabins never stood a chance. Trapped within the steel confines of *Lucona*'s hull, their deaths would have been mercifully swift but still terrifying and painful in the extreme. On the captain's official statement of the missing, he listed the names of the chief engineer, his fiancée and four of the Cape Verdean mates. Their bodies were never found or recovered.

The six who survived all did so because they were either working on the bridge or, like Captain Puister, were able to get to an upper deck immediately before the ship's final plunge. In addition to Puister and his wife, the chief mate, the cook and two assistant engineers survived, although several suffered broken ribs and other injuries. Having jumped from the ship in the nick of time, they all eventually made it to the same inflatable life raft and were picked up by a Turkish tanker early the following morning. When they recovered from the shock of the sinking and were able to give their accounts of what happened, they all concluded that an enormous explosion in one of the two cargo holds must have been the cause. Puister added that the damage to the sides and bottom of the ship must have been catastrophic for it to sink so quickly.

Two months after the *Lucona* sank, Udo Proksch, via Zapata AG, dutifully filed a claim with the Austrian insurers, Wiener Bundeslander Versicherung (the Viennese Provincial Insurance Company), for the total loss of his cargo in the amount of 31,360,725 Swiss francs, an unusually high figure for such a small cargo. The manifest listed it as an expensive uranium processing plant destined for export. Proksch had a few collaborators in Zapata, most notably his friend Hans Peter Daimler, of the Daimler-Benz fortune. The timing of the claim showed scant regard for the tragic loss of life caused by *Lucona*'s sinking. Proksch's sole interest was collecting on the insurance claim as soon as possible.

When faced with the claim, the Viennese insurers baulked. Knowing Proksch's reputation, they were immediately suspicious and requested documentation to prove the loss. When Zapata failed to comply, the insurers refused to pay, although one of their subsidiaries did compensate *Lucona*'s Dutch owners in full for the loss of their ship. Proksch's suit against the insurance company for non-payment was unsuccessful, and it led in turn to a counter-suit alleging that Proksch had made a fraudulent claim. This legal stalemate continued for six years. During this time the insurance company hired a Swiss private detective, who quickly compiled an impressive dossier on Proksch showing that part of the cargo had been worthless coal-mining equipment and that he had used a timed explosive to destroy the *Lucona*.

In 1983, the detective's file was handed over to the criminal prosecutors in Salzburg, who brought charges of murder and aggravated fraud against Proksch and Daimler. However, when the case was referred to Vienna for approval, the justice minister, Harald Ofner, had it blocked. The following year the judge assigned to continue the investigation ordered a search of Proksch and Daimler's premises, and a large number of incriminating documents were confiscated. It was the turn of the interior minister, Karl Blecha, to come to Proksch's rescue. His order for the police to immediately drop all inquiries into the *Lucona* affair lasted exactly one day before the outraged judiciary forced it to be rescinded.

In February 1985, the same judge had Proksch and Daimler arrested to prevent them from leaving the country and for obstruction of justice.

This time, Leopold Gratz, the foreign minister and co-founder of Club 45, produced documentation, never seen before, that appeared to certify what Proksch and Daimler had been unable to prove for the past eight years: that the uranium processing plant was real, it had originated in Romania and had been delivered to another middleman before being supplied to Proksch. It took Gratz just two days to get Proksch freed, but before that he demonstrated his personal concern for his friend by writing him a warm letter in jail: 'Dear Udo, don't get discouraged by this course of action; the truth will come to light.'

This ill-advised intervention on Proksch's behalf brought the *Lucona* affair into the wider public light and at the same time permanently damaged Gratz's political ambitions for higher office. He had been expected to be the socialist candidate for president in the 1986 election, but this was never going to happen after his close ties to Proksch were fully exposed. In addition to the Romanian document being revealed as a forgery, Gratz was forced to admit that he had actually been in Chiogga and inspected the storage sheds where Proksch kept his cargo, although he couldn't say what it consisted of.

In December 1987, the right-wing investigative journalist Hans Pretterebner published an explosive book which laid bare in painstaking detail Proksch and Daimler's criminal conspiracy and the network of political and social connections they relied on to cover their tracks and shield them from prosecution. Despite its serious subject, *Der Fall Lucona* (*The Lucona Case*) read like a thriller and became an instant bestseller throughout Austria. The accusations detailed in the book sparked a major public outcry and a parliamentary inquiry that poured even more pressure on Proksch. With his political friends hamstrung, the judiciary was no longer restrained and was able to go after him without fear of intervention. The endgame appeared near when on 8 February 1988, the federal attorney's office called for a full-scale probe into the affair. However, Proksch was one step ahead of the law again, and after yet another tip-off fled to the Far East the same day.

Despite international warrants for his arrest, the fugitive Proksch was able to travel freely using a passport that he forged himself. His favourite hideout was the Philippines, which conveniently didn't have an extradition treaty with Austria and where he was able to resurrect his friendship with Ferdinand and Imelda Marcos. One of the gems Pretterebner was able to turn up in his research was a photo of Proksch, dressed in white tie and tails, dancing with Imelda at the famous Vienna Opera Ball. Imelda must have been wearing a pair of the highest-heeled shoes from her huge collection, as she towered over Proksch's balding head by at least a foot. With Interpol actively hunting for him, Proksch realized the net was closing in and he would have to keep moving. He left the Philippines, and the comfort of the Marcoses' hospitality, but not before altering his appearance with plastic surgery. He may have sported a new nose, dimples sculpted into his cheeks, thicker eyebrows, a moustache, beard and hairpiece, but he was still a short, rotund figure who was not easily missed.

It was at Heathrow Airport in October 1989 that Proksch finally slipped up. His forged passport was spotted to contain a fatal flaw: a stamp showing passage through Heathrow's Terminal 2 on a date it was known to have been closed. The fact that he was carrying $400,000 in his suitcase, along with photographs of Hitler and naked women engaged in sexual acts with prominent Austrian politicians, didn't help his cause, nor did the fact that the man whose identity he had stolen – one Alfred Semrad – was also a serial criminal, with a rap sheet containing nineteen charges. Proksch and his girlfriend were placed on the next plane to Vienna, where the police were waiting to arrest him, but not before one last desperate attempt to escape was foiled when they caught him trying to board a flight to Germany after he had managed briefly to slip away from their custody.

Proksch's luck had truly run out. One by one his ministerial friends were forced from power, most because of their direct connection with him and/or the *Lucona* affair. In total, some sixteen politicians, lawyers and top officials were removed from their posts, prosecuted or convicted. Two of the most powerful, Gratz and interior minister Blecha, were forced to resign in disgrace when the parliamentary inquiry brought fresh attention to how they had protected Proksch. Others weren't so lucky.

Defense minister Karl Lütgendorf, who was a shareholder in Zapata and believed to be ultimately responsible for giving Proksch the explosives he used to blow up the *Lucona*, died in an apparent suicide in 1981. Few believe he took his own life. When he was found dead in his car, the door was open and the engine was running, while the gun he was holding in his left hand wasn't registered to him nor were there any fingerprints found on it. No suicide note was found either. The surest sign of foul play, however, was that the single shot through the mouth that killed him went through his clenched teeth.

An altogether different but equally unpleasant fate awaited Proksch now that he was under the control of the Vienna Regional Criminal Court and Judge Hans Christian Leiningen-Westerburg.

Our hired support vessel, the aptly named MV *Valiant Service*, got under way from Singapore for the search operation on 12 January 1991, four days later than scheduled. The basic plan was for us to complete the equipment mobilization and testing in Singapore and then steam to the island of Malé to pick up the judge and the other three experts waiting for us there before heading out to begin the search. However, nothing was really going to plan. The mobilization was turning into a nightmare and we were falling seriously behind schedule.

Generally, there are two types of problems that can crop up the first time a new spread of equipment is being mobilized on a ship or at sea. The first type is basically a routine one that can be solved with the resources you have on hand at the time. These can still be painful, expensive problems to solve but the key is that they *are* solvable. While we had plenty of those during the mobilization, causing our departure schedule to slip several days, they were all sorted out at the dock. The second type of problem is far more serious because it is potentially unfixable. These are the fundamental design or engineering flaws for which there is no short-term solution and which can bring a project crashing to a halt. Throughout the mobilization we were struggling with just such a problem, a real doozy, which jeopardized the entire search.

The problem was that our brand-spanking-new side-scan sonar wasn't communicating properly over the 10,000-metre cable we needed to reach the 4,000-metre depth of the wreck, and the sonar technicians trouble-shooting the problem for three straight days had run out of ideas. This was a seriously complicated problem and we had exhausted every option available to us, bar reducing the overall cable length (and thus increasing the electrical signal strength) by cutting short sections off – say 250 metres at a time – to see if reliable communication could be established over a shorter cable. I hated this idea, though, and frankly would have found it easier to cut off the ends of my fingers. Once a length of sonar tow cable is cut, it can never be spliced back together, and the last thing I wanted was for our cable to be too short to reach the wreck. No, it was clear to me that this was a fundamental problem that only the original engineers, Irv Bjorkheim and Larry Robinson, could solve.

It was a tough call to make, but I saw no other option but to ring Larry and Irv at home and demand they get on the next plane out from Seattle to join us. The sonar they had built for us, which I had christened *Ocean Explorer 6000*, was controlled by a mixture of digital and analogue circuitry, with Larry responsible for the digital design and Irv the analogue. Because the cable communication fault was essentially a digital problem, we really needed Larry more than Irv, but in the end Irv agreed to come too – in part because Larry had a heart condition and was generally a very nervous individual. Irv, in contrast, was completely unflappable and far better conditioned to step into the pressure-cooker environment that would greet them when they arrived.

Unfortunately, all the front-end testing we had planned to do in port of the new *Ocean Explorer 6000* and the ROV (Don Dean got the naming rights for this system and called it the *Magellan 725*) was now going to have to be performed en route during the transit to Malé. This was exactly the worst-case scenario we had all feared when the company accepted the five-month deadline set by the court. The risk that the project could end in failure had just increased dramatically. The only good news was that Larry and Irv were on the first available flight, while Judge Leiningen-Westerburg and the experts were relaxing in one of the

Maldive Islands resorts, completely oblivious to the fact that we were now operating by the seat of our pants. A simple summary of our status upon departure from Singapore was that the *Ocean Explorer* was inoperable and the *Magellan 725* was still only partly assembled. Generally, long ocean voyages following a mobilization are spent fine-tuning and relaxing, but not in this case. We were in for a very busy and anxious 2,000-nautical-mile voyage to Malé.

The first chance we had to test the *Ocean Explorer* was two days out of Singapore, at the northern end of the Malacca Straits, which separate the Malay peninsula from the Indonesian island of Sumatra. The Straits are a notorious area for pirate attacks, so we waited to pass through the narrowest section before attempting the first launch of the *Ocean Explorer* towfish,* in about 300 metres of water. Because pirates tend to prey on slow-moving ships with little freeboard, exactly like the *Valiant Service* when she was in sonar-towing mode, the marine crew was on high alert during the test. However, the watch didn't last very long, because the sonar failed after just forty-five minutes. Although this first test was mostly a failure, some useful sonar data was collected, which was a sign that Larry and Irv's efforts were starting to pay off.

The next test, three days later in deeper water just south of Sri Lanka, was more promising. This time the *Ocean Explorer* spent a few hours working at a depth of 4,200 metres before we pulled her back on board. This was the first real indication we had that the sonar was up to the job of finding the *Lucona*, and it was a huge relief to all of us. We were roughly 450 nautical miles out of Malé; less than two days from picking up our clients for a potentially make-or-break project. We weren't out of the woods just yet, but the anxiety levels on board started to diminish. Our spirits were also lifted by progress with the *Magellan*, which was essentially being built on board the *Valiant Service* by a few of our crew during the transit. Power had been switched on to the ROV – a major milestone for any new system – and while still not 100 per cent functional, it was closer than ever before.

* A torpedo-shaped platform containing the side-scan sonar and other electronics which is towed behind a ship.

Just a few miles before entering Malé harbour, we conducted one last test to see whether Larry and Irv's final tuning of the electrical circuits had had the desired effect. The proof in the pudding for side-scan sonars is whether they can produce good-looking sonar imagery, and I was finally pleased by what I was seeing, so I felt it was safe to allow the two tired engineers to disembark, their rescue work now complete. A couple of others from the team were getting off in Malé too, none more important than my boss, Don Dean. Don had been responsible for directing the overall project to date, and we all looked up to him as an inspired leader, especially when offshore. While I felt very fortunate to be taking the reins from him, I did wonder how I would handle the full responsibility of the project with him gone.

I knew I had the technical knowledge to run the search and the capacity to lead a team, but measuring up to Don was a very tall order. As we stood side by side on the fantail of the *Valiant Service*, recovering the *Ocean Explorer* after its last test, the late-afternoon sun was descending towards the horizon while surf from the Indian Ocean gently washed the breakwaters protecting Malé harbour. Even viewed from the deck of a noisy working boat, the setting seemed so amazingly serene and calm that I couldn't help but feel confident that things would turn out all right. Despite the frenetic, chaotic last five months, we were finally ready to begin the search for the *Lucona*.

As we left Malé harbour for the search area with Judge Leiningen-Westerburg and the other three experts on board, two additional things occupied our thoughts. One was the scores from the American football play-off games. The other, more seriously, was the Persian Gulf War, which had kicked off days before with the start of the aerial bombing campaign. There was no television on board and commercial Internet had yet to come of age, so we relied on faxes from our home office and the BBC World Service for all our news. Even though the war had just begun, we had already been put on notice that the US Navy might need us in case any aircraft went down near our location. With the

Ocean Explorer and *Magellan* mobilized, we were in control of the only combined search-and-recovery spread for deep-water operations to 6,000 metres anywhere in the world.

We arrived in the search area at six a.m. on 23 January 1991, precisely fourteen years to the day since the *Lucona* went down. As then, the seas were calm, which I took to be a good omen. Our pared-down operations team numbered ten in total. We split into two equal shifts of four crew each, with me supervising the midnight to noon shift and Bill Lawson taking the opposite twelve-hour shift. That left two of the crew free to concentrate on getting the *Magellan* ready if and when it was needed. In addition to Judge Leiningen-Westerburg and the naval architect Gerhard Strasser, two explosives experts, Major Ingo Wieser and Colonel Heinz Hemmer, had also joined the ship in Malé and were beginning to settle into a routine.

By the standards of the offshore industry in 1991, the *Valiant Service* was a small (57 metres long, 471 gross tons), run-down platform supply vessel that had the bare minimum of accommodation standards, which meant that the judge and three experts had to share a minuscule four-man cabin. It had double bunk beds but the floor space was so small that only one person could get in and out of them at a time. Even though this was the best cabin on the ship, bar the captain's, Leiningen-Westerburg claimed that it was worse than the jail cell Proksch was locked up in.

The area of sea floor designated for the search encompassed approximately 430 square nautical miles, more than double that originally estimated by the court. This increase came about as a result of a search probability analysis that Eastport had commissioned from Daniel H. Wagner Associates, an American firm who were experts at determining optimal search areas and probability maps based on proprietary computer simulations. Finding a single 75-metre-long ship within an area the size of London, Sydney or Bangkok might seem an impossible task, especially as the search was taking place below 4,000 metres of water and the ship's hull was probably shattered into pieces. My plan, however, was to conduct the search on a systematic basis to ensure that the entirety of the designated area was scanned. The cardinal sin of any search is leaving gaps, or 'holidays', in the coverage – expanses of sea floor, however small, effectively unsearched. To prevent this I'd have to make sure that each swath of seabed scanned by the sonar was overlapped by the adjacent swath, just like a gardener mowing his lawn. In fact, 'mowing the lawn' is what we call this method of searching.

I decided on a grid of thirteen equally spaced search track-lines that would each take about nine hours to run. With time for turns at the end of each, I estimated the search would take eight or nine days to complete. Whether the *Ocean Explorer* sonar could operate that long without failure was an obvious question mark in our minds, seeing that the longest time we had ever had it working at depth was barely three hours. Nevertheless, as the sonar was launched and began its long descent to towing depth, I was excited that this moment – one that I had thought about for years and that had occupied all my waking (and some sleeping) hours for the last five months – had finally arrived.

I chose line number 4, just on the northern edge of the central high-probability box, to begin the search, as Wagner's analysis indicated there was a good chance the wreck could be located somewhere along this line. While I wasn't counting on us being lucky enough to find it on the very first line, my plan was to start on the high-probability lines before progressing towards the south-eastern side of the box, which we felt was more promising than the north-western side. After line 4 was finished, we'd jump a line over to number 6, then number 8 and so on.

Searching every other line as opposed to adjacent lines in sequence would make the long turns at the end of each easier and safer to negotiate, especially given that the pilots driving the ship had no prior experience making the type of turns we required. We would also cover more of the sea floor right from the start of the search by leaving the intervening odd-numbered track-lines, and the overlaps they produced, until later. This meant the sequence of lines we'd be searching would be 4, 6, 8, 10 and 12, before heading in the opposite direction to search lines 9, 11, 7 and 5, with the outside low-probability lines (1, 2, 3 and 13) to be searched last if the wreck still hadn't been found. My search strategy was born out of a desire to be as efficient as possible by maximizing the rate at which the sea floor was being scanned, thus giving us a marginally greater chance of finding the wreck sooner rather than later. While this was mathematically sound, I could just as easily have started with line number 5, which would have made the next eight days of my life very different.

After the first two lines (numbers 4 and 6) were completed, we hadn't found the wreck but we had a pretty good idea of what the seabed terrain looked like. The dominant features were two rock ridges, roughly 200 metres high, situated at opposite ends of the SW–NE-trending search box. Thankfully the central area, coinciding with Wagner's highest probability, was essentially flat and rock-free. This was another extremely good omen, for as long as the wreck was located within the centre of the search box, we'd have no problem detecting it against the flat, homogenous sea floor. And even if we were unlucky and the wreckage was lost within the rock ridges, we'd still have a fighting chance of picking it out from the rocks because of the amazing sensitivity of the *Ocean Explorer* sonar.

One of the incredible things about towing a sonar in such deep water was how far behind the ship it was located. In order to get the sonar to the correct height off the seabed – we were aiming for an average altitude of 350 metres in water depths of 4,200 metres – over 9,000 metres of our 10,000-metre tow cable would need to be strung out behind the ship, with the sonar attached at the very end. This meant the sonar could be as far as 4.5 nautical miles behind the ship, making it very difficult to keep track of its precise position as it passed over potential targets. The US NAVSTAR Global Positioning System (GPS) had just become operational, in December 1990, but it wasn't yet fully functional, as 'blackout' periods where no satellites were visible was a common occurrence. One benefit of the start of the Gulf War, however, was that the US Department of Defense had stopped degrading the accuracy of GPS, because they needed it to be widely available to keep accurate track of military forces and equipment, whether on land, sea or in the air. This meant that our GPS was now ten times more accurate than before and we could determine the position of the *Valiant Service* to better than ten metres.

As we began the fifth day of the search without seeing any targets that could represent *Lucona*'s wreckage, Dr Strasser asked about the *Ocean Explorer*'s 'blind spot' and whether we could have passed directly over the wreckage without seeing it. Strasser had obviously been doing his homework to be aware of this particular trait of side-scan sonars, which means they are unable to see targets directly beneath them, or at nadir. The cause of the blind spot stems from the geometry of the sonar's transducers and the footprint of the acoustic beam, which is what creates the map-like 2-D images of the sea floor. As the transducers and resulting beams have to be pointed at an angle in order to sweep the seabed in a sideways direction, this leaves a physical gap at nadir, which the sonar cannot effectively scan. Our problem was that the size of this gap was unknown and very difficult to estimate with a reasonable degree of precision.

My sense was that the judge was growing nervous that nothing had been found so far, and this was what prompted the question from Strasser. Even though we had completed only 40 per cent of the designated search box, the lack of any solid targets had left Leiningen-Westerburg discouraged, and he must have been feeling the pressure of having taken the decision to conduct the search against the advice of those in Vienna who predicted nothing would be found. Because you never know when or where a wreck will be found, there is a definite psychological aspect to the search, where confidence can ebb and flow depending on your state of mind. One moment you can feel utterly convinced you are looking in the wrong place or that the wreck will never be found, and then literally in an instant it pops up on your screens, leaving you to laugh at your negative thoughts. To me it was a bad sign that the judge was showing his nerves so early in the game. His anxiety raised the tension levels on board and made me worry whether he'd have the stomach to see the search through to the end. Some good news was definitely needed very soon.

Day eight of the search, 30 January, started like all the preceding days. Fifteen minutes before midnight, I stepped up into the twenty-foot shipping container we had converted into our search operations command office and listened to Bill Lawson's summary of the previous twelve hours of operations while my shift relieved his. The news from Bill was that there was no news: no probable targets, no possible targets, not even the hint of a target. His shift had endured another tedious and uneventful twelve hours. I knew exactly how he felt as he trudged off to bed. Just an hour earlier, they had completed line 7 and were starting the turn for line 5, which would be the last within Wagner's high-probability area. If the wreck wasn't found on line 5, the prospect that we were either looking in the wrong place or had somehow missed the wreck would suddenly become very real. I had another five hours before the sonar towfish was back on line and searching again to come up with a plan B in case it was needed.

On the wall of our office I had pinned up a large plot of the search area with all the estimated sinking positions, search track-lines and zones of varying probability, from the highest to lowest. To keep people engaged, and for a bit of fun, I had created a pool at the start of the search, with everyone on board putting in $10 to pick the zone where they thought the wreck would be found, though I had had second thoughts about this when its existence had leaked out to the tabloid press in Austria. The last thing I wanted was for our harmless game to send the wrong message to the public. I could just imagine how the pool could be used to unfairly attack our professionalism, especially if the search ended in failure. As I studied this plot to work out where I would add new lines to fill in the blind spots or around the problematic rock ridges as part of a plan B, I noticed that the zone I had picked was about to come up on line 5.

The sonar towfish reached the start of line 5 at 5.01 a.m., but it wasn't all smooth sailing. For the next forty-five minutes we struggled with manoeuvring the ship to keep it on line. A southerly current was running, which meant that more speed and rudder was needed to maintain the correct track as the ship crabbed down the line. The weather and sea conditions thus far had been ideal, and I dreaded to think that this was about to end. The sonar was well into the flat, rock-free area of the search box when suddenly, and without warning, targets started to appear on our screens. These targets were unlike the few small, insignificant and thus easily discounted ones we had seen to date. They were hard, large objects grouped in a tight cluster, which told me we had either stumbled on a completely out-of-place field of rock boulders or we had just found the *Lucona*. In about twenty minutes' time I had my answer: the *Lucona* was lost no longer.

Although finding what we believed to be the wreck of the *Lucona* took the immediate pressure off the team, a new pressure quickly took its place. Whereas Judge Leiningen-Westerburg had been somewhat detached during the search operations, leaving Dr Strasser to monitor things whilst he mostly sunbathed on the upper deck of the ship, he was suddenly very anxious to view the wreck as soon as possible and terminate the operation. I didn't know what was driving this new sense of urgency, but it was clear to me that he had no idea or appreciation of what lay ahead of us when the ROV reached the seabed.

A second line over the wreck with the *Ocean Explorer* sonar in high-resolution mode revealed that the wreckage was more extensive and spread much further than we had originally thought. The magnified sonar image showed a single large target, which I interpreted to be the stern and bridge/superstructure of *Lucona*, with dozens of smaller targets scattered in every direction. Whatever had caused the ship to shatter this way, it had been impressively effective, and deadly, leaving us marvelling that anyone had survived such a violent sinking. Making sense of the wreckage, especially for a criminal trial where the defendant's freedom swung in the balance, was going to take a lot more time than the couple of days the judge appeared to have in mind.

To begin with, the *Magellan* still needed to be completed and tested. Although the basic vehicle was in pretty good shape, a number of important subsystems, like the scanning sonar, depth sensor and navigation transponder, were still not working. So much was riding on this next phase of the project, it was nerve-racking to think everything rested on a vehicle that had yet to pass even a basic wet test. My only comfort was that we had been in exactly the same situation with the *Ocean Explorer* sonar and had been amazed by its reliability to complete the entire search in a single dive lasting eight and a half days. I knew not to expect the same good fortune with the ROV, but getting the judge and experts to understand this wasn't going to be easy, as they were all primed for action and results.

There was plenty of action over the next four days but none of the results the judge expected or needed. The ROV was being worked on nonstop and even made one dive that resulted in ninety minutes of bottom time on the seabed. Frustratingly, we were unable to get either the scanning sonar or acoustic navigation system to work, making it impossible to navigate around the wreckage. This left us with only a standard colour video camera and a pair of lights to see, at best, six metres ahead of the vehicle as it was flown by the pilots. With the wreckage field measuring over 800 metres across, this meant it was like searching eight football pitches in complete darkness whilst armed with only a couple of flashlights. The two small pieces of twisted steel we found during the short dive provided no firm clue as to the identity of the wreckage and served only to tease the judge with what lay beyond the loom of our lights.

Even though it was strictly against best practice and potentially dangerous to dive in a wreckage field without a sonar or some form of subsea navigation, there wasn't any question in our minds about whether we were going to make another dive. The stakes for everyone involved were too high for us to back off now. Having come so far, to within touching distance of the court's objective, we were as keen as the judge to get the ROV back in the water and on the seabed. As *Magellan* touched down later that same day, it became frighteningly clear how difficult it was going to be to safely manoeuvre the vehicle in and around the dense field of wreckage, where any piece of exposed steel could sever our umbilical cable or damage the ROV irreparably. Without a working sonar to scan for objects beyond the short distance our lights could shine, we had no way of knowing what lay ahead of us. One moment there would be nothing ahead except emptiness; then within an instant a huge piece of wreckage would suddenly appear out of nowhere. Our ROV pilots,

Greg Gibson and Ron Schmidt, did exceptionally well not to be totally unnerved by the situation we found ourselves in.

Despite these dangerous conditions, the second dive started promisingly, and by its end, a remarkable nineteen hours later, we had answered most of the court's questions. Within twelve minutes of reaching bottom, at a depth of 4,192 metres, we found a section of derrick sticking vertically out of the seabed, confirming that the wreckage was indeed from a ship. A few hours later we came across a shipping container lying on its side with an identification code matching one listed on Zapata's manifest. The container's canvas top had been ripped away to reveal its true contents as scrap metal. The next significant object was a piece of coal-mining equipment that bore the German manufacturer's name in raised letters and the fresh coat of paint that Proksch had instructed a gang of Italian labourers to apply to the cargo while it was in storage before being loaded on *Lucona*.

To give his shipment an air of greater authenticity, Proksch had also had the labourers paint stencilled labels and code numbers on individual pieces of his cargo. These identification marks – B1, B10, XB19, the name ZAPATA SA and a four-digit manifest number – also appeared on numerous company documents discovered by the journalist Pretterebner, and were thus virtual fingerprints that tied Proksch and Zapata to the cargo. Each time one of the marks was found on a piece of wreckage, the judge and Strasser, looking on over the shoulders of our pilots, would excitedly ask them to stop and zoom in with the ROV's video camera. As the pieces of cargo found with the unique Zapata marking began to mount up, any doubts that we were looking at the exploded remains of *Lucona* faded away.

Twisted and torn pieces of hull structure littered the seabed like so many scraps of crumpled paper. The utter destruction of *Lucona*'s solidly built steel hull into so many relatively small and unrecognizable pieces shocked all of us, but none more than the explosives experts, Wieser and Hemmer, who began revising upwards the amount of explosives they believed Proksch had used. Based on the confession of an army explosives expert, Major Johann Edlemaier, who'd admitted giving Proksch a large amount of TNT, dynamite and plastic explosives in 1976 in exchange for making an army training video, they originally believed that 100 kg of explosives had been used, but now they felt the amount was closer to 150 kg and possibly as much as 200 kg. From what we could tell, the bomb must have been hidden in the forward cargo hold, as the one piece of destroyed hull structure we could still identify, based on visible draft marks, was the bow, while according to the sonar imagery the stern was intact, although we had yet to find it amongst the wreckage.

Finding *Lucona*'s stern became the new priority to answer the final and most critical question about how the ship came to grief. In the brief time filming the wreckage we had proved that it was from *Lucona* based on the cargo markings; that the cargo was mining equipment, scrap metal and other worthless items as opposed to the uranium processing plant expensively insured by Proksch; and that *Lucona* was sunk by a devastating explosion, as the damage we were documenting simply couldn't have occurred by any other means. However, Proksch's lawyers argued that if an explosion was the root cause of *Lucona* sinking, it could easily have originated outside the hull and not from a bomb hidden in the cargo hold. To give this idea some credence, they further suggested the ship might have been sunk by a Russian submarine ordered to torpedo the *Lucona* to prevent delivery of the uranium processing equipment to a possible Far Eastern enemy.

In the face of all the evidence stacked up against Proksch, this seemed an outrageous and desperate attempt by his defence team to make it more difficult for the prosecution to prove their client's guilt. The problem it posed for the judge and for us was that we had to find the seat of the explosion, which meant finding the stern and bridge/superstructure and whatever remains of the cargo holds were still attached to the hull. This was the only way we'd be able to determine whether the explosion originated from a bomb inside the ship, or from a torpedo detonating outside. It was the last piece of forensic evidence the judge needed, but it was proving the most difficult to obtain.

Three more dives to the wreck ended mostly in frustration before we had to admit defeat, at least temporarily, and head into port for reprovisioning. By systematically moving the *Valiant Service* to some forty different positions, we did eventually get the ROV on to *Lucona*'s stern, but only for a brief period before poor visibility and strong surface currents made us pull off the wreck without ever reaching the forward break. We desperately needed to get a replacement sonar shipped in to us and fitted to the ROV to have any chance of relocating the stern. After hours of phone consultations with the defence and prosecuting lawyers and the Austrian Ministry of Justice, the judge told me that he and his technical assistants wanted to come back out and complete the video documentation of the wreck now that it had been located. Colombo was the nearest major port, so we got under way and began planning the necessary improvements for the *Magellan* ROV. We had been at sea for a month and were so low on food that the only thing to eat during the two-day transit was cornflakes, tinned corned beef hash and, appropriately, Vienna sausage hotdogs. It wasn't exactly the celebration meal we were hoping for, but our spirits were high as we looked forward to the break in port.

I was aiming for a quick turnaround in Colombo, with just enough time to take on food and fuel and install a replacement sonar on the ROV, and for the judge to fly to Vienna for a mandatory court hearing to present the evidence we had collected to date. My plans were immediately thrown into turmoil, however, when I learned that because of shortages due to the Gulf War, the government fuel agency were limiting ships our size to 50 tons of fuel maximum, well short of the 166 tons we needed. The only way around the predicament was to get special authorization from the Secretary of State for Foreign Affairs, who for the next two days had me jumping through various bureaucratic hoops. He requested I get a letter from the Austrian consulate explaining our situation, but when I produced the letter the next day, he demanded additional confirmation direct from the Austrian government in Vienna. I was furious by his U-turn and worried that America's impending ground invasion to liberate Kuwait would make the fuel situation even worse.

Outside in the streets there was a slightly menacing anti-American feeling, with groups of Muslims staging demonstrations against the war and at one point even threatening to storm the US embassy. To make matters worse, Sri Lanka's protracted civil war with the Tamil Tigers had flared up and the word in Colombo was that eighty people had been killed in fighting up north. Although I found Colombo fascinating and the people bright and extraordinarily helpful, I didn't want to be stranded here. It took another day of urgent phone calls and faxes between the embassy in New Delhi and the foreign ministry in Vienna before the correct letters were dispatched to Colombo and the full 166 tons of fuel was approved for delivery the following day.

With the fuel drama behind us and the judge back on board in the nick of time, we left Colombo just before midnight and headed back out to the wreck site, hoping for one last perfect dive. Bill Lawson had successfully installed and tested the replacement sonar while we were in port, so there could be no excuses now about not being able to find the stern. The type of sonar we were using operated exactly like a radar, with the head rotating through a wide angle as it scanned for wreckage ahead of the ROV. Within minutes of the *Magellan* reaching bottom, the sonar was switched on, triggering sighs of relief all around as it immediately displayed a field of bright coloured targets ahead.

We picked out the largest target and started moving in that direction. The first small object we came to, however, was a white hard hat, reminding us of the human tragedy that resulted when *Lucona* sank. The next target was even more thought-provoking. It was the empty canister for one of the ship's inflatable life rafts. Whether this canister had held the raft used to save *Lucona*'s six survivors, or whether it was the one that sadly went unused, we could only guess. The ROV was now very close to the largest target. Although it wasn't visible yet, whatever it was filled the entire sonar display. The ROV pressed ahead another few metres and from the gloom the faint outline of *Lucona*'s port side finally came into view. I heard someone say quietly, 'Thank God for that'; otherwise the ROV hut was silent. We had been searching for only twenty-seven minutes.

Plummeting through over 4,000 metres of water at high speed, *Lucona*'s wreck had hit the seabed with such enormous force that it had driven the hull deep into the soft mud, obscuring the name on its transom. The violence of the impact had also badly buckled the stern and bent railings and platforms. It was tricky manoeuvring the ROV around the wreck, but ultimately Greg Gibson was able to make it forward of the bridge/superstructure, which was the main area of interest for the judge and the experts. What we saw next left no one in any doubt about what happened to the *Lucona* fourteen years ago.

While the ship's aft cargo hold was mostly intact, the forward hold, where the prosecutors suspected the bomb had been hidden in a large spherical container, was almost completely gone. The ship had been literally severed in half, with hardly anything left of the forward section. The break was roughly at the position of the middlemast and winch platform, situated between the two cargo holds. All that remained at the break were the double bottom tanks and remnants of the side shell plates still connected to the hull. Crucially, both the starboard and port side shell plates nearest to the break were very clearly splayed outwards.

Seeing this extraordinary level of destruction, we had absolutely no doubt that a massive explosion, originating from inside the forward cargo hold, had ripped the *Lucona* apart. The idea that a torpedo had caused the damage we documented was simply not believable. If a picture is worth a thousand words, we could only imagine the story the video we had collected of *Lucona* would tell when it was shown in court.

※※※※※※※※※※※※※※※

On 11 March 1991, less than three weeks after the conclusion of our investigation of *Lucona*'s wreck, Udo Proksch was found guilty of all charges against him, including the murder of six innocent people, and sentenced to twenty years in prison. When the sentence was read out, he shouted, '*Heil Hitler!*' to a stunned courtroom before being led away in handcuffs. Thankfully the video evidence we had produced spoke volumes by itself, and there was no call for me to testify during the trial as an expert witness.

Relying mostly on the protection of his political friends, Proksch had been able to remain a free man for some fourteen years after he conspired with Hans-Peter Daimler to commit their crimes in the coldest and most calculating way. During the trial it came out that Proksch personally gave Captain Puister one of Demel's famous chocolate tortes before the *Lucona* left port. In his summary of Proksch's motivations, the state's prosecutor maintained that 'The cake was not a travel provision but an executioner's meal.' The trial lasted thirteen months, and because of the costs connected with the expedition to find and film the wreck, it earned the distinction of being the most expensive criminal trial in Austrian history.

The following year, with Proksch serving his time in Graz-Karlau prison, an appeals court reviewed his sentence and increased it to life.

It took a further six years before Daimler was ultimately brought to justice. He was tried at the regional court in Kiel, Germany, on six counts of being an accessory to murder and of attempted murder. His defence, in an equally long and complicated trial, was that he was merely an innocent and unknowing pawn of the powerful Proksch. The jury disagreed and Daimler was sentenced to fourteen years in jail.

In Proksch's case, his life sentence really did mean life. On 27 June 2001, he died on a prison operating table during heart surgery.

II

MV *Derbyshire*

LOST WITHOUT TRACE

MV *Derbyshire*

SUNK 9 SEPTEMBER 1980

44 died
0 survived

The success of the *Lucona* project established the *Ocean Explorer* and *Magellan* as the premier search-and-recovery system in the world. Eastport International had become the go-to company for anyone needing something to be found or recovered from the deep ocean, and our services were suddenly in great demand. Even before the *Lucona* search was completed, we were lined up for another major investigation on behalf of a criminal court. This time the object was a commercial DC-9 aircraft that had mysteriously crashed just north of Sicily with the loss of all eighty-one passengers and crew.

Once again I found we'd be working directly for a judge, Rosario Priore, although this one had the good sense to remain in Rome and forgo the pleasures of joining us at sea. As we were told to expect the project to last several months, the *Valiant Service* got under way to Naples with just a skeleton crew on board while the rest of us took a much-needed break. The break didn't last very long, however, as the team was urgently called back into action by the US Navy to search for a pair of

Tomahawk cruise missiles that had failed to reach their targets in Iraq. Fired on the first day of the war by a surface ship, they malfunctioned shortly after launch and fell into the deep waters of the northern Red Sea. Seeing that the *Valiant Service* would be transiting close to their location en route to Italy, the navy scrambled us back to the ship to see if the missiles could be found, recovered and disarmed.

After that little classified adventure was completed, we spent the next eight months recovering hundreds of pieces of wreckage, including one of the DC-9's black-box flight data recorders, from the bottom of the Tyrrhenian Sea at a depth of 3,800 metres. The Itavia jet (IH-870) had been on a short internal flight from Bologna to Palermo on 27 June 1980 when it disappeared from radar screens as if it had just fallen out of the sky. The disaster, known as the Strage di Ustica, or Ustica Massacre, after the small island nearest to the crash site, is one of Italy's most enduring mysteries. Various theories had been put forward as to the cause of the crash, including that the plane was caught in the crossfire of a dogfight between a pair of MIG-23 jets, one of which was supposedly carrying the Libyan leader Muammar al-Gaddafi, and several NATO fighter aircraft attempting an audacious mid-air assassination of the Libyan leader. The true cause of the crash is still hotly debated and remains a mystery to this very day.

When the rough winter weather forced the recovery operations to end in late November, we decided to redeliver the *Valiant Service* back to her owners in Singapore and planned a route that allowed us to conduct two short shipwreck investigations in Oman on the journey east. I had thoroughly enjoyed managing the Itavia DC-9 project, especially as it gave me a chance to experience a culture towards which I felt a natural affinity, as my grandparents on my mother's side were both from the mountains just outside Naples, but I was keen to get back to working on shipwrecks. *Lucona* had given me a real taste for hunting down long-lost wrecks, and I had decided to focus on learning as much as I could about this fascinating field.

The two wreck projects in Oman couldn't have been more different. One involved mapping and filming the wreckage field of a US Liberty ship, the *John Barry*, sunk during World War II, whilst the other was the search for a modern cargo vessel, the *Demetra Beauty*, which had been lost only the year before. Of the two, the *Demetra Beauty* was the greater challenge, because in addition to finding the wreck, we needed to prove that she had been purposely scuttled. If we were able to save the insurer from having to pay out on a fraudulent claim worth millions of dollars, I knew this would open the doors to more work from the Lloyd's insurance market in London. The *John Barry*, on the other hand, was a straightforward salvage venture to help locate the large cargo of silver coins that our clients wished to recover.

After leaving Naples, the *Valiant Service* was battered by a severe storm that forced the ship to take shelter in the lee of Crete for several days. When she finally limped into Salalah, Oman, the ship was in a very sorry state and required urgent repairs. Despite this inauspicious beginning, the Oman projects went well and all our clients' objectives were achieved. The highlight for me was locating the *Demetra Beauty* in twenty-six hours based on my assessment of the navigational clues we were given. This was the first time I had handled this all-important responsibility myself, so I was chuffed we found the wreck so quickly in the face of some conflicting and confusing information. The experience made me appreciate what a crucial aspect the navigation assessment is when searching for a long-lost shipwreck. It seems an obvious thing to say, but a lost object, whether a downed aircraft or a ship, will never be found if you search in the wrong place. So the analysis and assessment of navigational clues, in context with oceanographic conditions at the time of the loss, was a skill I was determined to perfect before I led my next major project: the search for the British bulk carrier *Derbyshire*.

Very little is known about the last hours and minutes of the *Derbyshire* and her crew of forty-two, plus two wives, who were lost when the ship sank in the midst of Typhoon Orchid. What is known, from their last report, is that the ship was very close to the centre of Orchid's track and battling violent storm-force winds and seas of thirty feet. The ship was hove to, meaning that *Derbyshire*'s master, Captain Underhill, had pointed her bow directly into the mountainous seas to allow her to ride out the storm while maintaining little or no forward motion. Large ships like *Derbyshire* should be able to comfortably withstand severe conditions such as those posed by Orchid. But sometime on the evening of 9 September 1980, about 670 nautical miles short of her final destination in Kawasaki, Japan, *Derbyshire* succumbed to the extreme forces of nature and foundered.

No Mayday or distress call was ever received. No bodies or wreckage were spotted during the search and rescue effort that followed. More than six weeks later, one of *Derbyshire*'s lifeboats was sighted by a passing tanker some 650 nautical miles from her last known position. It was empty. To all intents and purposes, *Derbyshire* had been lost without trace. All that remained was the grief of forty-four families, devastated by the loss of their loved ones, who found it impossible to understand or accept how such an enormous ship could have inexplicably disappeared.

Losing someone at sea this way is the worst kind of bereavement, because there is no body to bury. The families can't even find comfort in the busy work of organizing a funeral. They are left in a hellish limbo until the last remaining hope of survivors being found is completely exhausted. Some never come to terms with their loss and forever hope that a knock at their front door will be followed by their loved one walking back in. Ultimately, the presumption of death becomes a formal pronouncement, and the families are advised to move on with their lives. In lieu of funerals, memorial services are held, at the conclusion of which the Naval Ode is invariably read. For the *Derbyshire* families, the last lines of the Ode – 'At the going down of the sun and in the morning, we will remember them. Lest we forget' – were especially resonant, as the unanswered questions they had about the vessel's sinking ensured they never would forget. In the face of a shipping industry blind to the evident problems with bulk carriers like *Derbyshire*, and government authorities unwilling to investigate further, getting to the bottom of the mystery was to dominate many of their lives for the next twenty years.

The *Derbyshire* was one of countless similar ships built in the 1970s to carry unpackaged bulk cargoes to industrial ports around the world. Although they ranged in size, many of the bulkers were absolutely massive ships built on a scale not seen before. The largest of them, called Capesizers because they were too big to fit through the Suez or Panama canals and thus had to sail the long way round past the southern capes of Africa and South America, were the workhorses upon which the global trade in bulk commodities was completely reliant. The economics of the shipping industry at the time dictated that the more flexible a ship could be in terms of the variety of cargoes it could carry, the greater the rate it could command, thus yielding increased profits for its owners.

The most flexible of this largest class of bulkers were combination carriers; specifically those that could carry ore, bulk commodities like grain, and oil (OBOs). Because of the different types of cargoes they were expected to transport, OBOs were built to a complicated design that meant they were subject to a higher order of forces, including sheer forces in the side shell structure; large hull girder bending moments; extremely high bending and shear loads in transverse bulkheads; and torsion of the upper wing tank structure that could lead to buckling of web frames and the distortion of hatch openings. In an age before computer analysis was routinely available to assist naval architects, they would have been unable to accurately predict the huge forces generated in such large structures, especially under dynamic sea conditions. A dangerous situation had developed, therefore, whereby the economics that drove the profitability of these enormous ships outpaced the ability of naval architects to ensure they were safe.

Derbyshire herself was the last and largest of the Bridge class of six OBOs built between 1970 and 1976 at the Haverton Hill (Teesside) shipyard of Swan Hunter. Originally named the *Liverpool Bridge* (renamed *Derbyshire* in 1978), she was built for the highly regarded Bibby Line – one of the oldest established family-owned companies in the UK – and registered at Lloyd's of London. Because of her size and her enviable pedigree as a 100 per cent British ship, there would have been no lack of seamen queuing up to sail on her. Her raw dimensions show that she was 294.2 metres long, 44.3 metres wide and 91,655 gross tons, with a deadweight capacity of 169,044 tons, meaning she could carry a maximum of 166,000 tons of cargo in her nine cavernous cargo holds. As impressive as these figures are, however, they cannot convey the immense scale of the ship, which was as long as a sixty-storey building was high. Nevertheless, all six of the Bridge-class ships suffered defects in a specific location that became the main focus of the families' questions when trying to determine what caused *Derbyshire* to sink.

The key defect was serious cracks in the deck plates above a transverse watertight bulkhead at frame 65, just forward of the bridge and accommodation superstructure. Although the cracks were found on the outside of the hull, where they could be repaired if discovered in time, their common location in each of the six ships implied a more worrying internal structural fault. Following the loss of *Derbyshire*, the second of the Bridge-class ships, the *Tyne Bridge*, suffered severe cracking that occurred so quickly that the crew had to be lifted off by helicopter. When the *Tyne Bridge* eventually made it to port, the severity of the deck plate cracks at frame 65 was evident; there was a nineteen-foot-long crack on the starboard side and an eleven-foot crack on the port side.

The *Tyne Bridge* incident finally prompted the Lloyd's Register to inspect a pair of the sister ships, the *Cast Kittiwake* and the *Sir Alexander Glen*, in the summer of 1982, which revealed that the likely problem involved two longitudinal bulkheads (girders) that ran nearly the length of the ship and were the main strength members for the hull. Contrary to the original design drawn up by the naval architects, the two bulkheads were butt-welded at frame 65 rather than continuing through this important transverse bulkhead, which marked the separation point between the superstructure and the line of cargo holds. Furthermore, they were misaligned on either side of frame 65 by 25–45 mm when to preserve continuity and maximum hull strength they needed to be aligned perfectly.

The discovery of this defect and the history of cracking in the other ships in the same place led some of the families to suspect that this was the root cause of *Derbyshire*'s loss. The ship was known to have had similar cracking problems at frame 65 and in conjunction with the enormous stresses her hull would have been experiencing during typhoon *Orchid* the belief that *Derbyshire* had suffered a catastrophic structural failure was born. Questions about the grade of steel used in building *Derbyshire*'s hull were also raised, suggesting that the deck plates could have been prone to brittle failure. The scenario arising from these theories was that when hove to against the thirty-foot waves of typhoon Orchid, the hull of *Derbyshire* cracked at frame 65 in a matter of seconds, causing the ship to snap in two and the superstructure – where everyone on board had been riding out the storm – to instantly capsize and sink.

To the families, this theory of total structural failure made a great deal of sense. It was an explanation they could accept when no others were forthcoming, and it provided comfort in that it confirmed the crew were not at fault for the loss of the ship. They were simply innocent victims of errors made by others years before they ever stepped on board. It also explained why no Mayday or distress call was made and why no survivors were found. With everyone trapped inside the superstructure when the ship suddenly snapped, they had no chance of escaping or of being saved. The only consolation for the families was that the end for those on board would have been mercifully quick.

A preliminary inquiry conducted by the UK Department of Trade in November 1980 was 'able to establish no evidence of any form of structural weakness in this type of vessel' and 'remain at a loss to know what caused the loss of the *Derbyshire*'. Without any material evidence to consider, the inspector in charge of the inquiry could only offer opinions as to the possible causes of the loss. Furthermore, he saw no reason to widen the scope of the inquiry to include other ships. The government minister responsible for merchant shipping, Lord Trefgarne, decided not to hold a formal inquiry as he felt 'the court could not reasonably be expected to establish the cause of the casualty and a formal investigation would serve no useful purpose and could not be justified'. He added, however, that he would 'be prepared to re-consider this decision should any new and material evidence come to light'.

The families were upset and disillusioned by Lord Trefgarne's decision. They could not understand how the loss of forty-four lives, along with the largest British-registered merchant ship ever to have been lost at sea (a sad distinction *Derbyshire* still holds today), did not merit a formal inquiry, especially as it was normal practice to hold one in the case of a very serious incident or one involving heavy loss of life. Certainly the sinking of *Derbyshire* justified an investigation on both these counts. The decision compounded the families' personal tragedy and left them feeling that the lives of their loved ones were expendable and unimportant because they were mere seafarers. Had forty-four innocent passengers been killed in a commercial airliner that crashed into the sea, surely the authorities would have investigated the disaster as thoroughly as possible. So why were they apparently satisfied to simply close the book on *Derbyshire*'s loss? It was the first of many setbacks the families would endure in their quest to get to the bottom of the mystery.

Although the government stopped short of conducting a formal inquiry, they didn't completely forget about *Derbyshire* as the families had initially feared. Quietly, the Department for Transport commissioned a series of technical studies from two independent teams to investigate the disaster, in particular whether it was caused by a catastrophic structural failure. The two teams, the British Research Ship Association (BRSA) and the academics Bishop, Price and Temarel, came up with essentially the same result despite approaching the problem in fundamentally different ways. Their common conclusion was that the likely point for a disastrous structural failure of the hull would have occurred at the forward extremity of the superstructure or in the region of frame 65. At the time they reported their initial findings, Bishop, Price and Temarel had no knowledge of the serious deck cracking suffered by the *Tyne Bridge*, or even of the existence of the other sister ships and the cracking they had experienced in this area, thus making their conclusions even more compelling.

Peter Ridyard, a highly experienced ship's surveyor and the father of David Ridyard, who was the fourth engineer on *Derbyshire*, had also begun his own investigations, and these too focused on the cracking in the sister ships at frame 65. He had seen for himself the damage in the *Cast Kittiwake*. He was also given reports by colleagues that documented the damage suffered by the *Tyne Bridge* and the *Sir Alexander Glen*. And it was Ridyard who made the key discovery that these ships weren't built to the original design with the longitudinal bulkheads continuing through the frame 65 transverse bulkhead. So no one was as keen as he was to read the Department's final report when it was eventually published five years later, in March 1986. However, when he did, he was stunned by its conclusions.

Pointing to the fact that it had no direct evidence regarding *Derbyshire* herself, the thirty-four-page final report was just as inconclusive as the Department of Trade's preliminary inquiry in 1980. Despite five years of internal investigation and several external studies by experts, it seemed that nothing new had been learned. The independent predictions made by BRSA and Bishop, Price and Temarel, identifying frame 65 as the likely location where major hull girder damage would have occurred, failed to get a single mention in the report's conclusions. All that was said about the theory of massive structural failure was that circumstantial evidence might be thought to support it 'but in the Department's view is not in itself conclusive'. And finally, 'In the last analysis the cause of the loss of *Derbyshire* is, and will almost certainly remain, a matter of speculation.'

The families, now loosely organized as the *Derbyshire* Families Association (DFA), based in Liverpool, where eighteen of the crew had made their homes, felt they were back at square one. Any hopes they had had of the government providing definitive answers to their questions were dashed. However, their anger at the report had a galvanizing effect on them and drove them to pile even more pressure on the government and the shipping industry, which they believed were complicit in turning a blind eye to the death of *Derbyshire*'s crew. Ridyard in particular was upset by the final report's conclusions because they read nothing like the conclusions of the draft report he was handed in July 1985. This earlier version had gone much further in documenting the known defects of the Bridge class of ships, and the author was clearly unafraid to suggest the logical cause for *Derbyshire*'s loss:

If *Derbyshire* was constructed in much the same way as the sister ships, and there is little reason to suppose that she was not, then in the severe tropical storm which occurred major cracking possibly developed as it did in the *Tyne Bridge*. In the case of the *Derbyshire* it is most likely that had cracking taken place it is probable that it took place so rapidly and extensively that total structural failure resulted. This was followed by the capsizal of the inhabited portion of the ship abaft frame 65. This probably accounts for the complete absence of any distress traffic.

As Ridyard and the DFA contemplated their next move, the news broke that another of the Bridge class of ships was in serious trouble off the coast of southern Ireland.

The thirteen-year-old *Kowloon Bridge* was carrying 160,000 tons of iron ore from Sept Iles, Canada, when she ran into trouble crossing the Atlantic in November 1986. Rough weather during the passage had given her a severe battering, including impact damage from waves breaking over her stern and a crack that was slowly expanding between the No. 9 cargo hatch and the pump room in the vicinity of frame 65. Knowing how *Derbyshire* was lost, and not wanting to suffer the same fate, the captain brought the ship into Bantry Bay on 18 November, informing the vessel's owners that she was unseaworthy. With the stricken ship in a safe anchorage, the crew began making preparations to temporarily repair her as per the Lloyd's surveyor's instructions. Unfortunately, the weather took a serious turn for the worse, and before they could complete the repairs it was blowing a Force 10–12 from the south-west straight up the entrance of the bay.

On the morning of 22 November, *Kowloon Bridge*'s starboard anchor cable parted, forcing the captain to flee the bay and whatever relative shelter it had been providing. The ship was out in the open ocean again, finding it very difficult to steer in the huge seas that had built up. Things suddenly went from bad to worse when the bows fell into the trough of a wave, eliciting a loud bang from the forward end of the vessel. Unable to see the bow through the driving snow and rain, the crew could never-theless tell that the ship was dangerously down by the head. Shortly afterwards, all steering was lost. It was now close to midnight and the situation had become life-threatening, with the *Kowloon Bridge* virtually uncontrollable on a rocky lee shoreline in near-hurricane conditions. Fearing the worst, the captain decided his only option was to abandon ship, and sent out a Mayday distress call requesting helicopter assistance. Luckily, a pair of RAF search-and-recovery helicopters out of Cork were fuelled and ready to go and were able to get everyone safely off in record time. But for the ship herself, there was no such luck.

Initially the abandoned and derelict *Kowloon Bridge* was about ten miles off the coast and out of harm's way but with her engine still run-ning. Much to the concern of the coastal residents, however, the storm turned the ship around and put her on a collision course with land. Anyone who was following her through binoculars at the time will tes-tify to the surreal sight of this unstoppable monster, lights blazing but without a soul on board, seemingly intent on slamming headlong onto the shore. That inevitability was only avoided when the engine stopped on the evening of 23 November, leaving the drifting hulk to the mercy of the wind and tide, the combination of which eventually brought her to ground on Stag Rocks, off Toe Head. Not long afterwards, the hull broke in two places, triggering the release of 2,000 tons of fuel oil that polluted miles of pristine coastline and shut down a centuries-old herring fishery. After two salvage tugs failed to pull the stern section off the rocks, it was left to the winter gales and pounding surf to reduce the *Kowloon Bridge* to a total loss. Months later, she finally slipped off the rocks into deeper waters, where she still lies today, of value only to the sea life that has colonized her hull and the scuba divers who visit regularly.

The sight of the *Kowloon Bridge* lying wrecked on Stag Rocks, her hull badly dislocated and streaming a constant flow of oil, became a front-page story that spread well beyond the narrow interests of the ship-ping industry. Obvious comparisons were drawn between her plight and that of *Derbyshire*. When divers began to inspect the remains of the *Kowloon Bridge*, however, the comparisons took on a completely different dimension. The aft break in the hull was found to extend all the way round frame 65. Added to the mounting evidence of a serious defect with the Bridge class of ships, this new information was impossible for the authorities to ignore. A formal investigation into the loss of *Derbyshire* was finally called to start in late 1987.

To say the investigation was a bitter disappointment to the DFA would be a serious understatement. The families who sat through even a fraction of the forty-six days of the hearings walked away bewildered by the technical complexity of the information discussed and upset that all the relevant evidence wasn't considered. For example, Peter Ridyard was prevented from offering his verbal testimony, as the commissioner doubted he would be of much help. Nor was Professor Bishop called to give evidence on the expert study his team had conducted for the Department for Transport, even though the inquiry was being held under their aegis. Strangely, the *Kowloon Bridge* wasn't mentioned in the final report, despite the fact that it was her loss that had triggered the inquiry in the first place. Ignoring the frame 65 issue with the *Kowloon Bridge* would have been difficult for anyone sitting in the courtroom on day forty, as divers supportive of the DFA had retrieved a huge section of it from the seabed and unloaded it onto the steps leading into the building. Its presence seemed to unnerve the commissioner, who declined the opportunity to go outside and inspect the steel structure that was central to one of the three theories he was investigating.

When the court's final report was published in January 1989, it only served to confirm what the DFA already feared: that another inconclusive decision would signal the end of any further attempts to determine the true cause of *Derbyshire*'s loss.

For the reasons stated in this report the court finds that the *Derbyshire* was probably overwhelmed by the forces of nature in Typhoon Orchid, possibly after getting beam on to wind and sea, in darkness on the night of 9th/10th September with the loss of 44 lives. The evidence available does not support any firmer conclusions.

By the start of the 1990s, it was readily apparent that the problem with bulk carriers being lost at sea was not confined to just the Bridge class of ships. In fact the rate of losses, including sunk or damaged ships, had been alarmingly high for more than a decade, and had experienced an upsurge in recent years. Whichever way you look at the figures, they are sickening. Between 1975 and 1990, 279 bulk carriers were lost, including combination carriers like *Derbyshire*, totalling nearly six million gross tons: an average of 17 bulk carriers per year. Estimates of the lives lost over this same period exceed 750.

This dirty little secret of the shipping industry finally broke in 1991, when the trade papers and magazines began to openly question the rate of bulk carrier casualties and their probable causes. The age of the vessels was seen as a primary factor, but poor management and maintenance and problems associated with loading and unloading were also cited. But where did the *Derbyshire* fit into this disturbing picture? She was only four years old when she was lost, and there was no question she was being well looked after by one of the UK's premier owner/operators. Moreover, she was entirely crewed by British seafarers, and thus far more effort had been expended to understand what had caused her to sink. As frustrated as the DFA were with the string of inconclusive investigations, no other bulk carrier casualty in the world had been investigated to the same extent, if at all. The shocking norm in other parts of the world, especially for ships registered under flags of convenience – where the owner registers the vessel in a different country to avoid strict regulations – is that merchant shipping losses are hardly investigated at all, and the loss of seafarers from poor countries like the Philippines, India and Indonesia, to name but a few, is seen by some as an acceptable cost of an inherently dangerous profession.

Against this backdrop of increased public awareness about bulk carrier losses, the DFA's campaign to have the government establish a new public inquiry started to gain wider support from politicians and the media. Along the way, various transport-related trade unions also offered support to the cause. In March 1993, there was sufficient confidence amongst the interested parties for an advertisement to be placed on the front page of the trade paper *Lloyd's List* seeking tenders from 'companies capable of deploying manned or remotely controlled camera vehicles at a depth of 3,250 metres' to investigate *Derbyshire*'s wreckage.

Within hours of the advertisement landing on my desk at Eastport International I had responded with details of our two ROVs that could reach this depth, and our track record on previous shipwreck investigations. I had no idea who I was writing to – only a fax number and an operation code name were provided – and it wasn't until three weeks later, at a meeting in a central London hotel, that I learned who exactly was driving this project. The list of attendees I was sent prior to the meeting seemed impressive: it contained the names of three trade union general secretaries; four members of Parliament; four legal representatives; two members of the clergy; an operations director named Shaun Kent; a metallurgist named John Jubb, and Paul Lambert of the DFA. I was naturally sceptical that everyone on the list was going to attend, but even if only half showed up, it had all the makings of being a fascinating meeting.

When Don Dean and I entered the conference room, it was immediately obvious that many of the high-profile names on the list were no-shows. Shaun Kent, a scrap metal merchant who had somehow acquired the rights to the hull and cargo of the *Kowloon Bridge* for the grand sum of £1, was acting as the lead representative and made the introductions. The sole DFA member was Paul Lambert. His younger brother Peter was just nineteen when *Derbyshire* sank, and Paul had dedicated much of his own life to finding the true cause of the loss. Like Shaun Kent, John Jubb was volunteering his time, as he had done when he testified at the 1987 public inquiry as an expert on welded structures. John had been extremely upset by the decision of the inquiry to rule the loss a result of the forces of nature and was determined to see this wrong righted. Also present were Eddie Loyden, a Liverpool MP who was working to build support within the House of Commons to press the government on reopening the inquiry; Belinda Bucknall QC, an Admiralty barrister who had represented the families at the 1987 inquiry, and who came to provide advice on the legal aspects of searching for the wreck; and the Reverends Peter McGrath and Canon Ken Peters, were there to support Paul and the families as they had done since virtually the day *Derbyshire* went missing.

Don and I were expecting a meeting focused on the technical and business aspects of finding and filming *Derbyshire*'s wreck, but given the people who were there, it seemed more like a community support group for the DFA. There was obviously a good case to be made that finding the wreck could finally provide the evidence needed to have the formal investigation reopened. What wasn't clear, however, was who in the room had the means to fund a search, which was going to be expensive given the extreme depth and remote location. When this question was raised towards the end of the meeting, all heads turned to a young man sitting at the back of the room who until that moment had remained silent.

Mark Dickinson was the personal assistant to David Cockroft, the acting general secretary of the International Transport Workers Federation (ITF), and was merely standing in at the meeting for his boss. He had no authority or mandate to commit ITF funds for a search, and had attended simply to listen to our presentation and the DFA's objectives and report back. So when everyone in the room looked towards him at the mention of money, he understandably began to squirm in his seat. No precise figures had been discussed yet, but everyone was aware we were talking about a significant sum in the order of several hundred thousand pounds. Nevertheless, it still seemed that some people in the room expected Mark to nod his head and approve the full cost of the search right then and there. As a seafarer and a Liverpudlian himself, there was no doubt that he personally empathized with Paul Lambert, but this wasn't his call to make. It was a major decision for the ITF that needed careful consideration.

In the end, it took eleven months for the executive board of the ITF to approve funding for the search and to award the contract to Eastport, which in the intervening period had been bought out by our closest competitor, Oceaneering International. The process of winning the competitive contract had been arduous, and included conducting a feasibility study to prove, amongst other things, that there was a reasonable chance of finding the wreck and bringing back the new evidence required. The ITF also hoped that the project would 'arouse world-wide public interest in maritime safety, in particular the plight of seafarers of all nationalities working on board bulk carriers and other large ships'. These objectives dovetailed well with the union's Flag of Convenience campaign, aimed at ensuring that its members around the world were paid proper wages and enjoyed decent working conditions. To achieve their publicity aims, the ITF agreed for an ITN journalist to be on board the survey vessel and file reports on the search as it unfolded.

It had taken fourteen painful and frustrating years for the *Derbyshire* families to get to this point, but now they were just a few short months away from having their questions answered.

<hr />

Of the numerous challenges posed by this project, the biggest had to be the relatively short period of search time the ITF's limited budget could afford. Excluding the time it would take to transit back and forth to the search area from the mobilization port of Yokohama, we'd have just eight days to find the wreckage, with no contingency for bad weather, equipment failures or any other losses of time. It was a huge gamble for everyone involved to accept the contract on this basis, as it left absolutely no margin for error or problems. Even if the search was conducted flawlessly, we could easily run out of time before finding the wreck and would be saddled with that failure and the inevitable damage it would cause to our reputation. Despite our successes in finding the *Lucona* and the *Demetra Beauty*, we knew we would only be considered as good as our last search. Although the ITF had already increased their budget from £320,000 to £360,000, that was as much as the executive board was willing to risk on a project with an uncertain chance of success. In the final analysis, however, we all felt it was a risk worth taking.

The other challenge, exacerbated by the limited budget, was the almost total lack of clues about where *Derbyshire* had sunk. The ship had sent no distress or Mayday calls, nor had any survivors or wreckage been picked up to allow an estimate of the sinking position, as was the case with the *Lucona*. It was also impossible to predict *Derbyshire*'s likely track beyond her last known position, as we didn't know the time the vessel sank or how far and in what direction it had travelled in the face of the extreme and changing conditions of Typhoon Orchid. There was, however, one set of possible clues, if only they could be trusted.

After *Derbyshire* was overdue arriving into port and Bibby couldn't raise the ship by radio, a request was made to the Japanese Maritime Safety Agency (JMSA) to mount a search. Because of a rule that a ship had to be overdue by at least twenty-four hours before a search could proceed, the JMSA didn't begin looking for *Derbyshire* until 15 September, a full six days after her last radio contact. The search-and-rescue (SAR) helicopter and aeroplane (no. 791) assigned to conduct the search that day found no signs of *Derbyshire* or any wreckage at the last known position (LKP) they were given: 25°18' north, 133°12' east. However, they did find oil, described as bubbling to the surface and causing a slick 1 km wide and 2 km long, at two different positions a good distance north-north-east of the LKP. Surprisingly, the two reported oil positions were quite far apart from each other: certainly too far apart to be coming from the same source. The oil sighted by aircraft no. 791 in position 25°50' north, 133°30' east was twenty-four nautical miles north-north-east of the helicopter's sighting at 25°30' north, 133°18' east. Later on, this distance between the two positions took on a much greater significance.

The next day, 16 September, the patrol vessel *Ohsumi* also found oil bubbling up to the surface at a more precise position (25°48'48' north, 133°37'18' east), considerably closer to the sighting made by aircraft 791. The search then had to be abandoned for a few days until Typhoon Sperry passed over the area, but was resumed on 19 September when another SAR aircraft (no. 811) found yet more dark brown oil bubbling up in a position (25°50' north, 133°33.5' east) more or less midway between the sightings made by airplane 791 and *Ohsumi*.

The oil sightings seemed like compelling clues to *Derbyshire*'s whereabouts, but it was still far from certain that they were reliable markers of the wreck's position on the seabed more than 4,000 metres below the surface. No one had ever used leaking oil to pinpoint a wreck's location in water this deep before. The key question was how far the oil would be displaced downstream by deep oceanic currents and surface tides as it slowly floated to the surface. At least one famous deep-sea shipwreck hunter believed the information wasn't sufficiently reliable. Dr Robert Ballard, who discovered the wrecks of the *Titanic* and the German battleship *Bismarck*, flatly ruled out a search for *Derbyshire*: 'There just isn't enough data available to locate the area the ship went down closely enough. I think that given the lack of data, the cost of finding the *Derbyshire* would just be prohibitive.'

Although one scientific study commissioned by the ITF estimated that the downstream displacement of upwelling oil could be as much as ten nautical miles, I intuitively felt it would be a lot less. What convinced me was the fact that the oil was sighted over a period of at least five days, and the description of it bubbling up to the surface. In my mind this meant it had to be coming up from the wreck in a steady, consistent stream that suggested little displacement. However, because we only had an eight-day budget to complete the search, I wasn't going to take any chances and decided to verify the JMSA data by visiting their offices in Okinawa and double-checking the SAR team reports.

The theory of catastrophic structural failure espoused by the DFA was based on the belief that *Derbyshire*'s hull separated completely and circumferentially at, or very near to, the transverse bulkhead at frame 65. One of the reasons for this long-held belief was the relatively large distance between the sighting of oil by the helicopter and the other three sightings twenty-four nautical miles to the north-north-east. To the DFA and other supporters of the frame 65 theory, this was proof that *Derbyshire*'s hull had indeed split and that oil was upwelling from the two separated sections of the wreck on the seabed. As *Derbyshire* was generally heading north at the time, they also believed that the stern section of the ship was probably located at the helicopter oil-sighting position while the forward section, including all nine cargo holds and the bow, was located where the oil was bubbling up twenty-four nautical miles away.

To cover this theory, I originally devised a three-phase plan in which a wide-area search for the forward section of the wreck would be followed by the high-resolution sonar imaging of any wreckage found. As long as there was time remaining in the allotted eight-day search budget, which couldn't be guaranteed, we would then attempt to locate the stern section of the wreck. This plan was accepted and approved by the ITF, but it changed completely after my visit to the JMSA in Okinawa.

I had specifically asked the JMSA to pull out all the original records, including the SAR team reports on the four oil sightings, and to have them ready when I arrived at their office. After I explained the importance of this information to our search, they handed me a single-page summary that cross-checked their records against the four positions I had given them, which came straight out of the court's investigation report published in 1989. All the information seemed to tally except for the sighting made by the SAR helicopter on 15 September in position 25°30' north, 133°18' east. Against this entry the JMSA summary had only two words: 'no record'. When I asked what this meant, they simply said they had nothing in their records about a helicopter being used in this SAR mission. I needed to be absolutely sure of this so I pressed them again and again in case we were having a language misunderstanding, but each time they assured me they had absolutely no information about a helicopter being involved.

I hadn't expected this at all, but when I got back to my hotel and pulled out a large-scale chart of the area, I could see why no helicopter had been used in the search: it was because *Derbyshire* was lost so far from land. The JMSA operation had been organized from their regional base at Naha, Okinawa, which is 310 nautical miles from the supposed helicopter oil sighting as the crow flies. This distance (620 nautical miles minimum round-trip) would have been at the far extreme of the longest-range SAR helicopters available in 1980, which meant that if one had been used, it would have had precious little time to search once it reached the location.

The helicopter oil sighting was an obvious piece of misinformation that had been accepted as fact for all these years simply because it was published in what people believed was an authoritative report. This was a hugely significant discovery because it meant we'd only have to search one location for the wreckage, not two. On the other hand, it went against what the DFA had believed about *Derbyshire* splitting and the two sections sinking in different locations, though I couldn't let that concern me. When, a little later, I discovered the source of the helicopter position, all the pieces of the puzzle fell into place.

At the time of the loss, Bibby Line had received a fax from their agent in Japan, Mr I. Yamada, which read: 'Now reported us that they heard from helicopter. Quote: found some oil spot afloating [*sic*] at the position of 25.5 north 133.3 east hence trying further research nearby sea water surface. Unquote.' As no helicopter was involved in the search, this report clearly refers to the sighting made by aircraft 791 in position 25°50' north, 133°30' east, except that an error was made when the position was quoted in decimal degrees of latitude and longitude rather than in degrees and minutes. As this information was relayed down the line, someone within Bibby Line naturally, and quite correctly, converted the 25.5 north, 133.3 east position into 25°30' north, 133°18' east without knowing of the earlier error. From this little piece of detective work I left Okinawa feeling 100 per cent certain that the helicopter oil sighting was essentially a phantom position and could be ignored. It also cleared up something else that hadn't made sense to me. I never understood how *Derbyshire*'s forward section, minus the engine and propulsion, was supposed to have been able to travel twenty-four nautical miles further north-north-east against the southerly winds and seas of Typhoon Orchid. There was still every chance that *Derbyshire* split at frame 65, but it was beginning to look as if the scenario presumed by the DFA was impossible and we'd find the two sections of the wreck in the same location, close to the three confirmed sightings of upwelling oil.

We got under way from Yokohama on the afternoon of 26 May, expecting it to take anywhere from two and a half to three days to complete the 660-nautical-mile journey to the search area. Unfortunately we set out in near-gale Force 7 conditions, which made it impossible to get any additional work done the first day of the transit. The cause of the poor weather was a tropical depression whose predicted track looked like it could be a real problem for us until thankfully it began to veer away. When you haven't been to sea for a while, the last thing you want on the first day out is rough weather, because it makes it very difficult to get your sea legs and become accustomed to the ship's movements. The constant pitching and rolling of the ship made the transit a bit unpleasant and laid low a couple of guys in our crew.

Overall there were eleven of us on board, in addition to the Japanese marine crew: Mark Dickinson of the ITF, Dr Chris Davies of the University of Wales, Rory Maclean, the ITN news reporter, and myself and seven others from Oceaneering. I was the search director and supervised one shift, while the project manager Craig Bagley supervised the other. We had been forced to keep our crew size small because of ITF's tight budget and the limited number of bunks on the *Shin Kai Maru*, the Japanese platform supply vessel we had hired for the project. However, the team was packed with some of the best and most experienced guys in our commercial group, and we had also brought with us a surprise the ITF was not counting on.

Our contract with the ITF called on us to provide only the *Ocean Explorer 6000* side-scan sonar. A lack of funds meant they could not afford any ROV dives to investigate or positively identify whatever was found during the search. Instead, they were counting on the high-resolution sonar images of the wreck to be distinctive enough to prove that *Derbyshire* had definitely been found and that this information met the criteria of 'new and important evidence' demanded by the government in order for the investigation to be reopened. From Oceaneering's point of view, this was a poor judgement. As the reading of side-scan sonar images is an interpretive skill, the conclusions that can be drawn from it are open to debate and thus are nowhere near as definitive as the visual images taken by an ROV. Our concern was that by relying on the side-scan sonar imagery alone, the ITF was running the risk of having the government reject out of hand any evidence they put to them because the sonar imagery was deemed too questionable to be accepted as definitive proof.

We felt so strongly about this that we decided to bring the *Magellan 725* to Japan with us, at no additional cost to the ITF, just in case it was needed. I stopped short of telling Mark about it until it was about to be loaded on board the *Shin Kai Maru* in Yokohama. My reasoning was that I didn't want to raise expectations too high, and that it was far better for the *Magellan* to be viewed as an unexpected bonus – in case it was needed to take pictures of something we found – rather than an unfulfilled promise. When I opened the container doors to reveal the ROV to Mark, our surprise had the desired effect. All he could say was 'Thank you' and 'Wow, that's great. Now I just hope we find something to dive her on.'

Despite the rough seas, we actually made very good time during the transit and arrived on site fourteen hours ahead of schedule. That was the good news. The bad news was that the weather conditions were still too poor to launch the *Ocean Explorer*, so the time we gained against our schedule during the transit was subsequently lost waiting for the weather to improve. Nothing could be done about it other than to use the down-time to train the ship's crew in running track-lines so they were ready when we needed them for the real thing. That occurred the following day, 29 May, when the *Ocean Explorer* was finally deployed in the early afternoon to begin the eight-hour descent towards the start of the first of seven potential track-lines. With the hopes of at least forty-four families in the UK and countless seafarers around the world riding on the outcome, the long-anticipated search for *Derbyshire* was finally under way.

When the first images of the sea floor started scrolling down my screen, I wasn't surprised to see it was a geologically active area with plenty of relief and outcropping rocks. If the upwelling oil was to be believed, *Derbyshire* had sunk somewhere along a steep slope that separated the Daito Ridge from an adjacent basin bearing the same name. Smack in the middle of the 200 square nautical miles I had designated for the search, the seabed dropped precipitously from 3,500 metres to well over

5,000 metres deep. Even with her enormous size, picking *Derbyshire's* wreck out from all this geology was going to be extremely difficult.

I purposely placed this first track-line (no. 4 in my grid of seven) just north of the three upwelling oil positions and parallel to the sloping sea-bed to make it easier for us to tow the sonar on an even plane. Although an oceanographer contracted by the ITF advised that the upwelling oil was probably drifting north with the Kuroshio current and that this indicated the wreck was located to the south, I had found more recent data during my research in Japan that indicated exactly the opposite: that the flow was actually south at about one knot, in keeping with a counter-current flowing against the main direction of the Kuroshio. This data, along with the JMSA SAR team descriptions of the oil bubbling up, convinced me that the wreck was close by and located probably just north of the reported sightings. My best guess was that the likely displacement of the oil was no more than three nautical miles, and this was the figure I used to establish the high-probability zone within the overall search box. If I was right, there was a good chance we'd find the wreck on one of the first three track-lines.

Four and a half hours into the search, we started picking up suspect targets at far range on the sonar's starboard channel. The targets were clustered and clearly hard, but it was impossible to discern any recognizable shape or structure because they were at such a distance. I liked the fact that they were located to the north and midway between the two aircraft positions, but it was much too early to start getting excited. Given the order in which I had decided to run the search lines (4, 6, 3, 5, 2, 7 and finally 1), we'd get a much better look at these targets on the next most northern track-line: no. 3 in my grid and also the third line in sequence to be searched. Before that, line 6 was searched, but when nothing of note was found, it simply heightened our expectations for what we might see on the next line. Although the search was only into its second day, it was progressing very well and there was a definite sense that we were closing in on *Derbyshire's* location.

The crew driving the *Shin Kai Maru* had already executed the most precarious part of the turn to line 3 and were steadying the ship on course when we suffered a disaster that jeopardized the whole project. The first sign that something was wrong was a spike of noise on the sonar's display. Within seconds, lights indicating an open circuit in the electrical core of the sonar tow cable flickered and then lit up brightly. There had been no warnings of any problems, but these symptoms meant only one thing: the tow cable had developed a fault, probably a minor ingress of water, and the sonar towfish would have to be recovered, meaning the search was over for now. As we trudged out of our operations van, disappointed at having to stop before learning what the targets on line 3 were but still thinking the problem was relatively minor, someone on the back deck shouted, 'Hurry, before the cable snaps!'

Short of snapping completely, the worst thing that can happen with a tow cable under tension is for some of the steel wire strands to break and 'birdcage' at the point where the cable exits the ship before entering the water. Seeing a cable in this condition will make your heart sink, especially when 8,000 metres of cable and your only sonar towfish are just one sudden heave of the ship's stern away from being lost forever. The only way to save a situation like this is to gingerly haul the damaged cable back on board until it is secure on a winch, thus releasing the huge loads on the damaged section. The real trick is getting the fat birdcage past the sheaves and narrow openings through which the normally thin cable passes without making things worse or snapping any more strands.

As everyone on my shift sprang into action, I ran up to the bridge to explain to the captain that I wanted him to maintain the ship's speed but to change the heading in order to reduce the rolling and heaving motion of the stern as much as possible. I knew the cable already had three to four tons of load on it, but that could easily double or treble with one bad heave of the stern, causing it to snap instantly. By the time I got back to the main deck, two of our most experienced crew, Greg Gibson and Ron Schmidt, had taken control and were easing the cable over the three large internal sheaves built within our deployment crane. Once the damaged section was inboard, I could see that the break was very bad, involving strands from both the outer and inner layers of steel wire that gave the cable its strength. I could hardly believe that so few unbroken strands – perhaps a quarter of the total – were keeping the full length of the cable and the *Ocean Explorer* from snapping free.

When a serious failure like this occurs at sea, it is hard not to panic. Though the potential for huge financial losses is staring you in the face, you must push it to the back of your mind and work quickly but calmly. A bad decision or action born out of panic can jeopardize not only your equipment, but someone's life. To avoid anyone getting hurt, we had to make sure everyone was forward of the failed section in case the cable snapped and whiplashed as it left the ship. The scariest part is not knowing when this might happen. You could be literally an inch away from saving the whole rig when just one more strand snaps, causing the cable to reach its breakpoint. Fortunately, we had two of our best crew feverishly trying to save the situation from becoming a complete disaster.

Greg and Ron had already done an amazing job, but the worst bit was still to come. In order to wind the cable up onto our traction winch, where it could be securely held, they needed to trim all the badly frayed strands of wire so that the cable could ride cleanly in the narrow grooves of the bull wheels. As Ron lay on his stomach to grind away each of the dozen or so broken strands, his hands and face were in constant peril should the cable suddenly snap and recoil back in his direction. This selfless and brave effort to save the rest of the cable and the *Ocean Explorer* sonar was typical of the man. In all my years at sea I can't recall being more relieved than I was to see what was left of the cable wind up on the winch and Ron stand up with all his fingers intact.

<hr>

Remarkably, less than nineteen hours of search time was lost while the *Ocean Explorer* sonar was recovered, the short end of the tow cable jettisoned and the long end wound back up on to its storage drum and electrically reconnected. We were left with about 8,000 metres of usable cable, which would be just enough to complete the search at slightly slower speeds. While we were dealing with the tow cable disaster, I had stopped thinking about the suspect targets we had found on line 4, but now that we were up and running again, I couldn't wait to see what they looked like from a better position on line 3. However, once the sonar had passed over the target area, I was left scratching my head about what the targets actually were.

Given that we were searching for the largest ship ever to have sunk, I was expecting to find at least some of its structure intact and easily recognizable. What we found instead could only be described as a very large patch of targets: some small, some large, but nothing on a scale that allowed me to be sure this was a shipwreck. In fact it looked more like the debris field from a plane crash, where the entire fuselage is obliterated after hitting the water surface at speed. If this was the remains of the *Derbyshire*, none of us could imagine what had caused the hull to be so completely shattered.

Mark Dickinson pressed me for a firm opinion about the targets. As much as I wanted to, I couldn't give him one. The sonar imagery was not conclusive enough, and with only a third of the search box covered, I wasn't about to change my plan. I told him we needed to run more lines to rule out other high-probability areas and to get a better feel for the seabed surface in case the patch of targets was typical of the geology for this area. In my mind, they could just as well be outcropping rocks as man-made wreckage of some type, but we needed more data to be sure. I decided the sensible thing to do was to continue the search as planned, but if no other suspect targets were found we'd return to this patch and have a closer look with a high-resolution pass.

Three lines and two days later, only 10 per cent of the search box had yet to be covered and there was less than three days remaining in the ITF budget. The patch of targets we'd detected on day one was still the only possibility we had for *Derbyshire*, especially as we found it looked less and less like geology the more sea floor we covered. Because time was becoming critical, I suggested we forgo the southernmost line in the box, which all the clues indicated was low-probability, and head straight to the suspect patch to see exactly what it was. We needed to resolve individual targets within the patch, so I took the gamble to narrow the sonar's swath setting more than I normally would, from 4.8 km to 1.2 km, thus increasing resolution by a factor of four. It meant, however, that the next sonar line had to be perfectly positioned otherwise we wouldn't be able to produce the detail I knew we needed. .

Getting the sonar towfish in the right position was completely dependent on how well the turn between the two adjacent lines was executed. It took full concentration for at least seven nerve-racking hours while the towfish, dangling from the end of roughly 7,000 metres of cable, was whipped around to the reciprocal course in a slingshot-like manoeuvre. Because a towfish will naturally drop during a turn, keeping a safe altitude above the seabed is paramount. Get the turn wrong and you can easily miss the target and then have to repeat the whole process over again with the loss of at least half a day. But get it badly wrong and you could watch helplessly as your towfish plummets and then crashes into the seabed to be destroyed or never seen again.

I had already had a very close call on an earlier turn when guiding the towfish over an uncharted submarine mountain on the eastern side of the search box. In that instance the seabed was already climbing faster than I could recover the towfish when the winch controller suddenly packed up and stopped working. As Ron worked feverishly to fix the fault, my eyes were glued to the sonar display showing the trend line of the seabed rising dangerously towards the sonar. A crash was inevitable unless I could do something in the next thirty minutes. With the winch still broken, my only option was to speed the ship up to get the towfish climbing in altitude itself.

I began by increasing the *Shin Kai Maru*'s speed in 0.2-knot increments, but with the seabed winning the race, I closed my eyes and increased speed a full knot, up to about 4.5 knots. I didn't dare go any faster, as drag forces had probably tripled by then and the last thing I wanted to do was overstress a tow cable that we suspected was already weakened following the earlier failure. I also knew that the faster I went, the harder the towfish would hit what was essentially the peak of a mountain. From a high of nearly a thousand metres, it was now just fifty metres off the seabed, and I started to brace myself for the inevitable impact. Just as I had given up hope, Ron called to say the winch controller was fixed and I could start hauling in the cable. How much those last few minutes of recovering cable as fast as I could mattered is hard to say, but when the towfish cleared the rocky peak by only fourteen metres, I was enormously relieved and knew we had dodged a second bullet.

While the turn to get the sonar on line to scan the patch of targets wasn't as dramatic, it still raised my pulse more than it needed to. In a bid to save time, I thought I had cut the turn too close and genuinely feared I was going to miss the target area. Sensing the towfish was off-line and was only going to sideswipe the patch, I ordered two last-minute course changes that I hoped would be enough to bring the sonar onto the track I wanted. Within minutes of the first targets appearing on my display, I knew I had hit the patch dead centre, exactly where I was aiming, and that it definitely wasn't geology but was actually an extremely dense field of wreckage. With the *Ocean Explorer* in exactly the right position, its 120 khz sonar proceeded to paint an incredibly impressive and devastating picture of destruction.

Our elation at finding *Derbyshire* soon gave way to shock and disbelief at the condition of her wreckage. This wasn't how anybody had expected the ship to look when found. Rather than a handful of pieces, her hull was obliterated into literally hundreds of individual fragments strewn across an area measuring 1,300 metres by 900. Although some larger pieces of wreckage stood out, the main story the sonar image told was of a ship that was completely shattered from stem to stern. After days spent discounting the patch because it didn't look right, there was no doubting any longer that it was *Derbyshire*, but how did she wind up in such a state? While the search appeared to be over, the sonar images instantly raised new and unforeseen questions about her demise that we hoped to attempt to answer in the few days we had left. Unfortunately, there was one more sting in the tail waiting for us.

After completing the initial sonar analysis that confirmed *Derbyshire*'s wreckage, I plotted out a second high-resolution line to collect even more detailed information about the wreck site. The aim of this second line, planned for 600 metres swath, was to collect better images of some of the larger pieces of wreckage so that they could be accurately measured. One huge piece in particular had produced a large acoustic shadow on the first high-res pass, and I suspected this was a major structural section of the ship. A second set of images and measurements might allow it to be identified. Pleased with the day's results, I went to bed leaving instructions with Craig Bagley to wake me once the next turn was completed and the sonar was on track for the line I'd plotted.

As the curtain concealing my bunk was drawn back, allowing the harsh glare of the ceiling lights to shake me from the deep sleep I was enjoying, my body clock instantly told me that I hadn't had the five hours of sleep I'd planned on. Something must be wrong. Craig quickly confirmed my fears, breaking the bad news in his usual direct manner: 'David, we lost the *Ocean Explorer*.'

It is one of the worst feelings ever to lose an important piece of equipment at sea. Old hands in our industry like to say that there are two kinds of sonar operators, 'those who have already lost sonars, and those who are about to lose one'. The gist being that if you do this job long enough, you will one day experience the embarrassed feeling that goes with leaving vital equipment behind on the seabed. Truth be told, this wasn't my first loss, but it was very personal because of my connection to the *Ocean Explorer*. While every piece of lost equipment can ultimately be replaced, it would be impossible to replace the history of success I'd shared with this amazing sonar.

The loss of the *Explorer* immediately changed the focus of the project from a shipwreck search to an emergency recovery operation. The decision to bring the *Magellan 725* with us now seemed like a masterstroke, as it would give us a real shot at recovering the sonar, thus sparing our blushes back at Oceaneering's head office. Given that we were planning to make at least one dive with the ROV to take pictures of *Derbyshire*'s wreckage, as promised to the ITF, it was possible that we could perform both tasks within the remaining time in our schedule, though only just. The main question now was how long it would take to locate the *Explorer* towfish, if it was even possible.

While everyone in the team kicked into high gear getting the ROV and associated tools ready for the recovery attempt, no one's job was more important than that of our lead navigator, Larry Ledet. Relying on a mixture of hard data and pure guesswork, it was left to Larry to calculate whereabouts on the seabed he thought we'd find the 4.2-metre-long torpedo-shaped towfish. Not an easy task considering the sonar was six kilometres behind the *Shin Kai Maru* when the cable suddenly snapped. Losing the *Explorer* was a cruel blow coming so soon after our triumph in finding the *Derbyshire*. It hardly seemed fair to be thrown into another pressure-filled search straight away.

The one thing you can count on at sea is that when everything starts to turn against you, bad weather will arrive to really compound the problems. True to form, as soon as we had the *Magellan* ready to dive, the wind kicked up to a Force 7, putting the recovery operation on indefinite hold. Waiting on weather is always frustrating, but never more so than when time is running out and your company's million-dollar towfish is still 4,000 metres below you on the seabed. Because of another project starting immediately after ours was scheduled to finish, there was a deadline for getting the *Shin Kai Maru* back to port, so an extension was out of the question.

As the hours ticked by and the weather showed no signs of improving, I began to fear the worst: that the *Ocean Explorer* was staying where it was and we had missed our chance to come home with definitive photographic proof of the *Derbyshire*. While I was absolutely certain we had found the wreck, I couldn't expect others to share my confidence in the sonar images. The government had already shown a willingness to either discount or ignore evidence about *Derbyshire*. I could easily see them doing the same when the ITF presented them with a handful of colourful but inconclusive sonar images without the backup of photographic proof.

Eventually the weather did improve, allowing us to dive the ROV, although two precious days were lost while we waited. The pressure was now squarely on the shoulders of the two *Magellan* pilots, Ron Schmidt and Greg Gibson, to find and recover the *Explorer* sonar as soon as possible. Fortunately, Larry's estimate of the sonar's position was spot on, so very little time was expended during the search. The recovery operation proved far more complicated. The main problem was that the buoyant sonar towfish was found floating in the water column while anchored to the seabed by its companion dead-weight depressor. This made cutting the still-connected tow cable and attaching lifting lines almost impossible. It was like trying to wrestle and control a huge bobbing cork. What should have taken minutes took hours, eating deeper into the time we had left. Nevertheless, when *Ocean Explorer*'s yellow body broke the water's surface and was craned on board, huge cheers and smiles of relief broke out throughout the ship.

After calculating how long it would take us to get back to port steaming at full speed, I told Mark we had just enough time to attempt one dive on *Derbyshire* but that we could spend no more than six hours on the seabed. I recommended we head straight for the largest sonar target in the wreckage field, which I suspected was the bow. I thought this would tell us more about the ship and how it broke up than any other piece of wreckage might. I also thought the biggest object would be the easiest to locate, especially as several other large pieces nearby could be used to navigate the ROV in the right direction.

Just before the *Magellan* was launched to begin the dive, I secured in one of its manipulators a small bronze memorial plaque that we were going to place on the wreck on behalf of the DFA, and in the other manipulator a sediment scoop to recover some of *Derbyshire*'s iron ore cargo, which we expected to find spread across the sea floor. I'd made the scoop out of an empty washing-up liquid bottle I found on the ship, cutting flaps into its base to create a one-way valve that would allow sediments to be forced in but prevent them from escaping afterwards. A lot more thought had gone into the plaque. I'd had it made weeks before we left home and it was inscribed with words of remembrance provided by the DFA. After all the ups and downs over the past seven days, I could scarcely believe we were just hours away from laying the plaque where it would mean so much to the families.

As the lights of *Magellan* began to illuminate the seabed at 1.40 a.m. on 8 June, a remarkable sight appeared. At a depth where absolutely no ambient light exists, thousands of tiny twinkling lights shone back at us from the seabed below like a galaxy of stars on a cloudless night. It took a minute or so before we realized what was causing this illusion. *Derbyshire*'s iron ore cargo was reflecting the lights of the ROV back at us like the facets of a diamond. It was the first sign that we had found the right wreck.

I purposely had us approach the wreckage field from the perimeter so as not to get confused by the huge amount of debris I knew we would find at the centre. Every sharp edge of fractured steel structure we encountered was a potential hazard that could instantly sever the ROV's umbilical. This was another reason for not descending straight into the greatest concentration of debris. An hour into the dive, we found our first piece of wreckage. Although it was too small to offer any clues for identification, it was a welded piece of steel plating that showed all the signs of coming from a modern shipwreck. At least we could rule out a wartime wreck, which would most likely be riveted and not welded.

The path I had plotted to the largest piece of wreckage would take us past three other large pieces, which we could use to vector the ROV the 500 metres it needed to cover along the eastern edge of the wreckage field. As Ron Schmidt flew *Magellan* along the vectors I'd given him, Larry Ledet and Craig Bagley were on the bridge of the *Shin Kai Maru* guiding the Japanese captain, with the help of a translator, to make the same movements in lockstep. It was a good plan but we weren't making the progress I had hoped. We were moving much too slowly; at this rate we were going to run out of time. Like a dog straining on its lead, the ROV was constantly tugging on its umbilical, waiting for the ship to catch up. Two and a half hours after touching bottom we had yet to reach the three large pieces, and everyone was growing increasingly frustrated. Rory Maclean, the ITN reporter, was on a live feed with Jon Snow, the Channel 4 news anchor in London, waiting for me to confirm the exclusive news that the wreck was indeed the *Derbyshire*. Rory was just about begging me to go on air during the broadcast, but I firmly refused. I wasn't about to jump the gun with a story of such important national interest before I was 100 per cent certain.

At 4.30 a.m., the first large piece finally came into view. It was twisted and torn beyond recognition but was clearly wreckage from a large, relatively modern ship. We desperately needed to pick up the pace, and now the ROV began making good progress over the ground. As the size and concentration of wreckage increased, we could all sense we were closing in on something big. Mark Wilson, running the ROV's sonar, was the first to see it. His screen was filled with an enormous target that was still beyond the sight of the ROV's video cameras. Everyone who was crammed into the small ROV control room leaned forward in anticipation as Mark rhythmically counted down the closing ranges to the target: 50, 40, 30, 20, 10 metres. Finally *Magellan*'s lights began to illuminate the enormous target rising out of the gloom. And then it appeared. A colossal wall of steel that could only mean one thing – the DFA's fourteen-year quest for definitive proof was finally over.

As Ron struggled to get the ROV closer, white lettering slowly came into focus. Not only had we come straight in on *Derbyshire*'s port bow, but we were staring at the last five letters of her name – *SHIRE*. Although the ROV's taut umbilical prevented us from filming the entire name, there could be no doubt about what was in front of our eyes. We had indeed found the wreck of the MV *Derbyshire*, 4,210 metres below the spot of angry sea where forty-four unlucky souls had been condemned to their deaths.

⁂

At first it was difficult making sense of the bow's orientation. It was clearly sitting upright, fairly level on the seabed, but an old photograph had to be consulted to show us which side was which. As Ron manoeuvred the ROV up over the wreck's side above the forecastle deck, the scene below us came into focus. The double drum mooring winch and spare anchor lashed near the port railing helped us get our bearings straight. The bow had obviously sheared off from the rest of the hull, straight across her entire 44.3-metre breadth from port to starboard. The important question was, at which frame number? To us it looked like the fracture was at frame 339. This was the forward area, along with frame 65, predicted by Bishop, Price and Temarel to experience peak dynamic stresses. What we had seen of the bow, with all the fractured surfaces around the circumference of frame 339, was enough to suggest that their conclusion about the location of dangerously large stresses in *Derbyshire*'s hull was valid and warranted further investigation.

Having satisfied the ITF's objectives by identifying the wreckage as that of *Derbyshire* and collecting potentially new and important evidence about what might have caused the ship to sink, it was time to end the dive, but not before completing two final tasks. I had been searching for a good place for us to lay the memorial plaque and spotted a flat open section of the forecastle deck near the spare anchor. As *Magellan* descended towards this spot, the plaque briefly spun in the current, revealing the writing on both sides. We had all signed our names on the back as a mark of respect for *Derbyshire*'s victims. On the front were the words requested by the DFA. As *Magellan* lingered above the deck, having carefully placed the plaque where it will remain as a permanent memorial to the forty-four who perished, the DFA's prayer stared back at us.

IN OUR THOUGHTS

IN OUR HEARTS

IN OUR AFFECTIONS CONSTANTLY

AT THE GOING DOWN OF THE SUN

AND IN THE MORNING

WE WILL REMEMBER

THE WHOLE *DERBYSHIRE* FAMILY

A long period of silence followed the laying of the plaque, everyone deep in their own thoughts about what we had seen over the past few hours. The horrific destruction of such an enormous ship, many times greater than our own support vessel, was a sobering sight to see. For me, the symbolism in laying the memorial plaque was very powerful.

Forty-four people had died on this spot, their lives tragically taken in circumstances beyond their knowledge and control. Although their remains were long gone, that didn't stop the wreck from serving as their grave site. And like any grave, especially one for so many people, it deserved to be marked and remembered eternally.

At 7.55 a.m., I announced over our communications system that the dive was over and for the recovery of *Magellan* to begin. In the ROV's left manipulator arm was my makeshift scoop, filled with a nice helping of deep-sea sediment. It was our very last act before leaving the bottom, and the first thing I checked when *Magellan* was safely back on deck: that its contents hadn't all washed away during the long ascent. To my relief there was still a good amount of the sticky greyish-green mud clinging to the inside of the plastic bottle when I slipped it from *Magellan*'s grip. I took the bottle to the nearest sink and began filtering away the mud to see what was left. Like a prospector panhandling for gold, my eyes widened when sparkles of iron ore fines began to appear in my hands: the first physical proof that *Derbyshire* was found. But there was more.

A pungent smell and a black film covering my hands was evidence of the most remarkable and unexpected discovery. Mixed with the iron ore fines were small clumps of *Derbyshire*'s fuel oil, carried to the seabed with her wreckage. As the oil stained my fingers, its significance was not lost on me. In addition to being the marker that had led us to *Derbyshire*'s resting place, it was now part of the proof we needed to establish that fact beyond doubt. There was also no doubt we had an amazing story to tell the DFA and all the interested parties when we got back to London.

The meeting organized by the ITF to brief representatives of the DFA and the government about the expedition's findings was an important moment in their campaign to get the formal investigation into *Derbyshire*'s loss reopened. It was scheduled for the afternoon of 5 July, which gave me plenty of time to get back to the States and to carefully analyse the sonar imagery and video we had taken of the wreck. In the meantime, the news of the wreck's discovery had been splashed across the front pages of all the national newspapers, and there was mounting political pressure on the government from the opposition Labour Party to reopen the investigation without further delay. The ITF's expedition had raised the level of public interest and debate about the appalling rate at which bulk carriers were sinking (nine had already been lost in the first half of 1994, with 120 seamen missing and presumed dead). However, no one was prepared for the bombshell I was about to drop at the briefing.

ITF's meeting room was full to capacity when I began my presentation of the sonar analysis. Only Mark Dickinson, who I had met the day before, had an inkling of what I was about to say. I began with some basic facts, chief amongst them my count of 115 individual sonar targets, which to qualify had to have at least one dimension greater than, or equal to, five metres. The large number underscored the surprisingly high fragmentation of *Derbyshire*'s hull. I spent some time talking about target #63, which was the bow fractured at frame 339, and how the video we had shot of it matched the sonar imagery. Then I passed around a close-up sonar image of target #9, which in my opinion was potentially the most significant of all. I firmly believed this target was the stern of *Derbyshire*, and that it appeared to be fractured at or near the notorious frame 65.

As everyone slowly took in what I was saying, I watched their faces to get a hint about how they were reacting to this new and unexpected information. Paul Lambert of the DFA was the first to ask me questions. As he did, I could see his eyes beginning to well up. He was visibly shaken by what I had just said, which surprised me at first, though on reflection after the meeting, the reason was obvious. In telling Paul that target #9 was the stern of *Derbyshire*, which of course incorporated the accommodation and working spaces of the ship, I was essentially telling him where his brother was buried. I felt stupid for not realizing how sensitive this information would be for a family member to hear, and regretted not sitting down with Paul beforehand. The other surprise of the meeting was that the government representatives present – the chief and deputy chief inspectors from the UK Marine Accident Investigation Branch (MAIB) and a pair of sonar experts they'd brought along as impartial advisers – readily agreed with my interpretation that target #9 was probably *Derbyshire*'s stern. Was this an indication that the government was finally interested in getting to the bottom of the vessel's loss?

The ball was now firmly in the court of the MAIB, who were tasked to provide recommendations to the Secretary of State for Transport about what to do next. They asked that I share the sonar images we had collected with another geophysical survey company in order for them to conduct an independent review of my interpretations, which I was more than happy to do. Meanwhile the DFA, the ITF and the shadow minister for shipping all called for an immediate return to the wreck site so that more evidence could be gathered. Cost estimates were prepared for the MAIB, but in the end no action was taken and the year ended with little progress being made.

In February of 1995, Mark Dickinson and I were invited to make a joint presentation to the all-party Parliamentary Maritime Group (PMG) in the House of Lords about the results of the ITF search expedition. Judging by the crowds of MPs in the oak-panelled room where our talk was held, the fate of the *Derbyshire* was still of great interest to politicians on both sides of the aisle, although it was no longer the front-page news it had been the year before. Mark began the presentation with an introduction of the ITF's objectives, which addressed a number of concerns about bulk carriers in general in addition to the specific aims of the *Derbyshire* search project. When he finished speaking, a video behind him showed the names of recent bulk carrier casualties involving large losses of life. For an audience concerned with shipping issues it had to be very uncomfortable viewing, and they sat in silence for the few minutes it took for the long list of names to be played out. I'm sure my presentation was equally sobering, as it would have been the first time any of them had had a chance to see the full set of images and video footage documenting *Derbyshire*'s fragmented state. How much influence the PMG and any individuals present might have in the process was hard to say, but it definitely felt like the long period of silence and inaction from the government was coming to an end.

I knew this better than most, as I had been asked to a secret lunch meeting the previous afternoon with Frank Wall, head of shipping policy at the Department for Transport. As the government's most senior civil servant dealing with shipping issues, Frank was responsible for formulating the department's overall plan regarding *Derbyshire*. He wanted to speak with me privately and confidentially about the various technical options and likely costs for a follow-up investigation of the wreckage. As if to impress upon me the confidential nature of our meeting, he chose a restaurant outside central London and made sure we sat in an out-of-the-way corner on the upper floor. Despite the unusual nature of our meeting, I was pleased to see that Frank had no secret agenda. His main question to me was perfectly reasonable and straightforward: what benefit would come from a second investigation of the wreckage? Essentially he wanted to know the likelihood that a second investigation would result in a definitive answer as to what happened to *Derbyshire*. My answer was that we couldn't give full assurances, but I believed there was a very good chance – in the order of 80 per cent – that the ultimate cause of the loss could indeed be determined.

Two weeks after my meeting with Frank Wall, his department acted. The announcement was left to Dr Brian Mawhinney, the Secretary of State for Transport, but the recommendation had clearly come from Frank and probably to some extent the MAIB. Former judge and Master of the Rolls Lord Donaldson of Lymington was to carry out an assessment into several questions, including 'What further work would be needed in order to learn more of and, if possible, make a judgement about the cause of the loss of the *Derbyshire*; the probability that the cause could be determined with reasonable confidence; and the likely costs to be incurred.' I wasn't surprised to see these were the exact same questions Frank had put to me, as they were the obvious ones to be asked in any assessment. However, there was an additional question, which I knew would be controversial and would upset the *Derbyshire* families: 'What benefit to ship safety would be secured if the cause of the loss of the *Derbyshire* was established?' and would the likely costs of determining this be justified. This last bit about justifying the costs infuriated the DFA, as it was essentially putting a price tag on the lives of their loved ones, and potentially on the lives of seafarers currently serving on bulk carriers.

In the meantime, the DFA were waging a public campaign to bring broader attention to their plight. Reorganized as the *Derbyshire* Action Group, they distributed 2,000 postcards for people to send to Dr Mawhinney urging him 'to authorize a new expedition and inquiry'. When Mawhinney's announcement about the new assessment didn't include a commitment to a return to the wreck site, they felt let down again. Their anger at the government's repeated mishandling of the case was evident in a Channel 4 television documentary entitled *Total Structural Failure* that aired two days later. Reported by Rory Maclean and produced by Rob White, the award-winning film gave relatives like Paul Lambert and Peter Ridyard a national platform to tell the story of their loss and their frustration with the government's inaction and perceived blocking of their efforts to determine the truth.

Rory and Rob's hard-hitting investigative report also uncovered questionable practices at the Haverton Hill shipyard where *Derbyshire* was built. These practices included forcing seriously misaligned parts of the ship together with jacks and large wedges that would have saddled the hull with inherently dangerous weak points. Finally, the documentary suggested that the government's reluctance to investigate *Derbyshire*'s loss in any greater detail was because they could face massive payouts to the insurers who had initially covered the £25 million loss, as in the process of denationalizing Swan Hunter in 1986, the government had retained liability for ships previously built by the company.

Lord Donaldson's report wasn't made public until December 1995, adding yet another year onto the long wait the DFA was made to endure in their fight for the truth. To be fair to Donaldson and the two technical assessors assisting him, Professor Douglas Faulkner and Mr Robin Williams, the assessment was wide-ranging and considered all possible causes of the loss including some not addressed during the 1989 investigation. The possibility that *Derbyshire* could have been hit by a 'rogue' or steep-sided abnormal ocean wave while she was hove to in the dangerous semicircle of Typhoon Orchid featured prominently in the report. In a separate paper Faulkner presented the scientific basis for such abnormal waves and calculated that there was a 60 per cent probability that the extreme wave height during the storm was in the range of 30–35 metres. A wave this high hitting *Derbyshire* would have resulted in green-water[*] loading of 11.6 metres over the eighteen MacGregor side-rolling hatch covers fitted to the ship's nine cargo holds. As if this wasn't a terrifying possibility by itself, the static design pressure of these same hatch covers was only 1.75 metres and the ultimate collapse load less than four metres. Donaldson found this information 'personally quite astonishing'. As for the frame 65 theory of failure, he concluded that this 'cannot be ignored and should be regarded as being a serious possibility'.

On the central question of whether the wreck site should be re-examined, and whether the expenditure of funds was justified – with respect to the probability that the cause of the loss could be determined with reasonable confidence – Donaldson was unequivocal, concluding that 'the case for expenditure is not only made out, but is compelling'. Nevertheless, he cautioned that there could only be one re-examination, 'once and for all', because of the large costs involved, which he estimated to be in the order of £2 million. In setting out the case for such a re-examination, Donaldson made an interesting argument comparing the way the aircraft industry designs new planes against the design of new ships. Whereas meticulous research and testing goes into the proving of each new prototype aircraft, at vast expense, no such degree of research goes into new ship designs. The shipbuilding industry is far less precise and doesn't concern itself with theoretical weaknesses. What concerns it most is proved failure. But how can failure be proved without a serious commitment of time and money to investigate accidents, as is routinely done in the aircraft industry? For these reasons Donaldson felt it was 'particularly important to make every possible effort to prove the cause of the loss of the *Derbyshire* at the very least to show that some of the theoretical causes are so likely to be the actual cause, whether singly or in

---

[*] The term 'green water' describes the actual solid seawater on the deck, as opposed to spray and foam.

combination, that the regulators who govern ship design will be obliged to take notice'. The DFA couldn't have put it any better.

〰〰〰〰〰〰〰〰〰〰〰〰〰〰〰〰〰〰

Throughout 1995 I continued to support both sides in their respective efforts regarding *Derbyshire*. It helped that despite my work for the ITF, I was seen as being independent, with no agenda other than wanting the true cause of *Derbyshire*'s loss to be determined. I advised Paul Lambert and John Jubb of the DFA and featured in the Channel 4 documentary, but equally I met with Professor Faulkner and Robin Williams to answer their questions and provide the cost estimates used in Lord Donaldson's assessment. However, by the end of the year I had left Oceaneering to join Blue Water Recoveries Ltd, a private salvage company formed to recover cargoes, mainly in the form of refined non-ferrous metals, lost in merchant ships sunk during both world wars. While I no longer had an active role in the investigation of *Derbyshire*'s loss, I watched with great professional and personal interest as the government came under serious pressure to provide definitive answers to the public and the DFA.

The first indication of how Faulkner and Williams were to conduct the government's investigation of the wreck was made public at a colloquium organized by the Royal Institution of Naval Architects in mid March 1996. The list of participants at this invitation-only meeting showed that interest in *Derbyshire*'s loss was at a much higher level since the Channel 4 documentary was broadcast. It included a wide spectrum of organizations and maritime experts with huge vested interests in the *Derbyshire* case specifically, and in naval construction and shipping safety in general. Amongst the audience of engineers, metallurgists, shipping companies, maritime lawyers, classification societies, MPs and civil servants from multiple ministries were the two staunchest members of the DFA: Paul Lambert and Marion Bayliss.

Marion's husband Curly, the ship's chief engineer, had already given notice that this was to be his last trip when the *Derbyshire* sank. Angry at the loss of the new life she had planned with him running a children's home together, and at what she perceived as the government's obfuscation of the truth, Marion sat through most of the initial formal investigation in 1987, but this made her even angrier and eventually quite ill. I hadn't met her before, but when Paul Lambert introduced us at the colloquium, the first thing she said was 'Can I give you a kiss?' She then told me how she had travelled to Japan after we had found the wreck in order to be as close as possible to Curly's grave. The visit allowed her to say goodbye and to rid herself of the oppressive guilt she had felt since 1980. 'I felt so much lighter suddenly,' she said. 'When I went to Japan I was part Marion, part Curly. I realized I had been carrying him on my back for fourteen years. Now I was Marion again, and Curly was part of my past.'

Marion's affectionate embrace was the one moment of emotion in an otherwise sobering day of technical discussions led by Faulkner and Williams after an opening address by Lord Donaldson. I was pleased to see that the technical assessors had decided to investigate a series of thirteen loss scenarios that, in theory at least, would cover every possible failure mechanism, including structural failure at frame 65. Their systematic and comprehensive approach indicated a level of commitment by the government that had been previously lacking. Lord Donaldson and the assessors also made a point of paying tribute to the DFA for their perseverance and persistence, which they believed had 'already made contributions to the cause of marine safety by maintaining the issues in the public eye'. Above all, it was Paul and Marion's presence that reminded everyone present of the human cost that needed to be at the forefront of the government's work into the loss of *Derbyshire*.

If the prevention of future losses was seen as a potential benefit of the investigation, a report that came out the following week from the colloquium on the latest bulk carrier losses underscored the scale of the problem. The numbers made for grim reading. Dry bulk carrier losses in 1994 were 50 per cent up on the previous year, with fifteen of these massive ships being actual or constructive total losses. More than 70 per cent of all the losses due to leaks, plate failures and outright disappearance, like *Derbyshire*, occurred in heavy weather. The number of seafarers killed was 141. An opinion piece that ran in *Lloyd's List* the same day summarized the sickening situation, declaring that the continuing bulk carrier losses 'remain a maritime scandal and a source of bewilderment to those who cannot understand how these very large ships can sink like stones'.

∞∞∞∞∞∞∞∞∞∞∞∞∞∞

Because of concerns about the seabed conditions where *Derbyshire*'s wreck was located, the government decided to conduct their investigation in two separate phases. The first phase was a short visual and sonar survey carried out in late July 1996 to ensure the conditions were suitable for the comprehensive photographic survey that was to be a major component of the second phase. Basically the government wanted to be sure that the wreckage wasn't buried in mud and that the water clarity was ideal for photography before proceeding with the second, far more costly phase. My former company, Oceaneering International, was again hired for this first survey, in large part because they already had a ship and deep-water ROV system in the region, thus saving the normally high mobilization costs.

Although I was told by one of the assessors that the survey was far from smooth, with disagreements between the various government representatives on board, the official report spoke of a successful outcome. Most importantly, they confirmed that the conditions for Phase II were suitable and that target #9 was indeed *Derbyshire*'s stern. They were also able to eliminate one of their thirteen loss scenarios, loss of the ship's rudder, as it was still bolted firmly in place. This scenario stemmed from another of *Derbyshire*'s sister ships – renamed the *Ocean Mandarin* – having to be sold for scrap in early 1994 after losing its rudder.

The small number of relatives who were shown extracts of the video footage were understandably upset by images of the stern, with the words *Derbyshire* and *Liverpool* (the ship's port of registry) still vibrantly painted on the transom. The DFA had campaigned long and hard for exactly this type of physical proof to be produced, but they would not have been prepared for the cold, hard reality of seeing what in effect was the grave site of their husbands, brothers and other family members.

Despite the passage of time, the emotions of many relatives were still palpable and raw, as without a body to bury, the grieving process had never really ended.

Their grief and anger was exacerbated by the treatment they felt they were receiving from their own government, even after the Department for Transport agreed to fund the new wreck investigations recommended by Lord Donaldson. The DFA considered that they themselves were responsible for a long list of evidence and technical achievements, including the actual location of the wreck, without which none of what the government was proposing in the form of an investigation would have been possible. Perhaps naively, they were expecting full transparency from the government and to be treated almost as technical partners during the investigation, invited to formally participate in the planning, execution and interpretation of the wreck surveys. They argued for DFA representation on board the survey ships, but their request, backed up by a Labour MP and the powerful Rail, Maritime and Transport Union, was rejected out of hand. The government's justification for their unsympathetic treatment of the DFA was that because the investigation results could have a direct bearing on legal proceedings in which the DFA had a vested interest, they could not be party to the investigative process. Of course, this position conveniently ignored the fact that the body with the greatest vested interest was the UK government itself, with a potential liability of upwards of £25 million.

Despite lobbying hard in Parliament and the press, the DFA did not get a representative on board the government's Phase I survey, nor would they be allowed on board for Phase II, which was scheduled to take place in the summer of 1997. Unbeknownst to anyone, the contractor for the Phase II investigation was selected even before the Phase I survey was conducted, and it was a surprising and clever choice. The Woods Hole Oceanographic Institution (WHOI), based in Massachusetts, was famous for having found the wreck of the *Titanic* and was a world leader in conducting science in the deep ocean. The engineers in WHOI's Deep Submergence Laboratory (DSL) were also pioneers in the development of innovative and cutting-edge underwater imaging and lighting technology. Having been a defence fellow at the nearby Massachusetts Institute of Technology, which shares joint programmes with WHOI, Douglas Faulkner was well aware of WHOI's capabilities and their previous work on deep-water shipwrecks under the leadership of Robert Ballard. Even though Ballard had left WHOI by the time of the Phase II survey and had earlier predicted the wreck would never be found, the team at DSL were clearly the best equipped to handle the key requirement of having to photograph the entire wreckage field and to digitally construct a single photomosaic from the many thousands of individual images.

There were other advantages to the selection of WHOI. First, they were the only organization that could take the project on effectively a lump-sum basis for the £2 million budget the government and the European Union had earmarked. Second, the direct award eliminated the need for a time-consuming and potentially controversial tendering process, which would have drawn fierce competition from several offshore contractors. Finally, and most importantly, with WHOI the government was getting an organization that would essentially be a technical partner operating under a memorandum of understanding with the US National Science Foundation (NSF), where the focus of both parties would be on achieving the best possible results and scientific analysis. Oceaneering might have felt the contract award to WHOI was a stitch-up – and if I'd still been working there I'd have felt exactly the same – but in truth no off-shore contractor could compete with the incredible value WHOI offered the government.

The Phase II survey took place during March and April 1997 from the US research vessel *Thomas G. Thompson*. It produced a huge amount of exceptional, high-quality data and imagery and was rightly claimed to be the 'most comprehensive photographic survey of a sunken vessel ever undertaken'. It set a new benchmark for wreck investigations and provided the three technical assessors (Dr Remo Torchio, an Italian naval architect, was added to the team as the European Commission had contributed a major portion of the £2 million budget) with virtually every piece of information they would need to determine the cause of *Derbyshire*'s loss. About the only disappointment was that WHOI failed to collect a physical sample of wreckage for the study of fracture surfaces because the hydraulic pump on their ROV could not produce sufficient power to run their cutting devices.

The major technical achievement of the survey was the photomosaic of high-definition electronic still images covering 98 per cent of the 1.5 km × 1 km wreckage field. A total of 135,774 individual images were used in the photomosaic, with resolution better than 5 cm; that is the equivalent of having a photograph of an area roughly half the size of the City of London but still being able to pick out an object as small as a quail's egg. Unlike modern-day aircraft investigations, where every piece of wreckage can be recovered and pieced back together on frames, the only way this could be achieved with the 2,000-plus pieces of *Derbyshire*'s wreckage spread out over an area the size of 165 football pitches was by taking the 135,774 photographs and electronically stitching them together into composite images. The navigational control and computer processing required to construct the 119 mosaics published in the UK/EC report was extremely impressive and completely justified the government's selection of WHOI. Today, such is the advance of technology, anyone with a decent digital camera and some free software can make their own mosaics with the push of a single button. However, that shouldn't take away what WHOI achieved nearly twenty years ago with their camera dangling from a cable suspended off the back of a heaving ship, 2.6 miles deep with zero ambient light.

Now that WHOI had delivered big-time, it was up to the three assessors to sift through the evidence and find answers. They got off to an awkward start, though, when in October 1997 Douglas Faulkner resigned his position. While personal reasons were cited, the reprimand he had received from the government for co-authoring a technical paper on hatch cover failure is probably closer to the truth. While it was no secret that both Faulkner and Williams believed that the failure of hatch covers played a central role in the loss of *Derbyshire*, for one of the assessors to publish a paper on the subject was seen as calling his impartiality into question. This perceived presupposition of the results of the investigation angered Faulkner's political masters and gave more ammunition to the DFA, who continued to complain bitterly about the assessors' lack of objectivity and not being allowed to conduct their own independent assessment of the Phase II survey results.

By the time the report was made public in March 1998, signed by both Williams and Torchio, the DFA had lost all hope of being told the truth about *Derbyshire*'s loss, which they still believed was precipitated by deck plate cracking in the area of the infamous frame 65. In the executive summary of the report, the two assessors went through each of the thirteen loss scenarios in order. First on their list was deck cracking at frame 65: 'Forms and patterns of plating failures and fractures, implosion/explosion damages to adjacent compartments fore and aft of Bulkhead 65 and the integrity of the cofferdam structures (bulkheads 64/65) show that this scenario was not the cause of the loss.'

Two more loss scenarios (deck cracking at mid-section and torsional weakness) were ruled out before the assessor's view about hatch cover collapse was summarized:

All the hatch covers were found within the wreckage field and indications are that all had been firstly driven into the holds by external pressure. No. 1 hold starboard was destroyed by dynamic impact.

Hatch cover failure was the final consequence which caused the loss. The slow filling of the bow prior to their failure indicates that this was not however the initiating event.

Despite their technical language, what the assessor was saying was crystal clear. *Derbyshire* sank not because of a break at frame 65, but because all the hatch covers failed, which would have filled the entire hull with water. No matter how large or strong the ship might have been, if its hull was flooded it would sink.

The report continued with elimination of the next two loss scenarios (hatch cover attachments and foredeck corrosion and fracture) before the crucial initiating event was finally revealed. As if the whole eighteen-year saga of *Derbyshire*'s loss wasn't controversial enough, the next two paragraphs in the report would really pour fuel on the naked flames.

> The bow was mostly flooded prior to the sinking which reduced the forward freeboard and made the vessel sluggish to rise to the oncoming waves. This resulted in considerably increased wave heights and dynamic pressures on the forward cargo hatch leading to its failure.
>
> This was the initiating event of the loss and this scenario describes the loss. The subsequent hatch cover failure was the final consequence causing the loss of the vessel.

The final six loss scenarios were all ruled out before the assessors provided a more descriptive explanation of what they believed happened in September 1980 as the captain and crew of the *Derbyshire* were riding out the extreme sea conditions caused by Typhoon Orchid. Without the benefit of surviving witnesses, they postulated a scenario in which the entire bow of the ship forward of no. 1 hold became flooded with seawater, starting first with the bosun's space and then the fore peak ballast tank. The mass of water caused *Derbyshire*'s bow to go down by some 2.5 metres, which in turn made the ship slower to rise to the massive waves during the storm. Because of the great length of the ship and the poor visibility caused by spray over the deck, the captain would not have been aware that the bow was riding lower, and in any event it would have been too dangerous for the crew to investigate further. At some point the ship was hit by a wave that was higher and steeper than all the others, made worse because the freeboard of the flooded bow was reduced by 2.5 metres. This wave destroyed the starboard hatch cover, causing the cavernous space of the no. 1 hold to fill with 10,000 tonnes of water in less than a minute. At this point, with all forward freeboard and reserve buoyancy lost, the ship was essentially doomed to sink.

The forward end of *Derbyshire* was now so low in the water that waves no longer broke on the ship; they simply rolled up the deck as far as the bridge, as if up a shallow beach. So much water was riding up the length of the ship that the hatch covers, starting with no. 2 and progressing all the way to no. 9, collapsed sequentially from the sheer weight of water on top of them. As the hatch covers collapsed one by one and each hold was filled with water, the bow was driven deeper and the ship started to sink. At certain depths all the confined watertight spaces integral to the design of *Derbyshire*, like wing tanks and the double bottoms, imploded due to the hydrostatic pressure of the surrounding water. Like a balloon being squeezed until it burst, the implosion of these spaces released shock waves that caused even more damage to nearby structures. The enormous forces generated by this implosion/explosion behaviour were what caused the near total fragmentation of the ship. One calculation in the report estimated the force that ripped the hull apart as equivalent to more than twenty tons of TNT. The report further emphasized that the implosion/explosion damage was a consequence of the sinking, not the cause.

As for the cause of the sinking, or what in their view was a key initiating event, the assessors left no doubt about who was at fault: 'The unsecured foredeck stores hatch cover is carried away and the Bosun's store space fills with every wave over the bow probably being full in less than 1 hour.'

With these thirty words, Williams and Torchio were clearly apportioning at least some of the blame for *Derbyshire*'s loss to her crew. That the foredeck stores hatch cover was knowingly unsecured – essentially left open by the crew – was an inflammatory and highly damaging accusation that implied serious professional negligence. The assessors must have realized that by concluding that 'the hatch cover was not securely closed and sealed by its toggles', which they presented in their report as a finding of fact, they were effectively impugning the reputation of *Derbyshire*'s officers and crew. That all the men were dead and unable to defend themselves only made matters worse for the DFA and upset the relatives to no end.

The many positive aspects of the assessors' report, in particular a lengthy list of recommendations for improved ship design and safety, were lost amongst recriminations by the DFA about a pattern of ignorance by the government and what was viewed as the patent bias of the assessors to the 'hatch cover loss scenario'. Captain Dave Ramwell, a master mariner who ran ships in and out of the Mersey and who co-authored an investigative book that first detailed the frame 65 structural issues with the Bridge class of ships, called the latest official conclusion 'predictable and wrong'. Despite these objections about the report, it and the Phase II survey results were deemed to be 'new and important information which could not be produced at the original investigation'; in accordance with the Merchant Shipping Act 1995, this allowed the deputy prime minister John Prescott to order the formal inquiry to be reopened. Prescott gave interested parties three months to submit representations on whether it should be reopened in full or in part, and whether it should be held before a wreck commissioner (as with the 1987 inquiry) or in the High Court. Once again the DFA were told to wait.

It took until 17 December 1997 before they got their answer. The formal inquiry would be reopened in full and held in the High Court. The attorney general would prepare the case and the government would contribute to the relatives' legal costs. At last the DFA were to have their day in the country's most senior Admiralty court, with proper legal representation.

∞∞∞∞∞∞∞∞∞∞∞∞∞∞∞∞∞∞∞

The one thing the DFA and other relatives had learned was that justice moved slowly. The formal inquiry (FI), to be heard by the Honourable Mr Justice Colman, didn't get under way until 5 April 2000. The parties to the inquiry included Swan Hunter, who built the Bridge class of ships; Bibby Line, who owned and operated *Derbyshire*; Lloyd's Register of Shipping, the classification society where the ship was entered; the Department for Environment, Transport and Regions, and the DFA. All parties had vested interests in the outcome of the case and thus were individually represented by their own Queen's Council barrister.

In his opening address Justice Colman set the tone for the rest of the hearing when he emphasized that the inquiry was about seeking the truth and was not a trial of anyone's civil liability. Part of the reason for the long delay before the hearing started was because the various parties and their expert witnesses had been instructed to cooperate in order to reduce technical disagreements. Colman encouraged this cooperation, and in the common purpose of seeking the truth he dispensed with the wigs and gowns normally worn by lawyers and judges in the High Court, saying that 'they have nothing to contribute to non-confrontational proceedings'. His decision to rid the court of its most symbolic garments would have immediately eased the tensions of the many family members who packed the public seating areas. Those who attended the 1987 inquiry would have remembered it as a frustrating affair during which important evidence was ignored and the frame 65 theory was under constant attack by the lawyers for Lloyd's and Swan Hunter.

The inquiry lasted for fifty-four days and heard oral testimony from twenty scientific and technical experts, six master mariners and five other witnesses. Robin Williams featured far more than any other witness, appearing for ten days of the hearing, and was at the centre of the most contentious and controversial testimony, involving the allegedly unsecured fore peak hatch cover cited by the assessors as the main reason for seawater flooding the bow spaces in the first place. Williams' opinion was that the hatch couldn't be secured in the normal way because someone had stuffed a mooring line from the port windlass down it, preventing the lid from being tightly closed. Instead of securing the lid with the eight hatch toggles, the crew, according to Williams' study of the wreck photographs and video, had attempted to lash it down with some heaving line tied between two toggles, with the line running over one corner of the lid. He also used the wreck photos to bolster his opinion that the toggles weren't tightened, as several appeared to be in a loosened position.

Williams didn't need to spell it out; it was crystal clear in his reports and testimony that he believed *Derbyshire*'s crew had been negligent in the extreme by knowingly leaving a hatch so poorly secured, especially in light of a typhoon being forecast. The DFA's barrister explained in his opening statement how much distress this had caused the families, especially the wives of the men whose job it had been to look after such things. It was particularly unfortunate that this accusation had been publicized in the press and on television from the release of the assessors' report in March 1998 all the way up to the present day. He wanted it to be clear from the outset of the hearing that 'The DFA's position is that to suggest such a course of conduct by experienced and competent seamen, with ample notice that they might be affected by typhoon conditions, is nothing short of preposterous.'

Throughout the course of the hearing Williams' dogged stance about the hatch lid grew weaker and weaker as other experts stepped up to give opposing testimony and factual evidence was submitted to prove that his reading of the wreck photographs was wrong. The first leg of his position was conclusively kicked away when it was shown by the DFA's technical experts that the mooring line emerging from the hatch was not attached to the port windlass. Several experts identified it as a four-square plaited polypropylene mooring line (only nylon mooring line was used on the ship's windlasses) that was stored in the bosun's space and floated out of the hatch when the ship sank. The second leg was lost when it was shown that the toggles were deformed when a heavy object struck the hatch during the sinking process, also causing the lid to be ripped away and one side of the coaming to be stoved in. Williams' reliance on the apparently loosened position of the toggles evaporated as more experts testified that they were prone to work free in severe weather and could even have rotated and unwound as the ship plummeted to the seabed.

The most authoritative and devastating repudiation of Williams' position came from Captain Nigel Malpass, a former chief officer on *Derbyshire* who also served on two of the other Bridge-class vessels and ultimately rose up the ranks to become the managing director and chairman of various Bibby Line companies. In an expert demonstration of his knowledge of seamanship, Mr Malpass (he spurned his title of captain) showed the court how Bibby Line crew normally applied additional lashings to hatch lids when severe weather was expected. The complicated lashing was called a 'cat's cradle', and once secured to all the tightened toggles and over the top of the closed hatch lid it acted to prevent the toggle nuts from working loose. He identified the fragment of line attached to two of the toggles as the heaving line they would use to make the cat's cradle. His demonstration of tying the cat's cradle on a half-scale model of the foredeck access hatch brought into the courtroom was termed 'very impressive' by Justice Colman. In his written statement, Mr Malpass left no doubt how absolutely wrong Williams was about the hatch lid being left unsecured, and about the professional conduct of *Derbyshire*'s crew.

> Mr Williams suggests that the mooring rope may have been draped over the coaming, and that the hatch lid was then dropped down on the rope and the hatch lid only secured by the loop of heaving line drawn across the corner of the hatch lid between the two toggles. I find it hard to believe that such a statement could possibly be made by anyone who has the remotest practical knowledge of sea-going life. I cannot accept that Mr Williams really believes that on the basis of a typhoon being forecast a crew member on any ship, let alone a well-run ship of this sort, would go forward to secure a hatch cover and leave a rope coming out over the coaming so that by definition the hatch could not be made watertight and thereafter just secure it on one corner.

Remarkably, Williams stuck fast to his implied position that the crew did not secure the foredeck hatch, even though he accepted he was wrong about the mooring line coming out of the hatch. It was left to the DFA's barrister, Andrew Moran QC, to finally get him to grudgingly agree, after repeated questioning, 'that it was more probable than not that the crew did fully secure the forward stores hatch'.

The final word on the matter was left to Justice Colman, whose report was made public on 8 November 2000. Some criticism was reserved for Oceanroutes, the routing agency hired to provide information to *Derbyshire*'s captain about the track of Typhoon Orchid and advice on the best route to avoid the worst weather. As to the cause of *Derbyshire*'s loss, Colman concluded with 'reasonable confidence that the initiating event was destruction of ventilators and air pipes on the foredeck' by waves breaking over the bow, allowing water to slowly flood the 'bosun's store, machinery spaces and probably the ballast tank in substantial quantities'. With the vessel trimmed by the bow, the sequence of events detailed in the assessors' report – including the destruction of hatch no. 1 by the overloading of green water, the flooding of hold no. 1 and then the successive collapse of hatch no. 2 and no. 3 and the flooding of these holds – caused it to sink. As for the conduct of *Derbyshire*'s crew, Colman was unequivocal in clearing them of any blame.

> The condition of the store hatch as found does not suggest that the lid was left unsecured by the crew or that the lid could not properly be closed because a rope was protruding from the hatch. On the whole of the evidence the lid was adequately secured both by properly tightened toggles and by a complicated roping device designed to prevent the lid coming loose because the wing nuts had ridden up the toggle shanks with the motion of the ship. The rope seen in the wreckage to emerge from the hatch was a mooring rope one end of which was originally attached to the inside of the hatch lid. This Report rejects the Assessors' conclusion that the crew had left the hatch lid inadequately secured prior to the *Derbyshire* entering the typhoon.

The DFA accepted the conclusions of the court.

During the twenty years the *Derbyshire* case rumbled on, the shipping industry was forced to look inwards and adopt some changes to improve the safety of ships, in particular bulk carriers. For example, in November 1997 the International Maritime Organization (IMO) adopted new rules for bulk carriers of 150 metres and upwards whereby the bulkhead and double bottom must be strong enough to allow the ship to survive flooding in no. 1 hold unless loading is restricted. A major portion of Colman's report was devoted to ship design regulations and standards; the most important concerned the relationship between cargo hatch cover strength (thought to be seriously deficient) and the height of bow freeboard, which Colman said should be totally re-evaluated. New amendments for specially strengthened hatch covers to be fitted to the forward holds of all new cargo ships (not only bulk carriers) came into force internationally in 2005. His report also included a list of twenty-two recommendations for the improvement of ship safety, nearly all of which have been adopted in new-build ships.

Despite resigning as an assessor, Professor Douglas Faulkner did contribute to the assessors' report and gave testimony at the inquiry. He also devoted considerable effort to studying *Derbyshire*'s loss and published numerous papers including his own independent assessment of the sinking of the ship. He was firmly of the opinion that hatch cover failure was the only cause of the loss and that *Derbyshire* would have sunk because of this whether the bow was flooded or not. His articles also criticize Robin Williams for his inflexible adherence to bow flooding as the initiating event in the overall sinking scenario.

The Derbyshire Families Association was awarded the Marine Society's prestigious Thomas Gray Silver Medal in 2004 for their tireless campaign to improve the safety of bulk carriers. Paul Lambert, who was chairman of the DFA and afterwards worked within the IMO on new safety regulations for bulk carriers, was awarded an MBE in the Queen's birthday list in 2010 for services to maritime safety.

Although bulk carriers and other large cargo ships still sink every year, and such are the inherent dangers in shipping that this will always be the case, the rate of losses has diminished over time, in large part due to the regulations and safety standards that have come into force both indirectly and directly as a result of the investigation of *Derbyshire*'s loss. I have been involved in finding arguably more famous ships, but for this simple reason the MV *Derbyshire* is the most significant.

III

HMS *Hood* and KTB *Bismarck*

SEARCH FOR AN EPIC BATTLE

HMS *HOOD*

HMS *Hood*	KTB *Bismarck*
SUNK 24 MAY 1941	SUNK 27 MAY 1941
1,415 died	*2,131 died*
3 survived	*115 survived*

One thing I have learned to expect after every wreck I successfully locate is for someone to ask me what I'd like to find next. When Rob White put this question to me at a party in March 1995 to celebrate the broadcast of the award-winning Channel 4 television documentary he'd produced about the loss of *Derbyshire*, my immediate answer was 'I'd like to find the wreck of HMS *Hood*.' I had no idea at the time how much this simple statement would change both our lives and bind us together in a twenty-year journey to honour the memory of this extraordinary ship and the 1,415 men who went down with her.

I will never forget Rob's reaction to my reply. His eyes widened and his mouth fell open but no words came out. When he finally regained the power to speak, he said, 'Is that possible, could you really find the *Hood*?' Having just spent the afternoon at the Guildhall Library in central London researching the loss of the battlecruiser, I did think it was possible and that there was a good reason for doing it.

The seed for finding the *Hood* had been placed in my mind by my boss and friend Don Dean, based on a casual conversation we had had about what wrecks would be interesting to find. The ship that was credited with sinking the *Hood*, the German battleship *Bismarck*, had been found some years earlier by Robert Ballard, but no one had ever attempted to search for the ship to which *Bismarck* basically owes its fame. Don and I both found it curious that of the central combatants in their epic battle, it was the *Bismarck* whose fame grew with the passing of time, whereas the *Hood*, a ship with a far more celebrated and eventful career, seemed to have been forgotten. Perhaps this was only the case in America, where there is a somewhat odd fascination with Nazi Germany, but I did get the sense that the average person in the street outside the UK would know the name of *Bismarck* before *Hood*.

I myself had little idea about the historical significance of *Hood* and what a tremendous shock it was to the country when they learned that her destruction had been so quick, so total and with such an unspeakable loss of life. Reading the contemporaneous newspaper accounts in the Guildhall was the start of my education. These reports emphasized that *Hood* – at 42,000 tons displacement – was the largest, most powerful warship afloat but that she was also an older ship completed just after the end of World War I, whereas the *Bismarck* was brand new and equipped with superior gunnery. There was also a lot of discussion about how *Hood* was sunk by an apparent lucky strike from *Bismarck* that penetrated the magazine, causing the ship to explode, and how magazine explosions also caused the loss of three British ships during the Battle of Jutland in 1916. I was especially drawn to the comments of two writers who, just a few days after *Hood*'s loss, already had diametrically opposed views about the real reason it happened.

The first writer suggested that perhaps the lessons of Jutland and the vulnerability of the magazines on British warships were not learned by the Admiralty. He surmised: 'Her loss therefore immediately raises the technical question whether a miscalculation was made, while probably leaving no evidence to assist in answering it.' This article drew an immediate and sharp rebuke in a subsequent letter to the editor, which argued that a lucky hit or technical miscalculation was not the reason for *Hood*'s sudden annihilation. Rather, claimed the writer, it was 'because she had to fight a ship twenty-two years more modern than herself'. So the debate had already begun about whether the disastrous loss of *Hood* was an unlucky or preventable event.

The instant attraction for me was whether evidence could be collected from *Hood*'s wreck to solve this debate. I wasn't to know at the time, but it didn't take me long to learn that no definitive answers to the central question about what had caused *Hood* to explode and why there were so few survivors had been put forward in the intervening fifty-four years. After finding the *Derbyshire*, I was keen to demonstrate to the shipping industry that it was feasible to investigate marine casualties to determine their cause and to use the lessons learned to prevent similar losses. It might have been an outlandish idea, but I thought if we could solve the mystery of the loss of *Hood*, perhaps we could convince more insurers and maritime authorities to take proactive action in the investigation of current shipping losses. While this motivation stayed with me right up to the search expedition I led six years later, in July 2001, to find and film the wrecks of both *Hood* and *Bismarck* on the sixtieth anniversary of their battle, it was meeting Ted Briggs and other former crew members that gave me a far better appreciation of how the search could be used to tell the human story of *Hood* through the experience of those who had sailed on her.

∞∞∞∞∞∞∞∞∞∞∞∞∞∞∞∞∞∞∞∞

The *Hood* had started life as a plan on paper for a new type of British capital ship that blurred the lines between the early-twentieth-century 'dreadnought' battleship, with its big guns and heavy plate armour, and the battlecruiser, which also had big guns but placed a priority on speed over protective armour. Her design stemmed from the transformational thinking of Admiral Sir John 'Jackie' Fisher, a modernizer and one of the most important and influential figures in British naval history. For Fisher, speed in battle was paramount and he consistently advocated it over armour as the best form of defence. His fondness for speed was readily apparent in this sharply worded letter to his successor as First Sea Lord, Winston Churchill, in 1911:

> The first desideratum is *speed*! Your fools don't see it. They are always running around to see where they can put on a little more armour! ... You hit him first, you hit him hard, and you keep hitting. *That's your safety!* You don't get hit back! Well! That's the improved 13.5 inch gun! But disassociated from *dominating speed*, that gun is futile.

The battlecruiser was thus the perfect embodiment of Fisher's maxim that 'speed is armour', and *Hood* was the evolutionary end point of this design philosophy, even though by the time the first designs were being formulated in late 1915, Fisher was no longer in charge of the Admiralty. After a change in plans, the final design was chosen in April 1916 and the Admiralty ultimately placed orders for four ships of the new Admiral class, designated *Hood*, *Howe*, *Rodney* and *Anson*. The *Hood* was the first of the four to be constructed, at the John Brown and Co. shipyard at Clydebank, but the catastrophic loss of three earlier vintage battlecruisers (HMS *Invincible*, *Queen Mary* and *Indefatigable*) at the Battle of Jutland on 31 May 1916 – including the combined loss of 3,309 men – brought the project to a shuddering halt.

Hearing of these losses, Admiral David Beatty, Commander of the 1st Battlecruiser Squadron at Jutland, is reported to have said: 'There seems to be something wrong with our bloody ships today.' While Beatty survived the numerous hits the German battlecruiser *Lützow* inflicted on his flagship HMS *Lion*, Rear Admiral Sir Horace Hood was not so lucky. As commander of the 3rd Battlecruiser Squadron, he was killed when his flagship HMS *Invincible* was hit by a German shell – also from *Lützow* – which penetrated one of its turrets, causing an explosion that literally blew the ship in half. One of the most evocative images from Jutland, which speaks to the high price the British paid in ships and men, is the photo of *Invincible*'s shattered bow and stern resting on the shallow seabed. Somewhere within her wreckage lie the remains of Sir Horace Hood KCB DSO MVO.

It is often said that the construction of *Hood* was started on the same day as the Battle of Jutland, but this is untrue. The reality is that the loss of three British battlecruisers to plunging shellfire that apparently penetrated their decks and turrets gave pause to the Admiralty's plans. An immediate redesign was necessary to correct several deficiencies that were fatally exposed at Jutland, including poor handling and flash-proofing of inherently unstable cordite, and inadequate armour protection around the vital parts of the ship. The question of how much protective armour a battlecruiser should have, and the part it played in the losses at Jutland, was still a vexing issue, and there was no unanimity of opinion within the Admiralty. The decision to potentially sacrifice speed for armour was clearly a struggle for the designers, judging from the number of changes that were made both before construction started and afterwards. By the time work began on *Hood*'s hull (designated ship no. 460) on 1 September 1916, there were increases in armour to her horizontal decks (1.5 inches to 3 inches), barbettes (8 inches to 12 inches) and angled main armour belt (also 8 inches to 12 inches), resulting in an additional 3,450 tons in weight.

When launched on 22 August 1918 by Lady Edith Hood, widow of Sir Horace, *Hood* was armed with eight 15-inch guns mounted in four turrets, twelve 5.5-inch secondary guns, four 4-inch high-angle anti-aircraft guns, and a total of six torpedo tubes placed above and below the waterline. Her design speed of 32 knots, intended to give her an advantage of several knots over any foreign capital ship, was achieved in part due to her extreme length, despite the additional weight she carried in armour and other structures. At 860 feet (262 metres), *Hood* was the longest ship in existence during her life. For those who witnessed her full-power trials at the Isle of Arran in March 1920, it must have been an extraordinary sight to see this 46,680-ton behemoth of a ship, with thick plumes of black smoke streaming from her twin stacks, ploughing her way through the mile course at a maximum speed of 32.07 knots.

In the end, whether she was really a 'fast battleship' or a 'super battlecruiser' – a debate that continues in certain circles today – mattered very little to the way she was viewed by anyone lucky enough to catch sight of her at sea or in the numerous ports she visited in the UK and the twelve countries that constituted her famous world tour in 1924. She was that rare beast: a fearsome warship bristling with deadly weaponry that also invoked visceral feelings of affection and respect. Her graceful, majestic lines were awe-inspiring and moved many a hardened man to tears. Loved by friends and feared by foes, the 'Mighty *Hood*' was quite simply the most famous and most powerful warship in the world for the next twenty-one years until the day her nemesis, the *Bismarck*, broke out into the Atlantic on 19 May 1941.

<hr />

Not long after the television documentary on *Derbyshire* was broadcast, I was back in the UK to accept a Seatrade 'Safety-At-Sea' Award for our work in locating the *Derbyshire*, and to meet with my new clients at Blue Water Recoveries (BWR). BWR was a brand-new consulting company founded to advise on the salvage of valuable metal cargoes from modern shipwrecks in deep water. It was formed off the back of the recovery of 1.3 million silver Saudi Arabian riyal coins from the wreck of the US Liberty ship *John Barry*, sunk off the coast of Oman during World War II. This groundbreaking salvage, from a depth of 2,600 metres, was the first to utilize the type of instrumented and controllable recovery tool that the CIA first developed for a highly secretive and daring attempt to retrieve the sunken Soviet ballistic missile submarine *K-129* some twenty years earlier. Like the CIA salvage, which was only partially successful, recovering the forward section of the submarine but none of its nuclear missiles, the *John Barry* project was a technical triumph but a financial loss. The team, led by the joint talents of Bob Hudson and Mark Cliff, had proved that seriously heavy cargoes could be extracted and lifted from wrecks in ultra-deep oceanic waters on a commercially viable basis. If only the 2,000 tons of silver bullion rumoured to be on board the ship had been recovered along with the 18 tons of numismatically unimportant coins, this one-off project would have been a roaring success.

Buoyed by their technical achievement, the various partners of the *John Barry* project decided to go into the salvage business on a full-time basis, and with that Blue Water Recoveries was born. For me, BWR was an important new client as they would be continuously on the hunt for multiple shipwrecks rather than looking for just one wreck at a time like other groups I had worked with in the past. I also knew the main players involved, and they knew me, as I was part of the team that had filmed the wreck of the *John Barry* and searched for its rumoured silver treasure during an earlier expedition. Although Mark and Bob were the named directors of BWR, the financial impetus came from a secretive UK-based businessman named Tim Landon. It was Tim who had primarily bankrolled and overseen the *John Barry* project on behalf of the Oman government, whose permission was required before the salvage could pro-ceed. He was an extremely close confidant and trusted adviser of Oman's ruler, Sultan Qaboos, owing to his involvement in a bloodless coup that had allowed Qaboos to overthrow his own father in 1970. He was also a natural risk-taker and liked the idea of being involved with a company working at the forefront of an exciting and innovative new industry.

This visit to the UK happily coincided with the ceremonies and events held annually to mark the loss of *Hood*. After I'd energized him with my idea to search for *Hood*'s wreck, Rob White had gone quickly to work and got us invited to both the annual dinner of the official HMS *Hood* Association and the commemoration service held the following morning in the Church of St John the Baptist in Boldre, near Lymington. The first surprise of the dinner was that Rob and I were at the top table; the second was that I was the main guest of honour, seated directly between the association's chairman, John R. Williams (JR), and its president, Ted Briggs. I wasn't expecting such special treatment but was relieved I had prepared well by visiting the National Archives in Kew the day before to consult the Official Boards of Inquiry and other records relating to where *Hood* was sunk. If I was to get the approval of the *Hood* Association for the search, especially from Ted Briggs, I needed to be super-confident that I knew where to look.

Ted's approval was crucial because he was the last living survivor of the three who miraculously escaped the ship when it sank with 1,415 of their shipmates. While Bill Dundas preferred not to revisit the tragic consequences of 24 May 1941, both Bob Tilburn and Ted joined the *Hood* Association and served as president. Sadly, Bob had died just three months before the dinner and I never got to meet him. Because Ted had served in the navy for thirty-five years and of the three lived the longest, he was the person most commonly associated with *Hood*. This meant that for every media request, church service, naval ceremony or event of any description connected directly or indirectly with *Hood*, he was called – and expected – to participate. The demands and pressure of carrying the memory of such a momentous and horrific event was at times a huge burden on Ted, and in the years to come I saw the toll it took on his health and his relationship with his wife Clare. Nevertheless, on the night of the dinner I was yet another person seeking his approval.

I felt it was fundamentally important that the *Hood* Association embraced my idea of the search and gave it their formal consent. In my mind it was the association, composed of men who had previously served on *Hood* and the relatives of the 1,415 who perished within her, who had the overriding moral authority to decide on any proposal connected with the shipwreck. Without their backing, I believed there was no way we could secure the wider level of government and public support I knew we would need to mount such a search, not to mention the large amount of funding that would be required. A lesson I had taken from *Derbyshire* was how much power the DFA families had in forming opinions and in determining the final course of events. The two searches would be for completely different reasons, but without the DFA the wreck of *Derbyshire* would never have been found, and I instinctively felt that the *Hood* Association would hold the same power somewhere down the line.

At the dinner I saw how Ted was a bit shy and reserved, whereas JR was a far more gregarious character, quick to tell a joke or make some pointed comment, generally at the expense of his ex-shipmates. There was a great camaraderie amongst the men and the evening was full of good-natured banter in all directions. The next morning it was left to JR, a senior figure amongst his shipmates in every way, to give Rob and me the first indication that the association was on our side. Standing in the sunlit churchyard at Boldre after the commemoration service, he announced that he thought our idea was a fitting way to pay tribute to *Hood* and those who died serving on her. Based on the number of heads nodding in agreement, I knew that other association members were in agreement and that we had a small but extremely significant measure of support for our plan.

It took a little longer to completely convince Ted Briggs; or more accurately, for Ted to become comfortable with the idea in his own mind. With all due respect to the others who served on *Hood*, Ted had a unique memory seared into him that made his perspective on the ship different. Whilst the others were able to remember the good times they'd had on board what was generally considered a happy ship, Ted could never forget his final view of *Hood*'s fractured bow sliding beneath the flame- and oil-covered waves at a sickening angle that told him all was lost, including the men he'd been standing next to just seconds before. His foremost concern, therefore, was a serious one born of his very personal perspective: that the wreck site was a massive war grave to be respected and left undisturbed. My own feeling was exactly the same, and I firmly rejected other people's ideas to recover artefacts from the wreck, as I knew this would be highly controversial and undermine whatever support we were beginning to build. I coined the phrase 'look but don't touch' for the policy of how we would approach any search, and I know Ted drew great comfort from my commitment to this pledge.

A month later, a fax arrived on my desk in America from Rob. It contained a letter on official HMS *Hood* Association letterhead signed by both J. R. Williams and Ted Briggs. Across the board there was approval for our plans to film the wreck for a TV documentary, conduct an investigation of the damaged hull and place a memorial plaque on the wreck on behalf of the association. The key paragraph stated: 'Having met these two gentlemen, who discussed their plans at length, members of the association wholeheartedly support the expedition to locate and film the wreck of HMS *Hood*, and are confident that this will be conducted with the utmost sensitivity and respect for the site where she lies as a war grave.'

Ted later explained his thinking in putting his name to this letter. While discussing his concerns a friend asked him, 'What's your objection about the wreck being a war grave? If you went into a graveyard and took a photograph of your mother's grave, would you regard that as a desecration?' This simple question got Ted thinking: what is the difference between a war grave on land and one at sea? Are they not fundamentally the same? People routinely visit war cemeteries and battlefields to pay their respects, and this is universally accepted as a good thing. If the technology is now available to find the wreck of *Hood*, and to use a camera to allow relatives the chance to make a connection – even in a virtual sense – to the last place their loved one was alive, surely that would be a good thing too. Once he had worked these issues through in his own mind, he became our biggest supporter.

<hr />

Looking back on the events of May 1941, it seems as though the *Hood* and the *Bismarck* were destined to meet in an epic encounter that would determine who would enjoy naval superiority throughout the rest of the war. Like two heavyweight boxers converging towards a title fight, the ships were being matched up beforehand in the minds and imaginations of their officers and crews. *Hood* had enjoyed the title of the world's most powerful capital ship for over two decades, but she was at the wrong end of her career to face *Bismarck*, the newest and most technologically advanced battleship afloat. *Hood* was still longer and just as fast, but *Bismarck* was bigger and heavier where it counted, her beamier hull making for an extremely stable gun platform, and her thicker armour (40 per cent of total weight compared to 33.5 per cent for *Hood*) able to shield her vitals from the most devastating blows. The name of the game in any big gun battle was to hit your opponent first while avoiding being hit in return, and *Bismarck* was blessed with the crucial advantage of targeting her fire very quickly and accurately.

The expectation that *Bismarck*, or her sister ship *Tirpitz*, was ready to commence operations was anxiously felt at all levels in the Royal Navy, from the chiefs and directors poring over confidential intelligence reports in Admiralty offices in Whitehall down to the ordinary ratings on ships stationed at Scapa Flow, where less informed opinions were openly discussed on the mess decks. The success of the *Scharnhorst* and the *Gneisenau* months earlier made the prospect of another German commerce raiding operation almost a certainty. Together these two battlecruisers, commanded by Admiral Günther Lütjens, had accounted for twenty-two ships and 122,000 tons of merchant shipping either destroyed or captured during Operation Berlin. Invigorated by his success, Lütjens tried to convince a reluctant Hitler that a repeat performance involving the more powerful *Bismarck*, which they had just inspected during a four-hour tour on 5 May, would inflict even bigger losses on the supply convoys criss-crossing the North Atlantic. Lütjens' clear desire was to attack the British 'wherever they showed themselves', but Hitler remained unconvinced and non-committal.

Hitler, a decorated army officer who reportedly had not seen the sea until he was over forty, was famously unsure about how to use the naval power within the Kriegsmarine (the navy of Nazi Germany). His feeble comment to Grossadmiral Erich Raeder, commander of the Seekriegsleitung (SKL), the overall maritime warfare command, when they abandoned plans to invade Britain by sea shortly after the successful blitzkrieg of Europe, summed up this lack of confidence: 'On land I am a hero, at sea I am a coward.' Although Lütjens was convinced that no single capital ship could trouble the *Bismarck*, Hitler was still wary of the numerical superiority of the British fleet. A great deal of Germany's prestige was invested in the building of the *Bismarck* and *Tirpitz*, and his fear of losing either ship appeared to paralyse him into inaction. Unbeknownst to Hitler, however, Lütjens had no such fear and had already issued orders on 22 April for Operation Rhine, the next exercise in commerce raiding. This time the newest German capital ships, *Bismarck* and the heavy cruiser *Prinz Eugen*, would be let loose into the North Atlantic to destroy merchant vessels and cause havoc to the convoy system upon which Britain relied so heavily.

Whilst the mission of Operation Berlin had been to attack unescorted convoys, the employment of *Bismarck* upped the stakes for Operation Rhine. Lütjens would now have a free hand to go after all convoys, including those escorted by British warships. The SKL directives still exhibited a high degree of caution about the use of *Bismarck*, however, warning that she should not be employed 'to defeat in an all-out engagement enemies of equal strength, but to tie them down in a delaying action, while preserving her own combat capability as much as possible, so as to allow the other ships to get at the merchant vessels in the convoy. The primary mission of this operation also is the destruction of the enemy's merchant shipping; enemy warships will be engaged only when that objective makes it necessary and it can be done without excessive risk.'

When the *Bismarck* and the *Prinz Eugen* departed Gotenhafen in the early hours of 19 May for Operation Rhine, it started the clock on the most momentous and dramatic nine days in the operational lifespan of any warship during World War II. By the time it was over, the two most feared and iconic warships in the world had been destroyed, 3,546 men were dead and the fate of big-gun capital ships had been ruthlessly exposed by a surprising adversary: the torpedo bomber. The breakout was meant to be secret – even Hitler was unaware until 22 May that the ships were at sea – in order to give the raiders the best chance of arriving in the North Atlantic undetected. But that wasn't to be, and a report by a Swedish ship that *Bismarck* was on the move was later confirmed when the German battleship was photographed at anchor in a Norwegian fjord. When Hitler was finally informed that Operation Rhine was under way, he wasted no time in asking Grossadmiral Raeder whether it was possible to recall the ships. When Raeder admitted that it wasn't, Hitler's reply was eerily prescient: 'Well perhaps now you have to leave things the way they are, but I have a very bad feeling.'

Despite Hitler's overt pessimism, the opening act of this three-part drama started extremely well; almost unbelievably so for Germany. Of the three possible routes into the North Atlantic, Lütjens chose the most northern, with his task force steaming into the Arctic Circle west of Norway, above and around Iceland and thereafter skirting the pack ice of Greenland through the Denmark Strait. Using the Norwegian fjord sighting to their advantage, the British Home Fleet laid on an ambush that relied on the speed of *Hood* and her consort, the battleship *Prince of Wales*, to intercept the German squadron before the breakout was complete. Two County-class heavy cruisers, HMS *Norfolk* and *Suffolk*, were also on the hunt, using their radars to track *Bismarck* and relay positions to the British commander, Vice Admiral Sir Lancelot Holland, who was flying his flag in *Hood*. Holland was intent on cutting off *Bismarck's* escape and forcing her into an engagement that – remembering the SKL directive – Lütjens was under orders to avoid. At 5.53 a.m. on Saturday 24 May, Holland caught up with his prey and put Lütjens to the sword.

Steaming into battle at 29 knots, the *Hood* and *Prince of Wales* were the first to fire their 15-inch and 14-inch guns, respectively, on the lead ship of the German pair, assumed to be *Bismarck*. Holland's battle plan was to concentrate their big-gun attack on the more dangerous battleship and leave the *Prinz Eugen* to the trailing *Norfolk* and *Suffolk*. At the opening range of 24,230 metres, it would have taken about 47 seconds for the British shells to strike *Bismarck*, but a misidentification meant that only the *Prince of Wales* had fired at the correct ship. What Holland didn't know was that a damaged forward radar set on *Bismarck* had forced Lütjens the day before to order a 'number change', which meant that the *Prinz Eugen*, and not the *Bismarck*, was in the lead position to scan for threats ahead.

Holland's other problem was his angle of attack. In going for an immediate strike rather than risk allowing the Germans room to slip away, he placed his ships in a distinctly unfavourable position whereby they were unable to fire all their guns in full salvos and also presented a bigger target for the enemy to hit. His only choice was to close the distance to *Bismarck* as quickly as possible, which he needed to do in any case as the *Hood*, because of her inadequate horizontal deck armour, was especially vulnerable to plunging fire at long ranges. If he could get close enough to turn on a parallel course with *Bismarck*, he would be able to fight a more even broadside battle with the full eighteen guns (eight on *Hood*, ten on *Prince of Wales*) at his disposal.

As the fire bellowed out from the British guns, and shells splashed all around his position, Lütjens seemed stunned by his predicament. No doubt he was thinking of Hitler's reluctance to risk *Bismarck* and the SKL's directive not to take on enemies of equal strength. Initially he had thought that the *Prince of Wales* was the better-equipped battleship *King George V*, which would have made his situation even worse, but there was no mistaking the *Hood*: the enemy they feared the most and the 'terror' they pictured during war games training, according to Baron von Müllenheim-Rechberg, the eldest surviving officer from *Bismarck*. No one will ever know whether Lütjens' despair was because Operation Rhine had been rumbled or because he was facing not one, but two enemies of equal strength. Whatever unpleasant thoughts were circulating in his mind caused him to freeze. Two interminably long minutes passed before it was left to Captain Lindemann to give the order to return fire, whilst muttering: 'I will not let my ship be shot out from under my ass.' Fortunately, for Lütjens' sake and the sake of his increasingly agitated crew waiting for this order, all the British shells missed. Holland wasn't so lucky.

Holland's oblique angle of attack meant that only the forward ten guns could be used (four on *Hood*, six on the *Prince of Wales*), which was further reduced to nine when one of the *Prince of Wales*' guns malfunctioned after the first salvo. The *Prince of Wales*, having been recently completed, was still suffering teething problems and had technicians on board from Vickers Armstrong during the action to help the inexperienced crew. Yet another disadvantage for Holland was that he was steaming into the weather, with sea spray obscuring *Hood*'s main rangefinders, while the Germans had the wind at their backs. The net effect of all this was that he was essentially fighting partially sighted and with one hand behind his back. Having realized his initial mistake in targeting the *Prinz Eugen*, he did give the order to switch fire over to *Bismarck*, but the crucial advantage of having fired first combined with Lütjens' inaction was squandered. In the end, none of *Hood*'s six salvos found their targets, though the *Prince of Wales* scored several hits on *Bismarck*, including one on her port bow that caused a troublesome fuel leak. In comparison, the German gunnery, once they did start firing, were deadly accurate.

Unlike Holland, Lütjens correctly concentrated all his gunfire on the lead British ship. Despite their delay in entering the fight, the German gunners in both *Bismarck* and *Prinz Eugen* found the range to *Hood* almost immediately. *Prinz Eugen* scored first with an 8-inch shell from her second salvo that struck *Hood*'s shelter deck near the mainmast, causing a fire that set off explosions of ammunition in nearby storage lockers. As chaos gripped *Hood*'s crew, who were struggling to douse the fire whilst crewmates lay dying and ammunition exploded around them, *Bismarck*'s first gunnery officer Adalbert Schneider saw that his aim was on target and ordered 'full salvos good rapid'.

By 6 a.m., the range between the ships was down to about 14,900 metres and *Hood* was making one last twenty-degree turn to port to bring her course near parallel to *Bismarck*'s and allow her gunners to fire full broadsides. She was no more than half a minute away from correcting her tactically unfavourable position and being able to fight on an even basis when a shell from *Bismarck*'s fifth salvo slammed into her starboard side, triggering a catastrophic explosion deep within her aft magazines. The entire structure of the ship shuddered violently and she slowed as if someone had slammed on the brakes. A second or two passed before a fierce pillar of flame shot from the ship, rising some 600 feet into the air. Inside the magazines, 112 tons of cordite ignited instantly, causing enormous explosive pressures that ripped the hull apart and fed the 'vast blowlamp' seen by those on the other ships watching in stunned silence. At the seat of the explosion decks lifted and debris as large as entire 15-inch gun turrets was flung high into the air before raining down in gigantic splashes.

Once stopped, *Hood* began a sickening roll to port from which recovery was impossible. Any of her men who were not killed outright knew instinctively that the ship they considered home was finished. No order to abandon ship was given, but none was needed. On the compass platform where Ted Briggs was on duty as a signalman, there was no panic or raised voices as the men made their way to the exits. As for Holland, he was last seen slumped defeated and dejected in his chair, fully aware of the horrific outcome of his decisions and seemingly resigned to going down with the ship. Remarkably, amidst all the carnage, there was a final act of gentlemanly conduct, as Commander John Warrand, the squadron navigating officer, politely gestured for Ted to pass ahead of him as they made their escape from the rapidly sinking vessel.

And with that the Mighty *Hood* was gone. Within scarcely a minute, 46,680 tons of machinery and steel disappeared from the surface of the Denmark Strait, dragging 1,415 men to the cold black depths below. The *Prince of Wales*, stationed four cables behind, was forced to alter course to avoid the remains of the ship, which included the three survivors, Briggs, Dundas and Tilburn, who apparently owed their good fortune to a large bubble of air escaping from the wreckage, propelling them back to the surface. Momentarily safe from the certain death inflicted on their shipmates, the three men, covered with oil, were able to find a three-foot-square Carley raft, which supported them during the desperately long two and a half hours they floated on the life-sucking seas until the destroyer HMS *Electra* arrived and was able to pluck their frigid bodies out of the water.

Having destroyed the *Hood*, *Bismarck* shifted her main and secondary armament fire to the *Prince of Wales*, and immediately scored several hits, including a 15-inch shell on the lightly armoured compass platform that killed nearly all personnel there except the commanding officer, Captain J. C. Leach. Without *Hood*, Leach was in a single-handed fight against two superior German ships that he knew he couldn't win. Assessing the desperate situation, he decided to break off the action and turn away.

Bismarck's ruthless destruction of *Hood* in an artillery fight that lasted just five minutes sent shockwaves throughout Great Britain and beyond. The 1,415 men killed represented the single largest loss of life in Royal Navy history and came during a particularly dark period of World War II, which counted May 1941 as the worst month ever for naval casualties, with 3,780 officers and ratings killed. The news that *Hood* was sunk genuinely caused the public to fear the war could be lost. Winston Churchill, now prime minister, who had been closely monitoring the battle with telephone updates from the duty captain in the Admiralty, was grief-stricken and deeply depressed by such a significant loss to the country. His mood wasn't helped when he was told that the *Prince of Wales* had broken off from the action and that contact with *Bismarck* had been lost. This was the nightmare scenario: the navy's most powerful warship seemingly easily destroyed in the same way as the Jutland battleships; two dangerous enemy raiders loose in the Atlantic, free to wreak havoc on British convoys; and the Admiralty, in Churchill's opinion, being overly timid in their actions to avenge the loss of *Hood*. If the *Bismarck* and *Prinz Eugen* were allowed to make it to a German or French port to be received as conquering heroes, the propaganda value to Hitler would be incalculable.

Despite Churchill's misgivings, the navy weren't about to let *Bismarck* escape their clutches. By the time the prime minister's message to 'Sink the *Bismarck*' was relayed to Admiral Sir John Tovey at 11.37 a.m. on 27 May, the British Home Force had already destroyed the German battleship and her remains had come to rest at the base of an extinct undersea volcano nearly 5,000 metres below the surface. *Bismarck* and her men had left their mark on naval history by defeating the mighty *Hood*, but she threatened no other ships during her last three days afloat, and in the end her dramatic victory led to her own cataclysmic destruction, which in turn caused Hitler to withdraw from committing surface ships of the Kriegsmarine to raiding actions in the Atlantic for the remainder of the war.

The pursuit and sinking of *Bismarck* was an epic operation that involved some twenty Royal Navy warships and waves of land- and carrier-based aircraft. Tovey, the Home Force fleet commander, who marshalled these overwhelming forces, saw to the killing of *Bismarck* once the wounded ship was finally caught, but it was the actions of two small aircraft and the fuel leak caused by the *Prince of Wales* that proved to be most significant.

Assessing the damage to his own ship, which included its bow being flooded with 2,000 tons of seawater, and 1,000 tons of fuel oil being put out of use, Lütjens knew that Operation Rhine was over before it had truly started. His new priority was to make it to the port of Saint-Nazaire in France for repairs, but first he would have to lose the British ships shadowing his every move. That was proving especially difficult because of the determination of the radar operators on *Suffolk* not to lose contact, and because wherever *Bismarck* turned, a long slick was trailing in her wake from the oil leaking out of her hull. Eventually Lütjens was able to lose his tail and make a beeline for Saint-Nazaire, leaving Churchill to fume once more at the apparent incompetence of the navy.

Tovey was completely flummoxed by *Bismarck*'s disappearing act, but he caught a huge break when on the morning of 26 May a Catalina reconnaissance plane piloted by a US Navy observer spotted the ship heading south-east towards the coast of France. This crucial sighting allowed Tovey to concentrate his forces onto *Bismarck*'s position, which set the stage for a second aircraft to write its page into history. This time it was a Swordfish torpedo bomber, launched from the aircraft carrier HMS *Ark Royal*, which scored a one-in-a-million strike on the one vulnerable part of *Bismarck*'s 251-metre-long hull. Flying at nearly wave-top height to avoid the Germans' intensive anti-aircraft fire, the Scots pilot John Moffat dropped his single torpedo into a wave trough and watched it run true towards *Bismarck*'s port side. Forced to turn away before it struck, he was unaware that the profound damage he had caused had probably changed the course of history.

Moffat's torpedo exploded against *Bismarck*'s unarmoured stern, causing severe structural damage that left the ship's twin rudders permanently jammed in a twelve-degree turn to port. Despite the herculean efforts of the crew, the damage couldn't be repaired nor the steering controlled. *Bismarck* was literally left turning in circles. By reducing speed to less than seven knots, some manoeuvrability was regained, but the course inexorably reverted to the north-west, away from the safety of France and directly towards the pursuing British attackers. In the evening of 26 May, Lütjens sent a series of radio messages that summed up the ship's dire predicament and his own resignation to their fate:

20.54   Attack by carrier aircraft.

21.05   Have torpedo hit aft.

21.15   Ship no longer steerable.

21.40 Ship unable to manoeuvre. We will fight to the last round. Long live the Führer.

Throughout the early hours of 27 May, the Kriegsmarine sent Lütjens promising messages about various aircraft and ships that were being dispatched to assist him. Although these messages bolstered crew morale, in reality *Bismarck* was too far away for any friendly forces to reach her in time. She was heading into battle, and certain defeat, with nothing more than the good wishes of Adolf Hitler to lift Lütjens and the crew:

01.53 All Germany is with you. What can be done, will be done. Your performance of duty will strengthen our people in the struggle for its destiny.

03.51 The Führer has bestowed Knight's Cross on you [first gunnery officer Adalbert Schneider] for sinking the battle cruiser *Hood*. Hearty congratulations.

On the morning of 27 May, *Bismarck*'s meandering course brought her right into the path of the British ships that had been chasing her relentlessly for the past two days. In her crippled state, she was effectively a sitting duck for the British guns that opened fire at 8.47 a.m. *Rodney*'s forward 16-inch guns rang out first, followed by the 14-inch guns on the *King George V* a minute later. Tovey, who was flying his flag on the *King George V*, was finally face to face with Lütjens and firmly pressed his advantage by closing the range in order to achieve maximum hits and damage. The heavy cruisers *Norfolk* and *Dorsetshire*, each with 8-inch guns, joined the attack, ensuring that every side of *Bismarck* was being hit.

After half an hour of action, Tovey observed that '*Bismarck* was on fire in several places and virtually out of control. Only one of her turrets remained in action and the fire of this and of her secondary armament was wild and erratic. But she was still steaming.' Although the four British ships fired a total of 2,876 shells, with an estimated 300 to 400 hits, *Bismarck*'s flag was still flying and she stubbornly refused to sink, much to Tovey's annoyance. Seeing the futility, and brutality, in continuing to shell the defeated and defenceless ship, the British resorted to torpedoes to finish her off.

On board *Bismarck*, the decision to scuttle the vessel had already been taken to prevent it from falling into British hands, and the few men who were still alive were preparing to abandon ship. Shortly after the last torpedo exploded against *Bismarck*'s port side at 10.36 a.m., she heeled over to port and started to sink by the stern. By 10.40, the 'Invincible' *Bismarck*, like *Hood*, was gone.

Of *Bismarck*'s 2,246 men, just 115 were rescued from the sea, making the loss of personnel an even greater blow to the Kriegsmarine than that suffered by the Royal Navy. Whether many more Germans could have been rescued before the British ships left the scene for fear of being attacked by U-boats was a question that didn't sit well with some of those who survived. Tragically, some several hundred men were left to perish in the unforgiving storm-tossed sea. The loss of *Hood* was avenged, and despite the great price paid by the navy and by all those who gave their lives, *Hood*'s actions in defeat, together with those of the *Prince of Wales*, had stopped Operation Rhine in its tracks and led to the removal of Germany's most powerful naval asset.

On a chilly but bright February morning in 2001, I had a remarkable meeting at Channel 4's offices on Horseferry Road in London that instantly made the previous six years of mostly frustrating inaction regarding the *Hood* search a thing of the past. Sitting across from me was Tim Gardam, at the time C4's director of programming, and one of the few television executives in the UK with the power to green-light programmes on his say-so alone. The meeting was only minutes old when Tim began laying out what he called his 'bona fides' for why he was prepared to back a search for *Hood*. Remarkably, his family had had relatives on board both HMS *Rodney* and the *Bismarck*, so he intuitively understood both sides of the story and the sensitivity the project required.

Having tried, and failed, to obtain funding for a search from numerous other sponsors over the years, I had come prepared to sell Tim on my ability as a shipwreck hunter, but was amazed to find that he was the one selling me. Even when I presented various budget options, ranging from £1 million to £1.5 million, figures that normally caused people to flinch, Tim was undeterred in his commitment and ambition to make the search a television event like nothing else seen in the UK since the live broadcast of raising the *Mary Rose* in 1982.

Four frenetic months after the C4 meeting, I was on a ship leaving Cork for the outer Bay of Biscay. In the intervening time I'd had to hire a survey ship, deep-water search and ROV contractors and navigation equipment, arrange for government permissions, commission memorial plaques and take care of an impossibly long list of other items. Normally twice as much time would be needed to get all this done, but our schedule was being driven by the tempestuous North Atlantic weather, which would only allow search and ROV operations during a short summer window, with July being the best month statistically. The other important driver was that 2001 was an important anniversary – the sixtieth – of the battle between *Hood* and *Bismarck*, and the producers at C4 were keen to take advantage of the increased public interest, which they knew would result in a much bigger television audience – but only if we were successful in finding and filming the shipwreck.

The other challenge the producers had set for me was that in addition to finding the *Hood*, they wanted me to find the *Bismarck* as well. They also wanted to broadcast live underwater video from the wreck sites for both television and the internet. At the time, this was a radical idea that had never been attempted before, relying on a novel satellite communication system called SeaCast and a huge dose of luck. The idea, in theory, was simple enough: to find and film both the *Hood* and the *Bismarck* in a single thirty-five-day expedition so that their respective stories could be independently told in a series of high-quality documentaries; and to broadcast special reports live from the expedition ship for the main evening news programme anchored by Jon Snow. In practice, however, what we were trying to pull off was devilishly complex.

Normally I would never guarantee that a wreck could be found, but that was exactly what I did with *Bismarck*. I told Tim Gardam that while there was a small chance we wouldn't find the *Hood*, there was no such risk with the *Bismarck*, for the simple reason that the wreck had already been located years before by Bob Ballard of *Titanic* fame. We'd still have to conduct a search, because Ballard and his investors had decided to keep the precise coordinates secret, but if I based my plans on the information that Ballard had used – basically the logs and radio messages of the British ships that had witnessed the *Bismarck* go down – I was confident that I would find the wreck. Because the *Ocean Explorer 6000* search system I planned on using was many times more efficient than the equipment Ballard had had to work with, I knew it was only a matter of time.

My guarantee to Tim was absolutely crucial to C4's decision to fund the search. In return for their million-pound-plus investment, they expected several hours of TV programming about the two famous shipwrecks, so as a minimum I had to guarantee that our expedition would deliver high-quality underwater footage of at least one of the wrecks, and ideally both. In the ultra-competitive world of TV broadcasting, a factual documentary covering a shipwreck search that didn't show any footage of the shipwrecks would be a sure ratings disaster. So *Bismarck* was my banker, but I was under no illusion that *Hood* was the real target, and thus the true measure of success would be whether I could find her or not.

There were thirty-five people on board our expedition vessel, the MV *Northern Horizon*, when it left Cork, but I was the one who carried the full weight of people's expectations and hopes. Everyone from Ted Briggs to Tim Gardam to all the people and organizations who had lent their support to my ambitious vision for the project were counting on me to deliver. My natural inclination was to turn that pressure into excitement, and what could be more exciting than going out to uncover history hidden away in the deepest depths of the ocean?

An essential part of our plan was to involve Ted Briggs in the expedition as much as possible. Ted had recently turned seventy-eight, and although his health was generally good, he did suffer from high blood pressure and could become anxious when recalling his experience of *Hood* blowing up, leaving him clinging for his life to a float. Although he had continued serving on ships during his thirty-five-year career in the navy, including a stint on HMS *Ceylon* during the Korean War, I personally suspected that he was still suffering from a form of post-traumatic stress disorder. He would sometimes talk about nightmares and flashbacks, especially of *Hood*'s upended bow sliding beneath the fiery sea, just one of several horrific images seared into his mind. While it was painful for him to relive these memories, I think he did it out of a sense of duty to his shipmates and the friends he had lost. More than anyone, Ted wanted to make sure their sacrifice was not forgotten, and for that reason he agreed to visit us in Cork to see the search equipment we'd be using and to wish us luck.

Having moved from America to England at the end of 1995 to permanently join Blue Water Recoveries as their director of research and survey, I had grown closer to Ted and was able to see him much more often. By now he trusted me to handle *Hood*'s legacy with care, but it was my wife Sarah who really gave him confidence that the wreck could be found. At a pub lunch months before the mobilization in Cork, perhaps sensing Ted's uncertainty about the search plans, she spoke directly to him in a way that reassured him far better than I ever could: 'David will find your ship, Ted; if anyone can find the *Hood*, it is David.'

The Bay of Biscay has a reputation amongst sailors for the fierce stormy weather that can turn it in an instant into a cauldron of treacherous wind-whipped waves. I had been through such a storm years before during a search for a different shipwreck, so I knew it could happen at any time of the year and had made a contingency in my schedule for just such an event. Unfortunately my bad luck was repeating itself, as the *Northern Horizon* was slammed by a Force 10 storm (winds 45–50 knots, six-metre seas) as soon as we arrived at *Bismarck*'s location. Standing down for bad weather is never good, but it is especially painful at the start of a project because it doesn't allow the crew, especially those who haven't been to sea before, to settle in and get accustomed to the movements and routines of the ship. No meaningful work could be done and the only form of amusement was watching the documentary film team run to the rails to be seasick, or visiting the bridge to watch the largest waves crash over the bow. Really bad Biscay storms have been known to last a week, but this time the blow was short and sharp and only two days were lost.

To make up the time, I really needed to find *Bismarck*'s wreck as quickly as possible. My problem, however, was that she had sunk in seriously deep water, over 4,600 metres, which meant everything would take longer than usual. My promise to C4 was that I'd find the wreck in no more than four search lines, but since every line and turn in this depth of water could take up to fifteen hours, I was aiming to find it on the very first line. I'd hired my old company, Oceaneering, as the search and ROV contractor for the project, which meant I was reunited with the *Ocean Explorer 6000* sonar search system and some of the guys I used to go to sea with. As nice as it was to be with old mates, I was most pleased to have the *Ocean Explorer* at my side again. I knew the system so well, and the best way to maximize its ability to search huge expanses of sea floor at a rapid rate. Finding shipwrecks was what the *Ocean Explorer* was built to do, and since its first success with *Lucona* nearly ten years before, I had personally used the system to find fourteen shipwrecks, including the *Rio Grande*, a World War II German blockade-runner officially certified by Guinness World Records as the deepest shipwreck ever found, at 5,762 metres.

The *Ocean Explorer*'s key feature was that she was equipped with two different frequency sonars optimized for two very different purposes: a 36 kHz unit for wide-range reconnaissance searching, and a 120 kHz unit for higher-resolution, close-in imaging. This was the absolute perfect blend of sonars to use when searching for wrecks, especially in ultra-deep water. It was akin to having a camera with both wide-angle and telephoto lenses without having to change out one for the other. I knew that if operated properly, the *Ocean Explorer* could initially detect a wreck at the extreme range of 2.4 kilometres (nearly 1.5 miles), but then zoom in during a second pass and resolve something as small as the ship's anchor. One revelation from Ballard's 1989 expedition was that all four of *Bismarck*'s 15-inch gun turrets had fallen out when the ship capsized and were now scattered around the hull. With so much heavy wreckage littering the seabed, there was every chance the debris field would give us the first indication of the wreck's location.

Sure enough, about three hours into line 1, the first sign of a definite hard target started to light up my screen, followed by a handful of bright yellow rectangular shapes characteristic of wreckage. In the place where I expected *Bismarck*'s 251-metre-long hull to be was a large crater at the head of a long curved chute, which I optimistically interpreted to be a slide scar representing the path the hull took as it slid down the steeply sloping seabed. Ballard had described *Bismarck*'s final resting place as the lower slope of an extinct undersea volcano, and the sonar image in front of me matched that description to a T. While others were thrown by the missing hull, I was confident we had found the wreck site and wasted no time in aborting the current search line and getting the ship and sonar to turn around to line up a second higher-resolution pass. There was a smattering of dissenting voices but they didn't understand the navigational clues like I did. *Bismarck*'s hull simply had to be near this debris field; it was just a matter of pinpointing its location.

Eleven hours later, the turn was completed and the sonar was at the start of the second line. Any doubts about whether we had found the wreck site soon evaporated. This second set of images matched the first but provided far more convincing detail. The huge crater *Bismarck* had caused when it impacted the seabed was now well defined, and there were fifteen large targets spread amongst numerous smaller hits. Once again there was no obvious sign of the hull, but I didn't think it was worth spending fifteen hours running another sonar line. In the same amount of time we could recover the sonar, reconfigure the deck for ROV operations and have the *Magellan 725* on the seabed taking actual pictures of the wreckage. I believed that was a better use of our dwindling schedule for the purposes of the C4 documentary. Although the *Bismarck* had lost a lot of her topside structure (turrets and bridge superstructure) in the sinking, we knew that the beautiful lines of the hull were undiminished, and this was what Gary Johnstone, the documentary director, was especially keen to film.

Shipwreck sites can be quite confusing, and *Bismarck*'s was an especially difficult one to decipher. What you finally see on the seabed is an amalgam of damage inflicted during a sequence of events, including the British gunfire and torpedo strikes; the scuttling charges set off by *Bismarck*'s crew; the violent capsizing and sinking; the crushing forces as the wreckage hit the seabed; the ripping of the hull's underside as it tobogganed 1,450 metres down the slope; and the corrosive actions of the sea over sixty years. There was no confusion, however, about the first objects we found when *Magellan* reached the seabed: a single pair of leather boots that foretold a scene of widespread death and destruction. Because leather holds up well in seawater, it is often the only visible material left behind where a human body once existed. *Magellan* didn't need to move very far from this spot before another pair of boots was spotted, and then another, along with a leather jacket indicating that the crewman who had worn them might have worked in the engine room. As we paused the ROV to take in this scene, the images were a chilling and poignant reminder that thousands of men had died on this spot.

We found the boots and jacket near the base of the slope where I wanted to begin the hunt for *Bismarck*'s hull, as it is easier to search moving up-slope rather than down. What I didn't know at the time was that we had landed *Magellan* roughly 210 metres from one end of the hull. At this extreme depth there is zero ambient light, so our cameras could only see as far as the lights on *Magellan* could penetrate, which was about ten metres. For longer-range searching and navigation beyond the limit of lights, we use a scanning sonar, but the unit we were using only had a maximum effective range of 100 metres. Although we would have been able to close in on the hull in less than an hour if we'd known in which direction to move, we didn't find it until nearly three and a half days later, having had to make three separate dives because of two equipment failures and another short spell of poor weather.

While searching for the elusive hull, we did locate and film a lot of other wreckage, like two of the 15-inch gun houses, the mainmast and the bridge superstructure, so the dives weren't wasted. However, the effort exhausted the entire team and my patience was stretched to the absolute limit. I believed the hull would be found at the bottom of the slide scar, but too many times the ROV pilots were getting side-tracked and chasing minor sonar targets outside the trench. To keep them focused, I told them they needed to be like bloodhounds and to follow the trail the trench laid straight to the wreck. The hut from which the ROV pilots operated was on a different deck from our control room and we communicated via an intercom system. As the ROV descended further down the trench, the tension ratcheted up a notch and became palpable. I spoke over the intercom in short, clipped bursts that reflected the tightness in my chest. The control room was packed, and anyone lucky enough to have a seat was perched on its edge.

The ROV pilots spotted the hull first in the monitor of the scanning sonar, which we couldn't see, but cheekily kept it to themselves, knowing full well what our reaction would be when it came into view. Suddenly, and without warning from my perspective, the picture in my video monitor changed and I was facing a curtain of black steel. 'What the hell is that, right in front of you?' I asked, not understanding how this enormous object hadn't been picked up by the scanning sonar long before it appeared in my monitor. Richard Daily, the big gregarious Texan who had played this little prank, calmly answered, 'That's the *Bismarck*.' Instantly, the room erupted in cheers. I dropped the intercom mic and raised both arms as the cheers turned to whoops and then laughter at Richard's trickery. Mainly, though, I was just massively relieved. It should have been easy, but finding *Bismarck* had turned out to be a real struggle, and we deserved our brief moment of celebration.

After the congratulations ended, we got down to business straight away and I asked the ROV pilots to start by conducting a full circumferential inspection of *Bismarck*'s fractured stern. In the sinking, or possibly when she crashed into the seabed, a section of her stern aft of the rudders detached from the rest of the hull. Whatever the cause, the fracture was remarkably clean, cutting straight across the full breadth of the ship in a way that exposed the welded construction of the after bulkhead. It was the one section of the ship where we could study its internal construction without having to make any penetrations with the ROV, a practice I had forbidden along with intentional touching of the wreckage.

Channel 4 had invited two experts – Bill Jurens, a Canadian consultant on battleship design and gunnery, and Eric Grove, a British historian – to join the expedition and assess the damage suffered by both ships in an attempt to answer some basic questions, such as: how did *Bismarck*'s armour hold up against the British onslaught; was the *Bismarck* sunk or scuttled; and what actually caused the catastrophic explosion that ripped *Hood* apart and led to the death of every man on board save three? This was going to be a tall order in view of our limited schedule for forensic-style examinations, but at the end of the day C4's sponsorship of the expedition was predicated on the making of television documentaries whose objectives were to inform, educate and entertain audiences. It was my job to balance all these different aspects, in addition to keeping control of the budget and schedule, so I had to take a pragmatic view about what was achievable and what was not.

My approach in managing the various objectives was to make the most of the fact that we were running a 24/7 expedition, and that except for equipment or weather downtime the *Magellan* would be in the water filming the wrecks on a round-the-clock basis. I gave preference to the documentary film team for the daylight hours, while Bill and Eric's detailed battle damage investigation was done mostly in the evenings, when activity on the *Northern Horizon* had quietened down. Our control room was completely blacked out from ambient sunlight, a necessity especially in the Denmark Strait, as we never had complete darkness there, with only about three hours of dusk each night. The rigid mealtimes on board were the best form of timekeeping, as nothing was more upsetting than missing your favourite meal of the day.

Devoid of her 15-inch gun turrets and bridge superstructure, *Bismarck* was still one of the most magnificent-looking shipwrecks I had ever seen. The lines of the hull, especially the way her mid-ship girth narrowed to the knife-edge of her stem, were especially pleasing. As we looked more closely, however, the extraordinary pounding she had absorbed at the hands of the British fleet was plain to see. Considering the great number of shells fired at her (2,876) and the large number estimated to have hit (300 to 400), relatively few appeared to have been aimed at her side or waterline, where the damage could have led to a sinking. It wasn't that the British shells didn't penetrate *Bismarck*'s stout armour, because they did; even the 320 mm main armour belt had been impressively pierced by a large-calibre shell. Yet far more hits were scored on the superstructure, guns and deck. Whatever their reasons, the gun directors of the four British ships that shelled *Bismarck* aimed high at her superstructure and gun turrets. Admiral Tovey was absolutely right when he radioed, 'Cannot get to sink her with guns', although this signal would later plant the seed that the British had not sunk her at all. He was also right to stop the brutal slaughter of the escaping Germans and order, 'Any ships still with torpedoes to use them on *Bismarck*.'

Although Tovey's unfortunate signal has led some to mistakenly believe that the British guns were not up to the job of defeating *Bismarck*, our expedition showed exactly the opposite. In fact, the surgical accuracy of the British gunnery was wholly remarkable. We found devastatingly destructive hits on *Bismarck*'s conning tower; on the barbettes for turrets 'Bruno' and 'Dora'; on the aft gun director; on one of the main rangefinders found in the debris field; and most impressively on five of the ship's six secondary 150 mm gun houses, which were all neutralized by a single well-placed shell. Tovey's view from the *King George V* 'after half an hour of action' was that as a fighting unit *Bismarck* was essentially defeated. The shelling continued – almost as a form of target practice, which some British officers considered wasteful and humanly unpleasant – for another twenty minutes before torpedoes were called on to finish *Bismarck* off.

Never before had anyone been able to document the damage caused by the torpedo strikes on *Bismarck*'s hull, which subsequently became the subject of a contentious debate about who could take the credit for sinking the ship. The fundamental question was whether *Bismarck* was sunk by the British or scuttled by her own crew. This question was first raised by Bob Ballard following his initial discovery of the wreck site in 1989. In an interview for the *New York Times*, he was quoted as saying he was satisfied that the Germans scuttled *Bismarck*, because 'only scuttled ships tend to make it to the bottom in one piece'. I knew this was untrue because Blue Water Recoveries had found many ships sunk during both world wars in one piece on the seabed even though they had been sunk by enemy action, whether it was gunfire or torpedoing. A ship sinks when it loses all the reserve buoyancy keeping it afloat because its internal spaces have become flooded. It matters not whether that flooding is due to holes opened up in the hull by the enemy or by scuttling actions.

I also knew that Ballard was unaware of the true extent of the torpedo damage, because in using a towed camera platform to photograph the wreck, he had only been able to look down on the main deck from above and couldn't see the sides of the hull below the armour belt where the torpedoes would have hit. My advantage in using a free-flying ROV like *Magellan* was that we were able to look at all surfaces of the wreck from every angle. I could ask Oceaneering's pilots to fly the ROV to any point on the wreck and basically have them hover in that position while we collected the best video footage possible. I also had the advantage of superior lighting and camera equipment that wasn't around in Ballard's day.

In order to see the torpedo damage, we needed to position *Magellan* near the lowest part of the hull, where it was buried deep in the muddy seabed, and fly along both the starboard and port sides looking for breaches. I wasn't surprised to find four gaping holes below the armour belt – two on the port side and two on the starboard – at locations that closely matched where eyewitnesses observed torpedoes to have exploded. Two of the holes were larger than normal but could have been caused by either multiple torpedo strikes or by being enlarged in the course of *Bismarck*'s slide down the volcano slope. The other two exhibited the unmistakable and classic signs of a torpedo explosion: hull plates splayed outwards by the internal explosion and the misalignment of heavy armour plates immediately above the hole. Despite my confidence in what I saw, others disagreed. Dr Alfred McLaren, a former US Navy submarine officer and ex-president of the Explorers Club – a club I have been a fellow of since 1991 – who dived to the wreck in a manned Russian submersible the month before us and again in 2002, appeared to completely discount the torpedo damage. In a second *New York Times* article to highlight the scuttling claims, he said: 'We conclusively proved there was no way the British sank that ship. It was scuttled.'

When McLaren's views were relayed to me by the journalist writing the article, I got the feeling that in addition to discounting the unmistakable torpedo damage, he was also overlooking the accounts of British and German eyewitnesses to the battle. For example, Baron von Müllenheim-Rechberg wrote in great detail about the extensive flooding *Bismarck* suffered before the final action; the two 14-inch shells from the *Prince of Wales* that struck the port bow caused the ship to be 'down 3 degrees by the bow and have a 9 degree list to port'. As for John Moffat's aerial torpedo that struck *Bismarck*'s stern on the evening of 26 May, Müllenheim-Rechberg wrote:

> The torpedo hit shook the ship so violently that the safety valve in the starboard engine closed and the engines shut down. Inspection by the damage-control parties revealed that the hole blown in the ship's hull was so big that all the steering rooms were flooded and the occupants had been forced to abandon their stations. Then the after depth finder tube broke and water rushed through to the main deck. The after transverse bulkheads had sprung leaks and the upper and lower passageways on the port side of Compartment III and the centreline shaft alley were making water.

By itself this information proves that an aerial torpedo was able to penetrate *Bismarck*'s defences and cause significant structural damage and flooding. Other reliable accounts reveal that *Bismarck* was hit by a minimum of six torpedoes and possibly as many as nine (three aerial and six ship-launched, with the ship-launched torpedoes carrying 90 per cent more TNT than the aerial types). On top of the flooding that resulted from the British attacks, German damage control parties were also flooding the ship to correct lists and to extinguish the fires that were raging throughout.

From the British perspective, the ship was in a sinking condition by about 10.15 a.m. Tovey observed that '*Bismarck* was a wreck, without a gun firing, on fire fore and aft and wallowing more heavily every moment.' Once again Müllenheim-Rechberg provides us with a vivid description from the German point of view; this time of the futility of Machinist Josef Statz trying to save a ship that was nearly fully flooded:

> In the damage control centre Statz remained standing at the table. His glance fell on the damage control board, which mercilessly revealed the bad situation of the ship. Red, the colour for 'taking water', covered almost the entire port side; green, for 'flooded', showed for the port shell and powder chambers and nearly the entire starboard side – the outboard list-control tanks there had been filled for a long time.

When the journalist asked for my thoughts about the scuttling claim, I told him I had no doubt that *Bismarck*'s crew initiated their standard scuttling procedures when all hope was lost. Müllenheim-Rechberg indicates that this took place at about 10 a.m. Taking into account the fact that nine-minute fuses were used on the charges and that more time passed waiting for the final order to scuttle, I think the earliest time the charges placed on the cooling water intakes would have gone off was 10.15. This fits with the time the crew were seen to be abandoning ship and jumping into the sea. Kapitänleutant Gerhard Junack, who set the scuttling charges and was the last man to leave the engine room, didn't hear them explode until he reached the battery deck just below the upper or main deck. By the time he reached the main deck, it was already awash. At 10.16, Captain Martin of the *Dorsetshire* described *Bismarck* as 'in extremis'; in other words, at the point of death.

Bismarck was a huge, brand-new and extremely well-built ship, which was proving more difficult to sink than the British had expected. While there was no single knockout blow, as with *Hood* when her magazine exploded, *Bismarck* was definitely on her way down from the accumulation of hits she had taken from all sides. Undoubtedly some of the seawater entering her hull from 10.15 onwards would have been coming in through the fractured cooling water intakes, but it would have been a minor amount compared to the massive volumes needed to sink – and already flooding – this Goliath of a ship. It was Captain Martin who decided to put an end to the carnage by firing three Mark VII torpedoes into *Bismarck*'s sides: two into her starboard side at 10.20 and the final one at 10.36 into her port side. All three hit and exploded. Barely two minutes later, she heeled over quickly to port, no doubt as a reaction to the final torpedo strike, and then sank by the stern, which by this time was fully flooded.

It is impossible for me or anyone else to be absolutely sure how much the scuttling hastened *Bismarck*'s demise. After carefully reading all the eyewitness accounts, and having conducted a 360-degree inspection of the hull, my personal belief is that it was only by a matter of minutes and not many hours like others have speculated. A debate about who actually sank *Bismarck* would make for a very interesting news article if all the historical information and facts could be presented. Unfortunately, the article that appeared left no space for such valuable content and instead read like a playground argument featuring the most provocative quotes possible. Who sank the *Bismarck* was a fair question, and one that I tried to answer, but I did not want it to overshadow the ship's historical significance and the effect her loss had on the naval war between Germany and Great Britain. Even though *Bismarck*'s operational lifespan lasted just nine days, I didn't think that time should be defined by the final twenty-five minutes, just as I didn't want *Hood*'s two-decade dominance of the high seas to be reduced to her pitiful five-minute defeat.

As we began to retrieve the *Magellan*, having laid a memorial plaque on the admiral's bridge on behalf of the Kameradschaft Schlachtschiff *Bismarck* (Comradeship Battleship *Bismarck*), I was already thinking about our next, and biggest, challenge: to find the *Hood*. This was a search starting from scratch, with no advantages and no guarantee of success. I had delivered on my promise to Tim Gardam about locating

Bismarck, but Ted Briggs and literally millions back in the UK who were following our expedition were expecting me to deliver once again. We needed to cross 1,100 nautical miles of the North Atlantic to reach the icy waters of the Denmark Strait, which gave me four and a half days to prepare the team and get ready for the hunt for *Hood*.

The good news from my research into the possible location for *Hood* was that there were plenty of positions to choose from; the bad news was that these were widely scattered, with as much as twenty nautical miles separating the furthest apart. If I simply accepted that all eight of these positions were valid, with an equal chance of being correct, it would leave me with a search box of 840 square nautical miles to cover after applying the same standard circle of error around each position. When you are out on the open ocean with nothing for your eye to use to judge distance, it is impossible to appreciate just how large an area that actually is. Even when standing on the monkey island – the highest deck of our expedition vessel – the distance to the horizon is only 6.7 nautical miles. So this was a truly huge area; four times larger than my search box for *Derbyshire* and nearly twice as large as that for *Lucona*. In terms that my sponsors could understand, I explained that it was an area greater than Paris, the largest city in Europe.

Searching for a 262-metre-long shipwreck within an area the size of urban Paris was already a daunting prospect, but it was made worse by the fact that I had lost nearly three days from my schedule due to the problems finding the hull of *Bismarck*. I'd started with funding from C4 for a forty-two-day expedition, including seven contingency days I could only use in case of bad weather and downtime, and arrived in the search area for *Hood* on the morning of day twenty-one. This meant that – excluding the seven-day contingency and the four-day transit back to Cork – I had exactly ten and a half days to find and film the wreck of *Hood* and to conduct the very special ceremony we had planned. My problem was that the fastest we could search the entire 840 square nautical miles, if that was necessary, was twelve days. I needed to find a way to bring the search time closer to what I had left in my schedule.

The main drawback of using a towed side-scan sonar in deep water is the excessively long time it takes to turn the ship and sonar at the end of each track-line before you can start the next. The problem is entirely due to the length of tow cable you need when searching in deep water and the correspondingly slow speed the ship has to move to minimize the drag forces on the cable. Unfortunately, speeding up is not an option because it just causes the towfish to fly higher, reducing the quality of the images, and in extreme situations can lead to the cable suddenly snapping like it did during the *Derbyshire* search. As I couldn't make the search go faster, the only realistic option I had was to reduce the size of the search box by discounting one or more of the sinking positions factored into the original area.

Of the eight possible sinking positions I had to work with, there was one obvious candidate for elimination. It came from a wireless transmission sent by Admiral Lütjens to Group North,* less than an hour after sinking *Hood*, a message that simply read: 'Have sunk battleship Qu. 73 AD.' I immediately recognized Qu. 73 AD as the 'Quadrate' code used by the German navy during World War II to secretly report their positions at sea. I had come across such coded positions many times before in U-boat logs and had learned the procedure for converting the alphanumeric code into latitude and longitude. Basically the code represented a grid system that divided the ocean into large squares designated by an alpha reference, medium squares fifty-four nautical miles on each side designated by the first two numerals, and smaller squares six nautical miles on each side that further divided the medium squares and were designated by a second pair of numerals. Qu. 73 AD therefore stood for Quadrant AD, which was located in the Denmark Strait, and medium square 73. However, because the second pair of numerals that would normally follow the 73 to identify the smaller square was missing, the

---

* A Navy Group Command within the German Kriegsmarine.

position lacked the degree of precision I needed. For some reason – probably because his ship was still being aggressively pursued by the *Prince of Wales* and the cruisers *Norfolk* and *Suffolk* – Lütjens did not report the full grid reference. As this meant that *Hood* could have sunk anywhere within a square that measured 54 × 54 nautical miles, it was of limited use to me and I could eliminate it without any worry.

Because Lütjens' position was in an outlying location, I could now cut the size of my search box from 840 to 591 square nautical miles. This was a very big reduction, which to my relief solved my immediate time problem in one fell swoop. I wasn't prepared to discount any of the remaining seven positions, as I had spent several weeks in the planning phase verifying and analysing them all to make sure in my own mind that I could rely on them as wholly credible pieces of information to factor into my search box. In my experience of searching for historic shipwrecks, these seven positions from five different sources constituted a veritable bonanza of information. Generally you are fortunate to find two or three sinking positions, and the most I'd ever had in all my previous searches was four. Most importantly, because the positions came from a variety of sources, they more or less corroborated each other. While it is entirely possible for one ship to be wrong when recording its position, it is extremely unlikely that five ships navigating independently will all be similarly wrong.

All my years of research and months of analysis were now captured on a single Quarter Million Plotting Sheet[*] that I had reworked so many times the creases were feathered and starting to tear. Written on it in pencil and highlighter pen was all the essential navigation information I had distilled from countless documents, as well as my calculations of bearings, dead-reckoning tracks, leeway drift and circles of error. A computer is arguably a better, and prettier, place to create such a 're-navigation', but I still preferred to work with pencil and paper. Old-fashioned manual plotting gives me a better feel for how the ships actually

---

[*] A plotting sheet on the scale of 1:250,000.

manoeuvred during the action, and it is also easier to spot errors. In the centre of the sheet was a rectangular box representing the overall search area where I was absolutely certain *Hood* would be found, and within it a much smaller box drawn in red. This was my high-probability box, which encircled the five sinking positions – three from *Norfolk*, one from *Prince of Wales* and one from *Hood* herself – that I trusted the most. If my analysis was correct, *Hood*'s wreck would be found somewhere in this high-probability box early during the search and all my concerns about the schedule could be forgotten.

The trick to finding the wreck quickly was to search the central high-probability box first before moving outwards to cover areas of lower probability in sequential order. This approach was based on the common-sense principle of looking in the place you think you've lost something first before moving on to less likely places. The concept is exceedingly simple, but when the stakes or the costs involved are very high – as in our case – sophisticated computer programs are sometimes used to determine the absolute optimum search plan with the objective of finding the target in the shortest time possible with minimal effort expended. There is even a whole field of mathematical theory on the problem that began with the research of an English statistician and Presbyterian minister called Thomas Bayes. Bayes developed the mathematical equations for his theorem in the mid eighteenth century, but they were not published until after his death and not applied in an underwater search until 1968, when John Craven of the US Navy used the methodology to locate the sunken nuclear submarine USS *Scorpion*.

By assigning subjective probabilities to various sinking positions or loss clues, you can use Bayesian statistics to develop a distribution map of probabilities to show you where to search first. If you fail to find the object in that location, the distribution map is updated – manually or by computer – to show you where to search next, and so on. We had used computer programs to help design the search plans for both *Lucona* and *Derbyshire*, so I had a good working knowledge of what they could do, but for *Hood* I opted to design my own plan. It essentially boiled down to me picking the sequence of track-lines I thought gave me the best chance of finding the wreck the quickest. As there were only ten lines in total across the entire search box, with four covering my high-probability box, it wouldn't be hard to assign new probabilities and adjust the order of lines to run as the search progressed.

We arrived on location in the Denmark Strait at 4 a.m. on a fine, clear day with light winds and small seas: perfect weather for deep-tow operations. The temperature was noticeably colder, but the biggest change we had to get used to was the amount of daylight. The darkest it ever got was civil twilight for about four hours. I had arranged for a small transfer boat out of Iceland to meet us when we arrived, but was surprised to find a fleet of twelve fishing boats trawling the waters near the eastern edge of the search box where I wanted to begin the search. As they, like us, were going to have restricted manoeuvrability whilst towing their nets, we needed to keep a close watch on their movements to avoid the danger of them cutting across our course – or vice versa.

The boat from Iceland arrived on time and in the ship-to-ship transfer we lost the documentary director but gained a two-person Channel 4 news team, including the journalist Lindsay Taylor. Channel 4 were so pleased with the public's response to the footage we had shot of *Bismarck* that they'd decided to attempt live coverage of our hunt for *Hood* for the nightly news, using the SeaCast satellite communication system. Having spent in excess of £100,000 to hire the SeaCast system and have it mobilized on board the *Northern Horizon*, they'd decided it was time to capitalize on their investment.

A few hours later, Lindsay was able to file his first story, about the deployment of the *Ocean Explorer* sonar officially starting the search for *Hood*. The fishing boats had kindly moved south, away from our location, which gave us a free run-in to the track-line I judged to have the highest probability. There was a new energy on the vessel, in part because of our new arrivals and the change in location, but mainly because of the excitement that builds with the start of every new search. Once the sonar had reached the correct depth, I had my first look at the type of terrain we'd have to contend with while trying to pick out possible targets. I'd expected the seabed to be flat, which it was, but the surface geology was more active than I had hoped. There were large areas of outcropping rock scattered about, and what appeared to be mobile sand-waves travelling across the seabed in response to a bottom current. We were mainly searching in the Irminger Basin, but this side of the search box was close enough to the lower slope of the seismically active Reykjanes Ridge for such geology to be normal. I wasn't overly concerned by what I was seeing, but my initial thought was that it was going to be easier to look for the wreck on the western side of the box, even though the probability of that side was considerably less than the eastern side.

The reason for favouring the eastern side of the box was that four of my five most trusted sinking positions were located there in a relatively close cluster. One was based on the dead-reckoning* track of the *Prince of Wales*, fighting alongside *Hood* in the battle, while the other three came from *Norfolk*, trailing the action by fourteen nautical miles but first on the scene to report that *Hood* had sunk. It was the second of these wireless transmissions from Admiral Wake-Walker, commanding the First Cruiser Squadron from HMS *Norfolk*, that most excited me.

At 6.15 a.m., as *Norfolk* was steaming towards *Hood*'s last position, Wake-Walker sent his first transmission with what I believe was an immediate but estimated position: 'HMS *Hood* has blown up in position 63° 20' N, 31° 50' W.' Some twenty minutes later, *Hood*'s remains – some balsa rafts, charred wood and hammocks bobbing about in a spreading slick of oil – could be seen from *Norfolk*'s bridge as she steamed past in pursuit of the enemy.

At 6.37 a.m., a second transmission was sent by Wake-Walker to the destroyers and the Admiralty: 'HMS *Hood* sunk in position 63° 21' N, 31° 47' W. Proceed search survivors.' This position was clearly a revision of the first, and there can be no doubt about Wake-Walker's intentions. He was instructing the destroyers to steam at once to the reported position to rescue any survivors. As a cruiser that could keep up with the German ships, *Norfolk* could not afford to stop. Along with the *Prince*

* Dead reckoning is a simple method of navigating, using course (or heading) and speed to calculate a current position from a previously determined position.

of Wales and *Suffolk*, she gave the Admiralty its only chance of catching the fleeing *Bismarck*. This responsibility was made absolutely clear some hours later in an urgent message from the Admiralty to 'continue shadowing *Bismarck*, even if you run out of fuel'. The search for survivors would be left for the destroyers, but by transmitting a second, more precise sinking position, Wake-Walker was giving anyone still alive a better chance. When I found a pink copy of this message in the archives, stamped MOST SECRET in bold red letters, I knew this was about the best pointer I could ever find to where *Hood*'s wreck might lie.

As the *Northern Horizon* progressed down the first track-line into the slightly deeper waters of the basin, the rugged geology seen in the eastern side of the box transitioned into a carpet of soft, muddy sediment. The resulting sonar images had the ideal background that shipwreck hunters like me pray for. The seabed had become a uniform tableau of burgundy against which any hard sonar targets, depicted as bright yellow, green and blue angular shapes, would readily stand out. However, other than one not terribly exciting sonar target around the halfway mark of the line, nothing else appeared. There was a moderate increase in water depth from east to west of roughly 300 metres, but that would have no real effect on the search. As long as the fine weather conditions held and I had placed my search box in the right location, I could see no impediment to finding the wreck.

In order to be awake for every moment the sonar was actively searching, my plan was to catch short sleep breaks during the turns. At the speed we were operating, the track-lines were taking nine to ten hours and the turns slightly more than five. With all the other responsibilities I had managing the expedition and doing interviews with Lindsay Taylor, I was hoping to get three hours of sleep per turn. Maintaining such a punishing schedule over the duration of a long-drawn-out search would be tough, but I was determined to be on duty when *Hood* popped onto my screen.

The second track-line was slightly more interesting than the first, with a few more targets logged but nothing remotely close to being *Hood*, or even part of *Hood*. During the turn to the start of the third line, Lindsay wanted to conduct an interview with me that would be broadcast live on the main evening news. Because what we were attempting was so technologically novel – streaming live real-time video from our ship in the North Atlantic to a news studio in London via a combination of satellite and fibre-optic cable links – it would represent a first in the history of UK television broadcasting. Lindsay understandably wanted the interview to go well and for the timing to be spot on, so we practised several run-throughs before waiting for the cue from the producer in his earpiece. We were just seconds away from going live when Lindsay shouted out in language more in keeping with a sailor than a journalist. The video link had dropped out during the producer's countdown, leaving him with no option but to kill the segment entirely. Lindsay was furious at our bad luck, while I was left ruing the hour of sleep I had just lost.

Lindsay had met Ted during the mobilization in Cork and was keen for me to speak in the interview about Ted's hopes for the search. The story I would have told had the interview not been killed was how Ted wanted me to find the wreck close to the position I'd derived from the last wireless message sent from *Hood*. The significance of this was that the position had been worked out by the squadron's navigating officer, Commander John Warrand. Ted could never forget Warrand stepping aside to let him leave the compass platform first as *Hood* sank, and believed he owed him his life as a result. Warrand's position was the last one I found during my research, and there was a long, laborious process before I could make any sense of it. I even had to break a British navigational code similar to the German Quadrate system that I had never seen before in order to figure out Warrand's position, which was sent at 5.43 a.m., about seventeen minutes before *Hood* blew up and sank. Having determined the lat/long coordinates and time of the position, I used them to project where *Hood* had sunk, based on the same dead-reckoning track I'd calculated for the *Prince of Wales*. In the end I was left with a sinking position that came directly from the most senior navigator plotting his ship's movements during the battle.

Deriving a sinking position from Warrand's information was a major victory in my research effort, but it did leave me with a new problem.

The position was a long way – about eight nautical miles – west of where the other four positions were clustered. Incorporating it into my search box, which I felt I had to do, increased the area by about 40 per cent. I had another important piece of information, however, that bolstered my confidence in Warrand's navigational skills. It came from HMS *Malcolm*, one of the destroyers that had raced to the scene after receiving Wake-Walker's order. When *Malcolm* arrived at Wake-Walker's revised sinking position nine hours after *Hood* had sunk, they found nothing. They proceeded to another position reported by a Sunderland flying boat that had witnessed the battle, and again found nothing. At 9.45 that evening, after conducting broad sweeps of the area within ten nautical miles of *Norfolk*'s position, *Malcolm*'s persistence finally paid off, and the crew found 'a large patch of oil, 1 mile in diameter, covered in small bits of wreckage'. Joined by *Antelope*, another destroyer, *Malcolm* searched the area through the night but sadly found no survivors. In their finely observed report of how they scoured the sea, however, they did provide me with another vital clue.

In the time that transpired after *Hood*'s loss – nearly sixteen hours – the oil and wreckage found by *Malcolm* would have drifted away from the sinking position, carried and pushed by a combination of current and wind. I knew that if enough information was available, I could calculate this total drift and backtrack the movement of the oil and wreckage from *Malcolm*'s position to where it originated –in other words, to where *Hood* sank. The process is called reverse drift analysis, and I learned how to do it years before finding shipwrecks for Blue Water Recoveries. Fortunately, I had the information I needed, which was good data on the average speeds of the Irminger current in the month of May and winds at the location recorded by four of the British ships, including *Malcolm*. When I calculated and plotted the current and leeway (the correct term for wind drift) components of the drift relative to the position where *Malcolm* sighted the wreckage, I was surprised and more than pleased with what I saw. The result covered a range of outcomes defined by a trapezoid on the plot, with Warrand's position for *Hood* very close to the centre. The correlation between the two independent positions could hardly have been better. Commander Warrand's position was no longer on an unexplainable outlier and there was every chance Ted would get his wish.

The third track-line, like the first, ran east to west through the heart of my red high-probability box. We were scanning it overnight, which meant our control room would be quieter than normal. I had always preferred to work the night shift while at sea: there were fewer distractions and it seemed that the best results were always achieved then. The sonar had passed the halfway point of the line and was now imaging the flat and featureless Irminger Basin, and I was struggling to keep my eyes open as the long hours and lack of sleep caught up with me. I caught a second wind, however, and my eyes were glued to the screen as the first few pixels of a hard target emerged from its top. Another eight seconds passed, producing two more scan lines that revealed a yellow-green angular shape with a sharply defined acoustic shadow behind it. In that instant, sitting alone watching this stunning picture reveal itself in super-slow-motion, I knew I was looking at the wreck of HMS *Hood*.

There were no whoops of joy or back-slaps as with *Bismarck* and every other wreck I had found before. It was such an emotional and sad moment for me, I didn't feel like celebrating in any way. With other shipwrecks I avoided connecting my emotions to the objective of the search, but *Hood* was different. In part it was because I had grown so close to the veterans and families, and of course to Ted. I knew they would all be pleased that the wreck was now found, but I was sure their first emotion too would be sadness. Even though I had no relative to mourn, I was hit hard by knowing I was looking at the grave site of so many boys and men.

As the full devastating picture of *Hood*'s destruction scrolled down my screen, I knew it would be difficult for Ted to see his ship in this condition. We would need to collect high-resolution sonar images of the wreckage and debris fields, which were scattered across two kilometres of the seabed. And I also needed to get visual proof with the *Magellan* of what the sonar was telling me before I broke the news to him. I did have some news that would please him, though. His wish had come true: on 24 May 1941, Commander John Warrand was without question the best navigator.

∞∞∞∞∞∞∞∞∞∞∞∞∞∞∞∞∞∞

Our speedy location of *Hood* – thirty-nine hours from the time the sonar was launched – meant that instead of being three days behind schedule, we had caught up and were suddenly one day ahead. The extra time allowed me to collect two stunning high-resolution sonar images of *Hood*'s extensive wreckage field and to plan for the arrival of a very special visitor. From our earliest meetings with Ted Briggs in 1995, it was plain to both Rob White and me that Ted really wanted the chance to come out to the wreck site himself. He had carried the burden, and probably some element of guilt, of being a *Hood* survivor his entire adult life. It was important for him to say goodbye to his shipmates, and knowing his selfless devotion to the *Hood* Association, I believed he felt it was his duty to make the trip. However, it was also his chance to personally lay the ghost of *Hood* to rest.

For the first ROV dive on the site, my sole objective was to film some distinctive part of the ship as conclusive proof that the wreck was indeed HMS *Hood*. If we were going to announce our discovery to the world, and to the many thousands of people related to *Hood*'s crew, I wanted visual evidence, even though in my own mind the sonar images were already 100 per cent conclusive. As the largest piece of wreckage generally provides the best features for identifying a shipwreck, I chose a path for *Magellan* directly to the very large sonar target that had first alerted me to the wreck and that I believed was a major intact section of *Hood*'s hull. What I'd thought would be a fairly straightforward dive, however, turned out to be an eighteen-hour mini-marathon that ended in frustration, disappointment and nearly disaster.

The problem we faced was that there was a very strong subsea current flowing past the wreck site that made visibility poor and flying the ROV extremely difficult for the pilots. Even when applying full thruster power they were struggling to make any forward progress against the current. The biggest upset, though, was that the large section of hull was sitting upside down on the seabed. With *Hood*'s main and upper decks buried in the mud, all we were able to see was the bottom of the hull and a couple of large square vents covered by a heavy grating. While these vents were a perfect match for the ones I found on a docking plan for the ship, and thus gave me the proof of identification I was after, we weren't going to learn anything else from the underside of the ship, and were starting to reposition the *Northern Horizon* and the ROV when without warning our video screens flickered once and then went dark.

Recovering an ROV when power and control has been lost is called a 'dead vehicle recovery', and in the best of circumstances it can be a hairy operation. In the nasty weather that had kicked up during the dive, however, it might have been downright dangerous if not for the experience of Oceaneering's team. Like a handful of cowboys roping a runaway bull by the horns, they essentially manhandled the 1.5-ton vehicle into a safe recovery position while managing to prevent it from being sucked under the *Northern Horizon* and into its spinning propellers. It was only when *Magellan* was safely on deck that we realized just how precarious the situation had actually been. A threaded mechanical termination that physically connects the long steel cable coming off the ship to the depressor platform and the ROV, which together make up the *Magellan* system, had very nearly unscrewed and sent the whole rig crashing to the seabed. Unbeknownst to us, the depressor was spinning around in the strong current, causing the threaded termination to unscrew and the central cable containing the power and signal conductors to twist and ultimately snap. Of the fifteen threads holding the two ends of the termination together, twelve had unscrewed, meaning we were only three rotations from an unmitigated disaster that would have brought the expedition to an abrupt and undignified halt.

Despite our problems, I had seen enough of the wreck to fulfil a promise I had made to Ted that he would be the first person I called once I was sure we had found the *Hood*. He was waiting at home, having been primed by C4 that momentous news was about to break. Despite the fact that this was a moment we had dreamt about for years, our conversation lasted a couple of minutes at most. Ted was so choked with emotion he could hardly speak. He did, however, ask me where the wreck was found, and was pleased to hear it was close to Commander Warrand's estimated position. We were both being filmed as we spoke, but it was many months before I saw the completed documentary and the impact this news had on Ted. After saying goodbye to me, he got up from his chair and walked past the documentary team into his kitchen saying 'yep, he's done it, well and truly,' his hands visibly trembling as he pulled a handkerchief from his pocket to wipe the tears from his eyes. As he stood staring out of the kitchen window into his garden, I imagined that my news had taken him back sixty years to the Denmark Strait and that haunting view of *Hood*'s upended bow sliding beneath the fiery sea.

The next few days filming the wreck and preparing for Ted's arrival went by in a blur. With the ROV diving on a continuous basis and no darkness to break up the days, it was easy to lose track of time. At one point I walked out of our darkened control room into the bright sunlight and asked one of the documentary producers how long until breakfast was served, only to be told that the next meal was supper at 6.30 p.m. I had already been up for more than a day and a half and was so absorbed in our investigation that I continued working for another twelve hours without any conscious awareness of time. I find that the concentration required to navigate around the wreckage field and to reassemble in your mind ship structures that have been blown apart forces you into a purely mental zone.

After my initial disappointment at the state we found *Hood* in, my mood lightened as more recognizable sections of the ship were discovered and we began to make sense of the widely scattered fields of wreckage. Our discovery was also publicly announced on C4, and congratulations started trickling in to the ship. Julian Ware, the executive producer in charge of making the documentaries for C4, had been instrumental from a very early stage in securing Tim Gardam's support for the project. For this and other reasons I was most proud of his brief message, especially in the way that it harked back to one of the most famous quotes in British naval history: 'England expects, but it took an American to deliver.'

When I was not glued to the video monitors directing the wreck investigation, I was on the phone finalizing plans for Ted's journey out to the *Northern Horizon*. Because our location was too far for a helicopter, the safest and surest way for him to travel was on a reliable and sturdy Icelandic tugboat. The 500-nautical-mile round trip would take at least two days – the longest time Ted would have been at sea since retiring from the Royal Navy in 1973 – and would be an arduous journey for a man half his seventy-eight years. We were relying on good weather for his visit, otherwise it would be called off, but were blessed with the absolute best day we had had since leaving Cork.

After completing the delicate business of transferring Ted from the tugboat to the *Northern Horizon*, he joined me in the lounge for a detailed review of the video we had shot of *Hood*'s wreckage over the previous four days. Back in London, he had caught snippets of the video during an interview with Jon Snow, the C4 news anchor. Now, however, he was about to see for the first time the full-scale destruction that had been visited on his former ship. I started by explaining the overall spread of wreckage revealed in the high-resolution sonar images. The bird's-eye view provided by this imagery was key to understanding what had happened to *Hood*, and in my opinion answered the fundamental questions of why the ship sank so fast and why the loss of life was so great.

The surprise discovery of our expedition, which hadn't been forecast by any of the eyewitness accounts, was that *Hood* had apparently suffered not one but two explosions, ripping the hull apart and sending her to the bottom of the Denmark Strait so quickly that the men were unable to get clear. The main evidence for that conclusion was the two large teardrop-shaped debris fields covered with dense concentrations of wreckage. These separate and distinct fields marked the exact spots where *Hood*'s hull, while still afloat, broke apart. Their relative position on the seabed, approximately in line with *Hood*'s heading when she was hit by the shell from *Bismarck*'s fifth salvo, further reflected the close sequence of two separate events. The first, easily seen by anyone with their eyes on *Hood* at that moment, was when the 112 tons of cordite in the aft 15-inch magazine detonated within a second or two of being hit by the shell. We estimated that this explosion obliterated a seventy-metre section of the hull in the area of the aft magazines, the remains of which, including *Hood*'s two forward propellers with parts of the shaft still attached, and two complete stern tubes, formed the eastern debris field.

What was left of *Hood*'s stern, a forty-metre section that was blown clean away from the rest of the hull when the aft magazine exploded, was also found lying in this eastern debris field. The damage to its forward end, nearest to the seat of the explosion, reflected the enormous power of the blast. I wondered what Ted would make of seeing familiar parts of his ship so utterly destroyed. As a signalman he would have mustered near the ensign staff on *Hood*'s quarterdeck – the longest of any warship afloat – every morning at 8 a.m. for Colours and at dusk for Sunset. He would also have remembered its bleached teak decking on more joyous occasions, like dances and receptions when *Hood* was in port. Now, fractured clean across from one side of the ship to the other, it bore no resemblance to how it had once looked. We could see no structure below this part of the quarterdeck, as the bottoms of the ship had been blown away and the contents strewn throughout the eastern debris field. Near the seabed on the port side a section of shell plating I estimated to be at least fifteen metres long was peeled back like a banana skin. Everywhere we looked there was evidence of the tremendous forces that had ripped the hull apart.

One of the only parts of *Hood*'s stern that had escaped damage was her port rudder, which I especially wanted Ted to see. We had found it frozen in a twenty-degree turn to port, proving that Holland had ordered this final manoeuvre, and that *Hood*'s crew had executed it. This was a remarkable and important discovery for our expedition. The video I was showing Ted was literally a snapshot in time of a historic moment that had been discussed and debated since the day *Hood* sank. Ted, who had been in earshot of Holland ordering this final turn, was able to finally see how unlucky his admiral had been that day. Had the turn been completed just thirty seconds sooner, *Hood* would have moved out of the shot pattern of *Bismarck*'s fifth salvo and probably not been hit.

Once it was determined that the eastern debris field marked where the aft 15-inch magazine had exploded, it was possible to put into context the western debris field, whose centre was roughly 500 metres further west, or in the approximate direction *Hood* was travelling. In terms of time, the western debris field had to have been created as a secondary event after the initial explosion of the aft magazine. It was also about twice as large as the eastern debris field and much more densely concentrated with wreckage, which indicated that whatever cataclysm destroyed the forward part of *Hood* must have been on a scale greater than, or certainly no less than, the aft magazine explosion. Like the stern, *Hood's* bow appeared to be just an exterior shell, with its internal structures and contents missing. We found the bow lying port side down with the anchor cable draped along the upper starboard side, its heavy links cascading over the railing to the muddy seabed below. This was where I thought Ted should lay the memorial plaque, and he readily agreed.

The other remarkable and surprising discovery we made was of *Hood's* conning tower, which had it remained in its place on the ship just forward of the bridge superstructure would have been hidden from our view, buried beneath the upturned hull. Instead, this heavily armoured cylindrical structure weighing some 650 tons was found 1,100 metres north-west of the western debris field and 2,100 metres away from the mid-ship hull section. In my experience documenting deep-water shipwrecks I had never seen an individual piece of wreckage like this so far from where it had broken away from a ship. Some objects, especially those that have kept their hydrodynamic shape, can take a horizontal glide path as they fall underwater. The mid-ship hull section had obviously done this, which was how it was found over 800 metres south of the two debris fields. In the case of the conning tower, however, I didn't believe it had the right shape or mass to glide such a long distance. If I was right, the only way it could have got there was if a second explosion of *Hood's* forward magazines had caused it to be blown free and propelled away from the ship.

A second explosion, which could have been hidden behind the plume of smoke that enveloped *Hood*, or even occurred partially underwater, would also account for the larger western debris field. A likely trigger was the sudden shifting of cordite in the magazines as *Hood's* bow rapidly reared up as the ship slid underwater. The cordite used by the navy was notoriously unstable and had the propensity to spontaneously explode in spectacular fashion. Even Ted, when he saw the damage to the bow, wondered without prompting whether a second explosion had occurred.

As much as we learned in the few days we spent documenting *Hood*'s wreckage, I knew we had only scratched the surface in understanding how the ship had broken apart. A full forensic-style examination of the wreckage would take weeks, and our time on site was nearly over. One of the questions we wanted to answer was why, out of a ship's company of 1,418 men, only three survived. Having seen the terrible state of *Hood*'s wreck, that question had changed to how anyone had survived at all. It made me appreciate what a truly lucky man Ted was, and how fortunate and privileged I was to have him sitting beside me to share this experience. It would have been completely understandable for him to be upset by the images of the fractured and mangled wreckage, but mainly he was relieved that the deaths of his shipmates had been so mercifully quick.

For sixty years Ted had carried a burden of responsibility as one of *Hood*'s miraculous survivors. This was the reason for his pilgrimage to the Denmark Strait. Because the wreck of *Hood* represented a massive war grave, I also felt a strong personal responsibility to make sure that it was appropriately marked. My idea was to create a bronze memorial plaque that included the names and details of every one of the 1,415 brave men and boys who had gone down with the ship. I had an identical plaque made for *Bismarck*'s men and placed it on the wreck on behalf of the Kameradschaft Schlachtschiff *Bismarck*. As president of the *Hood* Association, Ted was given the honour of physically releasing the plaque from the ROV's manipulator at the spot against the anchor cable we had chosen together. As he pressed the switch to open the jaws and let the plaque go, he paid his last respects to his shipmates: 'Farewell, good friends, I have never forgotten you.' *Hood*'s bow, which had been the enduring symbol of her beauty in life, now represented the entombment of her crew in death.

When I got back home to the UK after having demobilized the *Northern Horizon* in Cork, the question everyone was asking was why hadn't we tried to recover a bell we had spotted amongst a large pile of wreckage. The subject was being openly debated all around the country in newspapers and amongst veterans' groups, which caught me completely off guard because at sea we hadn't given it a second thought, such was our commitment to the policy of 'look but don't touch'. I didn't even know if recovering the bell was possible, as it was in a precarious location nestled in a corner with possible obstructions on all sides. The only reason we'd even spotted it was because of the quality of the underwater camera we were using, and sheer luck. We had been scanning the wreckage, looking for something recognizable, when I saw a curved shape that looked out of place and asked for the camera to be zoomed in on it. We had no idea which bell it was, as large capital ships like *Hood* often had more than one, but it did seem a decent size and its condition looked good.

Even after I got home and saw the commotion the bell had caused, I didn't give the idea of a possible recovery attempt much consideration. We had achieved nearly everything we'd set out to do with respect to the wreck and *Hood*'s historical legacy, and I felt more than satisfied with that. However, the images of the bell published after the expedition and shown in the C4 documentaries had clearly struck a chord with the public, and there was much discussion of the symbolic importance it might have if it could be used as a focal point for some future land-based memorial. Sailors who had formerly served on *Hood* also began to make their feelings known. Having heard the bell rung throughout their period of service on the ship, they thought it embodied her spirit and soul. Their sentiments were unanimous: if any one part of *Hood* was ever to be raised from the wreckage, it should be the bell.

As the years passed, the subject of recovering the bell would crop up from time to time. I was regularly approached at the annual memorial services at Boldre church and asked whether I thought it could be retrieved, and whether that would be a good thing. My answer to the first question was always the same. I thought there was a fair chance we could recover the bell, but it wasn't guaranteed and it would be an expensive operation requiring significant sponsorship. However, on the question of whether it would be a good thing, I demurred. I was a strong advocate of the 'look but don't touch' policy towards naval war graves, and felt that you could only breach that sanctity if there were exceptional reasons. In my mind such a decision would have to start with those who held the moral authority on questions about *Hood*'s wreck.

It wasn't until September 2004 that a conversation with Ted Briggs made it clear to me that he was in favour of having the bell recovered. The occasion was the opening ceremony at the Royal Navy training base HMS *Collingwood* of a new building named after *Hood*. Ted and I had been invited to the ceremony by Commodore Philip Wilcocks, the former commanding officer of *Collingwood*, whose uncle had gone down with *Hood*. I think it is fair to say that Ted's opinion about the wreck was an evolving one. Initially he was against the idea of cameras filming it, then he changed his mind and gave his full support to our expedition, and it was a similar situation with the bell. To begin with he was resistant to the idea, knowing that a number of people would object to any disturbance of the wreck site. However, three years on from our expedition, he saw how the discovery of the wreck and the resulting documentaries had made *Hood*'s story more widely known. There were more new members of the *Hood* Association than ever before, and the building at *Collingwood* was yet another example of the ship's enduring legacy.

From that day onwards, Ted made it clear whenever we met that if I was ever able to find a sponsor to fund a recovery attempt of the bell, he would gladly give us his support. He never applied any pressure on me to actively make it happen; he wasn't that type of person. He just wanted me to know that he and others in the association would get behind a recovery attempt if I could organize one. Sadly, Ted's health began to fail, and he died in Portsmouth hospital on 4 October 2008. He was eighty-five years old. His funeral was held in St George's Church at HMS *Collingwood*, not far from his home in Fareham. Such was his popularity, the small church was overflowing, and a marquee had to be set up outside for people to listen to the service through speakers. The navy made sure he received an impressive send-off, with scores of uniformed sailors lining the long road up to the church. There was a lot of love shown for Ted that day, but I couldn't help but feeling profoundly sad about losing my friend.

In 2009, I began an association with a well-known American philanthropist, who was ultimately able to grant Ted's wish of having *Hood*'s bell recovered. Paul G. Allen, the co-founder of Microsoft, was open to the idea of providing his expedition yacht *Octopus*, which was equipped with the necessary deep-diving ROV and specialist equipment, for a recovery attempt at some future date. He had a long-standing enthusiasm for preserving and sharing World War II artefacts and had previously donated *Octopus* to film the wreck of the aircraft carrier HMS *Ark Royal*. His offer, subject to the Ministry of Defence (MOD) providing me with an official licence to conduct the recovery, was a unique and extraordinarily generous one that I felt I had to take to the *Hood* Association for consideration by its membership. The association reacted with unanimous support, no doubt helped by knowing Ted's earlier stance. With the help of Philip Wilcocks, who had retired from the navy as a rear admiral and taken the reins of the association as its president, I was able to make an exceptional case to the MOD for the recovery licence, which was finally granted in June 2012 just in time for an expedition to the site in late August.

Although this attempt ended in huge disappointment when a recovery hook failed to penetrate the twin lugs on the crown of the bell by a margin of less than half a centimetre, it gave us all the information we needed to return three years later with a much better method. This time a specially engineered suction device called a 'sticky foot' was developed by the *Octopus* team to physically capture the bell and lift it out of the wreckage. Instead of trying to thread the needle of the small openings on the lugs, which was impossibly difficult to achieve at a depth of 2,800 metres, we simply placed the sticky foot over the waist of the bell and applied suction. I had to endure a slightly nervous wait of ninety minutes before the bell broke the surface and was hauled safely on deck, but eventually I had fulfilled Ted's last wish.

The bell was a magnificent prize: eighteen inches high, weighing seventy-five kilograms and with two raised lines of writing that encircled the rim. As soon as it was released from the ROV, I began cleaning away the mud and surface rust that was obscuring the inscription. A crowd of people watched over me as I scrubbed away with a soft towel. Once I got enough of the lettering clear, I began reading aloud for the benefit of the huddle of spectators: 'THIS BELL WAS PRESERVED FROM HMS 'HOOD' BATTLESHIP 1891–1914 BY THE LATE REAR-ADMIRAL, THE HONOURABLE SIR HORACE HOOD, KCB, DSO, MVO, KILLED AT JUTLAND ON 31ST MAY 1916.' The word 'preserved' was the key, because it literally meant that this bell had been used on board the battle-ship *Hood*, but had been taken off at the end of that ship's life (she was intentionally sunk on 4 November 1914 to block the southern entrance to Portland harbour from U-boat attack) and personally saved by Sir Horace. The timing of the decision to reissue the bell to the new battle-cruiser *Hood* is not known, but as the order to build this *Hood* was made in April 1916, at least a month before Jutland, there is every chance that Sir Horace himself was behind the plan. Although neither *Hood* was involved in the Jutland action, the connection with Sir Horace, who was killed at the battle aboard HMS *Invincible*, is another significant historical reference point.

We had moved the bell, which was heavier than I'd expected, to an upper deck on *Octopus*, where there was more room for it to be cleaned and inspected. While carrying it with another crew member, I noticed more writing on its waist. It was a shallow engraving that was hard to read, so I had to rub a bit harder with the towel to get this area clean. I was holding the crown of the bell for leverage when suddenly one of the heavy lugs snapped off, causing my hand to slip and my palm to be sliced open by the sharp edge that was created. As blood oozed out of the cut, I couldn't understand how such a thick piece of bronze had broken away with so little force. Later I learned that the lugs had been cracked in the original explosion, and thought how lucky it was that we had not used them to try and lift the bell. With all the excitement and adrenalin that was flowing through my body, I hardly noticed the cut and continued my cleaning, which eventually revealed a second inscription showing the personal importance of this bell to the Hood family: 'IN ACCORDANCE WITH THE WISHES OF LADY HOOD IT WAS PRESENTED IN MEMORY OF HER HUSBAND TO HMS 'HOOD' BATTLE CRUISER WHICH SHIP SHE LAUNCHED ON 22ND AUGUST 1918.'

I could hardly believe our luck in finding and recovering a bell marked with so much history. In addition to representing the soul of *Hood*, it was essentially a piece of physical history that documented the most significant naval events of the first half of the twentieth century. It had served in the Royal Navy on two capital ships for a period of fifty years, and had connections to the battles of both Jutland and the Denmark Strait. The dedication from Lady Hood to her late husband was also symbolic of the great sacrifice made by all of *Hood*'s men, starting with Sir Horace in 1916 and ending with the 1,415 who had perished below us. I could scarcely imagine a more important bell in the entire history of the Royal Navy, and I was proud and overwhelmed with emotion to have played my part in its recovery for the nation.

The last act of the *Octopus* ROV team before we were chased to port by yet another North Atlantic storm was to leave a White Ensign on the wreck site, as requested by the Royal Navy. The flag was carefully unfurled next to the memorial plaque at the bow and was gently waving in the current when the call from the bridge came that the weather was deteriorating, causing the ROV to be immediately recovered. The chance to explore more of *Hood*'s debris fields had been lost, but we left the site knowing that our principal objective had been achieved. While running for the shelter of Reykjavik harbour, we received the copy of a statement made by the First Sea Lord, Admiral Sir George Zambellas, which perfectly summed up why going back to *Hood* to retrieve her bell was so important: 'Her story, her sacrifice, continues to inspire the Royal Navy today. The recovery of the ship's bell will help ensure that the 1,415 men lost, and the name *Hood*, will always be remembered by a grateful nation.'

IV

TSS *Athenia*

THE FIRST CASUALTY OF WORLD WAR II

TSS *Athenia*

SUNK 3 SEPTEMBER 1939

112 died (69 women, 8 men, 16 children, 19 crew)
1,306 survived

Of the two dozen major shipwrecks I have found in my career to date, more than half were sunk in conflicts during World War II. Most of these were either enemy combatants like *Hood* and *Bismarck* or *Sydney* and *Kormoran*, or merchant vessels carrying raw materials and war supplies that were sunk before reaching their intended destinations. It wasn't my plan to specialize in finding World War II shipwrecks, although I do find the stories of the ships and the events leading to their loss fascinating. It helped that I was able to interview and get to know some of the people involved, whether they were survivors like Ted Briggs or relatives of those who weren't as lucky as Ted. Hearing their experiences first-hand allowed me to imagine more vividly the historical accounts I would come across while conducting my research in archives. It was the experiences of my own family, however, that gave me the closest connection to the war and how such events shaped people's lives.

I was born long after World War II had ended, at a time when America was thriving off the back of the baby boom and when the main threat was the Cold War with Russia. My mother was one of ten children of Italian immigrant parents who settled in the small rural town of Honesdale in north-eastern Pennsylvania. My father's childhood was far less stable, as his mother ran off to be an actress while he was still in grade school, leaving him to be raised mainly by a stepmother. Despite their Italian heritage, my mother's family was very civic-minded. During the war, her brother Louis served as a signalman on a PT boat in the navy; Bob was billeted to a naval tanker in the Pacific, while Vince served with the 28th Division, 109th Infantry Regiment of Pennsylvania and was one of the first troops since Napoleon to invade Germany, where he was subsequently captured on a reconnaissance mission behind enemy lines. He spent over a year in a Nazi prisoner-of-war camp somewhere in East Germany before his family found out he was still alive. Through a mixture of good fortune and sheer bravery, he was able to escape the camp when Russian forces moved in, using the confusion to flee to Russia, then Poland and eventually to Italy, where he was put on a ship back to America. When he returned to Honesdale he was given a hero's parade down Main Street on the same fire truck he'd driven as a volunteer fireman before the war. The family dog, a Dalmatian named Bivouac, rode alongside him.

My father served stateside, working as a draftsman for the army, while my mother trained as a cadet in the US Army Nurse Corps, looking after wounded veterans. Had the war continued until her training was complete, she would have become a commissioned officer with the military rank of second lieutenant. Many of the men she cared for were amputees, and she remembers them being very bitter about the pain and personal loss they'd been forced to endure.

After the war, my parents met and married in New York City, but their lives, and mine, might have been very different if a Japanese destroyer named *Harukaze* hadn't sunk the US submarine *Shark II* on 24 October 1944. One of *Shark*'s crew of eighty-seven who died that day was William 'Dewey' Wall, a torpedoman third class and my mother's hometown boyfriend. Dewey and his brother Art were keen sportsmen at Honesdale High School, where my mother cheered them on whether the game was basketball or baseball. When they graduated, both boys went off to war, but only Art returned home.

Shark's story is so incredibly tragic you might expect it to be better known, but among all the other tragic events of the war, it barely rates a single mention in the history books. All that is known for certain is that on the day she was sunk, the submarine made contact with a Japanese freighter west of Luzon Strait between the Philippines and Taiwan and radioed her intention to attack. That was the last message ever received from her. When submarines in enemy waters fail to surface or report in after a period of time, it can only mean one thing: that they've been sunk. Several weeks later, the US Navy obtained information about a Japanese freighter, the *Arisan Maru*, transporting some 1,800 American POWs and 100 civilians from Manila to Japan, that was sunk on 24 October near to *Shark*'s last position. The navy's conclusion was that *Shark* had torpedoed the *Arisan Maru* and was then sunk herself. When all attempts to contact her by radio failed, the navy announced on 27 November – my mother's nineteenth birthday – that the submarine was presumed to be lost.

The lack of solid information about the loss of both the *Shark* and the *Arisan Maru* barely hides the gruesome truth of what almost certainly happened on 24 October 1944. According to the few Americans who survived and made it home to tell their stories, the conditions on board the Japanese freighter, with some 1,900 men crammed into the ship's holds sardine-style without sufficient water, food or air, were so intolerable that the prisoners prayed for deliverance from their misery by a torpedo or a bomb. These 'hell ships' served as bait for the Japanese, for when they were attacked – the *Shark* duly obliging in this case – the Imperial Navy could count on the enemy submarine to surface and search the area for Allied survivors, whereupon they themselves were easy targets to be destroyed. So as the *Shark* moved in towards the cries of American voices calling for help, so did the *Harukaze* to take her revenge. The *Shark* must have endured a tremendous hammering as seventeen depth charges were dropped before heavy oil, clothes and other debris bubbling up to the surface told the *Harukaze* that the pesky submarine was destroyed.

There is no telling what might have happened if Dewey, like his brother Art, had survived the war and been able to continue his relationship with my mother. Other than my father, Dewey was the only man she ever talked about from her youth, so his importance to her at the time was undeniable. For his part, Art was lucky enough to return home, attend and graduate university and fulfil his athletic promise as a professional golfer of some renown on the PGA tour. He went on to win fourteen PGA tournaments, including the 1959 Masters by birdying five of the last six holes in what is arguably the greatest ever final-round finish in a major. That year he won the Vardon Trophy and was named PGA Player of the Year, eclipsing even the great Arnold Palmer. Until he stopped playing golf many years later, Art Wall was the first name I'd look for on the regular PGA and senior tour leaderboards.

I don't come from a military family and all my relatives who did serve during World War II led normal civilian lives after the war ended. Nevertheless, it was a major event for all of them, during which they were thrust into remarkable and sometimes extraordinary situations the like of which they would never experience again. The war had the same impact on millions of people around the world, and this is what I find fascinating: how otherwise ordinary individuals found the courage and mental fortitude to deal with extraordinary and challenging moments. When deciding what shipwrecks are worth pursuing, I specifically look for such human stories hiding within the overall drama of the shipwreck itself. The stories are not always positive; indeed some, like the *Lucona*, demonstrate unconscionably murderous behaviour as you might expect in the case of ships being purposely sunk. The best ones, however, are those that are the most relatable and memorable, and that can certainly be said about the story of the first ship sunk in World War II, the passenger liner TSS *Athenia*.

In late November 2006, I was contacted by a BBC production team who had the idea of making a shipwreck search the centrepiece of a week-long series of documentaries they wanted to broadcast the following summer. The series was about the natural and social history of the British coastline and was simply entitled *Coast*. By the end of 2006, *Coast* was already a very popular series shown on BBC2 averaging more than four million viewers an episode over its first two years, which in TV terms are terrific figures. The producers, however, wanted more. Their idea in a nutshell was to make a more entertaining, less sobering version of *Coast* for broadcast on BBC1 to pull in an even larger audience. They referred to it as '*Coast*-light'. Their original plan was to have four half-hour episodes during a week in late July, culminating in a one-hour special on the Friday. The catch was that the special was to be a live broadcast and for that reason they needed me to find the shipwreck well in advance. It was a ballsy, even wild idea that was another leap beyond what I had pulled off with C4 on the *Hood* and *Bismarck* expedition in 2001. However, when they told me that their first choice for the shipwreck was the TSS *Athenia*, I wasted no time in telling them to count me in.

As stories go, *Athenia*'s ticked all the boxes that attract me to a shipwreck, and for this reason it was on my personal wish list long before I took the BBC's call. It starts with a highly controversial sinking on the very first day of the war – 3 September 1939 – and ends with one of the most significant convictions for war crimes at the Nuremburg trials. It involves tragedy, with 112 people, mostly women and children, being killed, but also a daring and heroic high-seas rescue by six other ships that saved the lives of 1,306. And despite the fact that she was sunk by the first shots fired in World War II, the full story wasn't revealed until the end of the war in 1945. Her sinking thus became one of the most guarded secrets of the German navy. The political ramifications of her sinking were so huge that if the truth had got out, the course of the war would likely have been very different, with the probability of an earlier defeat for Hitler and Nazi Germany. Few shipwrecks are as historically interesting and significant as *Athenia*.

There was no question in my mind that *Athenia* was worth finding, but the next question – and this was what the BBC was hiring me to determine – was whether she *could* be found. My gut instinct, for several reasons, was that she definitely could. Firstly, knowing how many ships had been involved in the rescue, I was almost certain there would be at least one – and possibly multiple – reported position of where she had sunk. So I expected there to be a good starting point and a reasonably sized area in which to conduct the search. Secondly, the sinking was such a major incident that I expected there to be a large volume of primary source documents in the archives upon which I would base my analysis. Controversial events tend to generate lots of correspondence, reports and other documents, which is the lifeblood of my research. Thirdly, I already knew the general area where the ship had sunk, and there was nothing geologically complicated about the seabed there that worried me. The depth of the wreck might be an issue if we were unlucky, but the geology wasn't. And finally, *Athenia* was a big ship: a 160-metre-long, 13,465-ton passenger liner, which made her wreck a large and distinctive target easily distinguishable with sonar.

I needed to get started on the research straight away because the BBC producer, Jane Merkin, was typically in a rush and I was preparing to leave for Antarctica on 1 January. I had been successful in getting time on board the Royal Navy's icebreaker HMS *Endurance* to attempt to locate Otto Nordenskjöld's research vessel *Antarctic*, which sank in the Erebus and Terror Gulf in 1903, and was finalizing my travel arrangements. The *Antarctic* search was a sort of dry run to prove it was possible to find a wooden ship in Antarctic waters, which at the time was a necessary step in order to get backing for the shipwreck I really wanted to find: Sir Ernest Shackleton's *Endurance* (see Chapter VIII).

Over the space of two weeks I found everything I needed on *Athenia* in just three visits to the Naval Historical Branch in Portsmouth and the UK Public Record Office in Kew (now the National Archives) and delivered a six-page feasibility report with supporting documents to Jane. In short, although there were fewer reported sinking positions than I had hoped, my opinion was that 'there was a good starting point for the search and that the wreck could be found within the usual parameters and risks for wreck searches of this type'. My confidence stemmed from the careful action taken by *Athenia*'s chief radio officer, David Don, in communicating the SOS message and ship's position (56° 42' N 14°

05' W) multiple times from almost the minute the ship was torpedoed and throughout the next three hours until he was instructed to abandon ship. The fact that all the ships steaming to this position were able to locate *Athenia* without difficulty is a further indication that the position he transmitted was reasonably accurate. To answer the BBC's question I focused on navigational details in my research, but in the process I uncovered more than forty documents laying out the whole sordid story of how Oberleutnant zur See Fritz-Julius Lemp of *U-30*, in attacking and sinking an unarmed, unescorted transatlantic passenger liner carrying mostly North American women and children, made one of the biggest naval blunders of World War II.

<hr />

The passengers who had been able to book late passage on the Donaldson Atlantic Line's *Athenia* would have breathed a huge sigh of relief as the ship pulled away from where they had boarded, whether it was Princes Dock in Glasgow, Belfast Lough in Northern Ireland or the River Mersey in Liverpool, believing they were making a safe escape from imminent war in Europe. Since late 1938, the prospect of widespread hostilities would have been a worrisome possibility to all as it became increasingly evident that the appeasement of Adolf Hitler by Neville Chamberlain and others had failed in thwarting Hitler's European ambitions. When the unified armed forces of Germany marched virtually unopposed into Bohemia and Moravia on 15 March 1939 to complete the conquest of Czechoslovakia, the threat of war moved one step closer to the doorstep of Western Europe. Finally, news that the Soviets had signed a non-aggression treaty with Germany on 23 August made the invasion of Poland a near certainty. In fact, Hitler had planned to invade Poland on 26 August but postponed it until 1st September after learning that Britain had signed its own treaty with Poland which promised to provide mutual military assistance in case of attack by a European country.

As the crisis worsened, the rush to leave Britain in the final days of August grew to a flood. Most of those wishing to flee were Canadians and Americans, and women and children lucky enough to have relatives in North America where they could bide their time until the hostilities had ended. The problem, however, was finding a ship with empty berths available. In normal times there were several transatlantic passenger ships to choose from, depending on the class of travel one could afford, but these weren't normal times. The Royal Navy was already on a war alert and the government moved quickly to requisition commercial passenger liners for conversion into troop transports and hospital ships. The removal of these liners from the transatlantic trade greatly exacerbated the problem, as scores of passengers holding existing reservations on ships including the *California*, *Aurania* and *Britannic* were told that their reservations were cancelled, sending them scrambling to find alternative vessels. One such alternative was the *Athenia*, and Donaldson's were happy to accommodate the overload.

As morning broke on 1 September over the River Clyde in Glasgow, where the captain and crew of *Athenia* were readying the ship for the arrival of their first passengers, 1.5 million German troops had already begun the invasion of Poland along its western border with Germany, while the Luftwaffe and Kriegsmarine were simultaneously bombing Polish forces on land and at sea. For those who had already taken the decision to leave Britain, the events in Poland that morning placed the urgency of *Athenia*'s voyage in sharp relief. Over the next two days, the ship was scheduled to embark passengers and 900 tons of general cargo, including bricks, curling rocks, textbooks and cars, in Glasgow, Belfast and Liverpool before getting under way to Montreal and Quebec City, Canada. Knowing the ship would be full beyond capacity, Donaldson's made provisions for the increased number of passengers by building temporary berthing for 200 and taking on extra crew. By noon, the vessel was loaded and a pair of tugs were called to guide *Athenia* away from Princes Dock and down the Clyde. As she slowly moved past other ships being prepared for naval service, shipyard workers apparently taunted the passengers standing along the railings with shouts of 'Cowards!' and 'What are ye running from?'

The loading of the remaining passengers and cargo in Belfast and Liverpool took place without incident, and at 4.30 p.m. on Saturday

2 September, *Athenia* weighed anchor from Liverpool. Before the hour was out, with the ship still in the Mersey, a boat drill for the Belfast and Liverpool passengers was held; the Glasgow passengers had had their boat drill on Friday afternoon. All told, there were 1,418 people on board, made up of 316 crew and 1,102 passengers. Of the passengers, 469 were Canadians, 311 Americans, 172 British or Irish, and 150 European refugees of various nationalities. Fully three quarters of the passengers were women and children, which made the gibes of the Glasgow shipyard workers particularly offensive.

Athenia's master for this voyage, Captain James Cook, was a highly experienced officer who had made fourteen previous trips with *Athenia* and served in the Royal Navy as a lieutenant on destroyers and mine-sweepers until 1919. In Liverpool Cook had received secret instructions from Naval Control to steer a course well off the normal track he usually took, and to adopt a defensive zigzag pattern at night-time when submarines normally attacked. The advantage of zigzagging, as opposed to running in a straight line, was that the irregular course and speed changes made it harder for an attacking submarine to predict where the ship would be going next and thus where torpedoes should be aimed. Zigzagging was a standard tactic used by both merchant and naval warships.

The other precaution taken by Donaldson's was to completely black out the ship by painting both sides of the glass in all the portholes and windows. The running lights would be dimmed in the evening, smoking on deck was prohibited, and officers would patrol the ship to ensure there was no visible light to attract submarines. Captain Cook had hoped he'd be well clear of danger if and when war was declared, but in the meantime he was taking every precaution possible. As *Athenia* steamed into the Atlantic at fifteen knots in the early hours of 3 September 1939, he might even have believed they were safely away and that nothing was on the horizon in front of them except the open expanse of the Atlantic all the way to the coast of Canada. In a matter of hours, that belief was suddenly and violently shattered.

At the outbreak of war, the Kriegsmarine was no match for the strength of the Royal Navy in terms of the number and composition of warships in their respective fleets. As dictated by the Anglo-German Naval Agreement of 18 June 1935, the total tonnage of the Kriegsmarine could not exceed 35 per cent of the total tonnage of the Royal Navy, while the ratio for submarines was limited to 45 per cent as long as the 35 per cent total tonnage ratio was maintained. Even though Hitler renounced this agreement in April 1939, Germany was unable to expand the Kriegsmarine because the rapid build-up of the German army and air force had consumed all available resources. Thus, despite being allowed seventy-two submarines under the 45 per cent ratio, there were only fifty-six commissioned boats available to the submarine force on 19 August 1939, when U-boat officers were secretly recalled to prepare for war on an urgent basis. Of these, forty-six were in an operational state, with eighteen of them designated to cover a sparse but wide network in the Atlantic west of England and the Iberian peninsula. The boat assigned to the sector north-west of Ireland was *U-30*, a Type VIIA submarine commanded by twenty-six-year-old Oberleutnant zur See Fritz-Julius Lemp.

Hitler wanted the Atlantic U-boat screen in position well in advance to give Germany the advantage at sea if Britain declared war. It was expected to take the British several weeks to organize a convoy system, and in any case there would be plenty of single vessels already at sea for the U-boats to attack without fear of running into naval escorts. So as *Athenia* ploughed steadily into the Atlantic, seemingly leaving the prospect of war behind her, Lemp and *U-30* were already in their designated patrol area 250 miles out to sea. Lemp's standing order was simply to await an urgent signal and be ready for immediate action. His boat carried ten torpedoes and had a surface speed of 16.5 knots, more than enough to match the speed of *Athenia*.

The British response to Germany's invasion of Poland was to issue an ultimatum for German forces to withdraw from Poland or face war. The deadline for a reply was timed at 11 a.m. on Sunday 3 September. As the hour was reached, and passed, without reply, Prime Minister Neville Chamberlain announced war against Germany at 11.15. Chamberlain's

Me (third from left) and my siblings Susan, Gina and Bobby in our back yard in
Union City, New Jersey.

This green moray eel would attack me every day I worked on an artificial
reef in St. Croix. As he refused to relocate to a nearby reef, my only choice
was to be bitten or kill him.

Right: Standing watch over the EG&G 259-4 side-scan sonar recorder during a survey in the Bahamas.

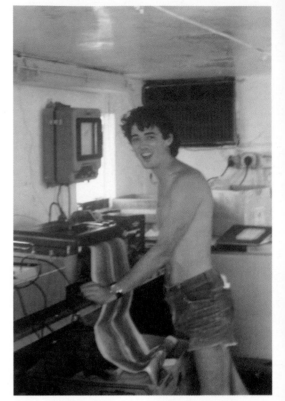

Below: Udo Proksch surrounded by his staff at der Demel pastry shop in Vienna. Proksch gave the captain of *Lucona* one of Demel's famous torte cakes, knowing that the bomb he loaded on board would kill the ship's crew.

The crowded back deck of the *Valiant Service*. The ship was home for the *Ocean Explorer 6000* and *Magellan 725* systems for 15 months.

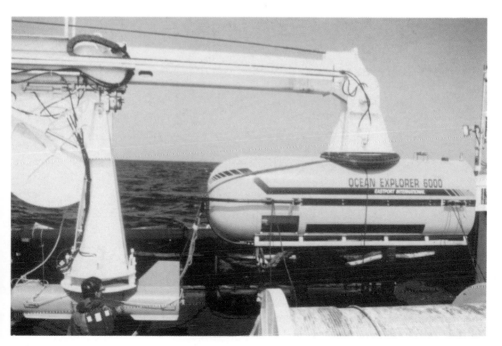

Lifting the *Ocean Explorer* off the deck prior to launch. I'm wearing a radio headset that allowed me to direct the crane and ship drivers during the launch.

Recovering the *Ocean Explorer* sonar in the Mediterranean Sea. The back deck was the most dangerous place to work on the *Valiant Service*.

This innocuous box, resting upside down, contained all the evidence to prove the ship we found was the *Lucona*. Stencilled on its side was the codes (XB 19 and B10) manifest number (02354) and company name (ZAPATA SA) connecting the cargo to Udo Proksch.

The investigative journalist Hans Pretterebner managed to find this photograph of the actual cargo Udo Proksch had shipped in *Lucona*. It helped us identify the exact same pieces of machinery we found lying on the seabed in the debris field. Compare the object circled in both photographs.

The search for *Derbyshire*, and the DFA's quest to learn the truth about her loss, frequently made front-page news in the UK.

Every bright yellowish target in this sonar image represents a piece of *Derbyshire*'s shattered hull.

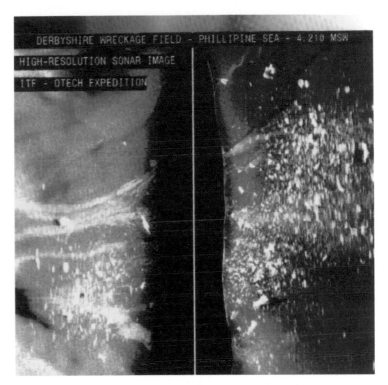

This diagram shows the location of the controversial Frame 65 section of *Derbyshire* just forward of the bridge.

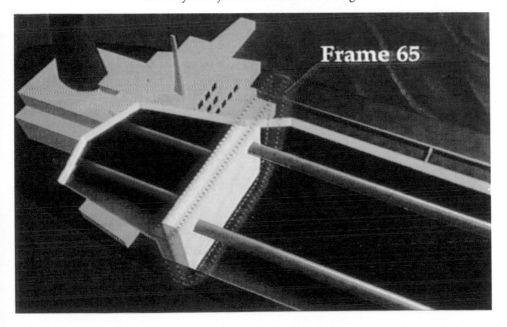

Mark Dickinson bravely backed ITF's search for *Derbyshire*. He is now the General Secretary of Nautilus UK, the Maritime Officers' union.

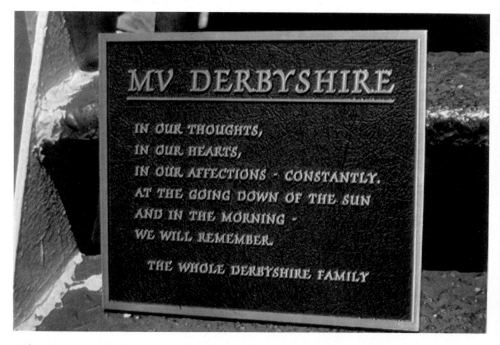

The plaque we laid on *Derbyshire*'s bow meant so much to the families it made me determined to so the same with *Hood*, *Bismark* and *Centaur*.

At nearly 50,000 tons fully loaded, *Bismark* (along with *Tirpitz*) was the biggest and most powerful warship ever built by the German Navy.

This picture of the open hatch door (U 145) through which Baron von Mullenheim-Rechberg escaped demonstrates the quality of pictures we could take at a depth of 4,900 meters.

Others have said that *Bismark*'s amour belt was impenetrable. Here is visual proof that the British guns did put holes in the side of the German battleship.

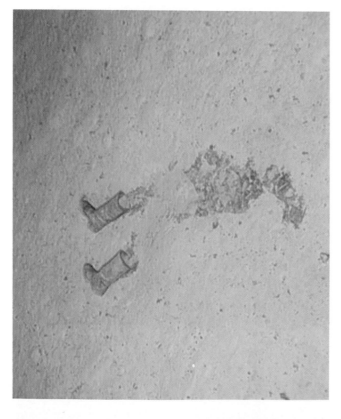

The leather boots and jacket identify this as the remains of one of *Bismark*'s engine room crew.

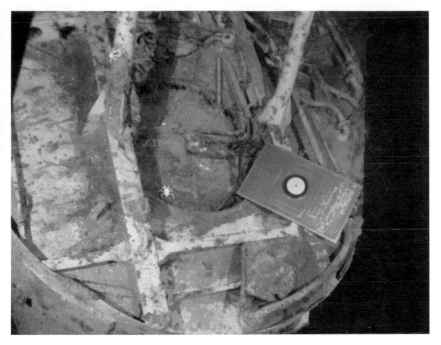

I wanted to lay the plaque commemorating *Bismark*'s men on the bow, but I had to settle with placing it on the upturned Admiral's Bridge as it took us so long to finally find the hull.

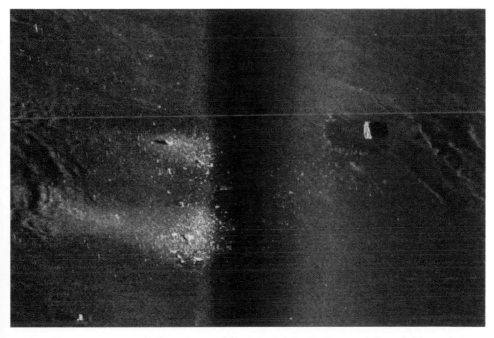

Hood's wreckage, including the middle part of the hull, two debris fields and the conning tower (lower left sonar target) was scattered over a distance of 2.1 kilometers.

Greeting Ted Briggs with a bear hug when I finally got him safely on board our survey vessel over the wreck of his ship — HMS *Hood*.

Enjoying a toast and a cheeky whisky with Ted Briggs and other members of the production team after our successful mission to HMS *Hood*.

The bell of HMS Hood is now on permanent display in the National Museum of the Royal Navy in Portsmouth. When the Princess Royal rang the bell eight times to signify the change of watches, it was the first time the bell had been heard in seventy-five years.

The plaque that Ted Briggs laid on the anchor cable of HMS *Hood*. The gold CD contains the Roll of Honour of all 1,415 men lost in *Hood*.

A cartoon that appeared in *Punch* 10 days after *Athenia* was attacked by U-30, drawn by E. H. Shepard, who once lived in my village in West Sussex.

THE MURDERER

" Remember, women and children first ! "

Below: Karl Dönitz (right) with Fritz-Julius Lemp, Captain of U-30, in August 1940 after Lemp was awarded the Knights Cross.

The twenty-two-year-old John F. Kennedy was sent by his father to meet the
SS *Athenia* survivors that were brought into Glasgow.

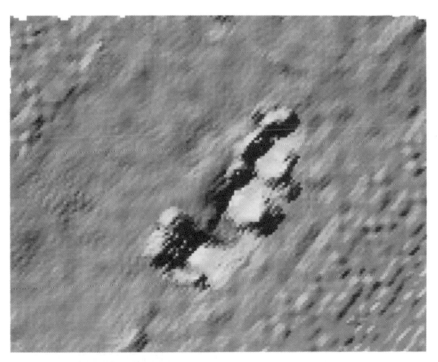

It doesn't look like much, but this multi-beam sonar image contained all
the information I need to know it was the wreck *Athenia*.

THE ILLUSTRATED LONDON NEWS.

The World Copyright of all the Editorial Matter, both Illustrations and Letterpress, is Strictly Reserved in Great Britain, the British Dominions and Colonies, Europe, and the United States of America.

SATURDAY, SEPTEMBER 16, 1939.

A CHILD VICTIM OF AS CRIMINAL A DEED AS THE SINKING OF THE "LUSITANIA": A LITTLE BOY CARRIED ASHORE AT GALWAY FROM THE "KNUTE NELSON," WHICH RESCUED MANY "ATHENIA" SURVIVORS.

More than anything, the idea of the sufferings of the children in the "Athenia" horrified people in every civilised country in the world. Anguished passengers told of how they lost sight of their families below decks when the lights went out, and of sons and daughters who were never seen again. Even more dreadful was the experience of those who were forced to look on while their children drowned before their eyes, or saw them drifting away in lifeboats as they themselves stood on the decks of sinking ships. These were pathetic scenes when the little ones were brought ashore, dressed in darker borrowed from men in the ships that rescued them, or wrapped in blankets. Drawings made by our special artist from the accounts of witnesses of the highest standing, giving an impression more vivid than anything words are capable of, of one of the most terrible, as well as one of the most futile of German crimes, appear on later pages. [Central Press.]

One of the many child survivors from the SS *Athenia* being brought
ashore in Galway, Ireland.

Troops disembarking HMAS *Sydney* in Suda Bay, Crete in early
November 1940. By this time *Sydney* had already earned multiple battle
honours, which included sinking the Italian cruiser *Bartolomeo Colleoni*.

The launch of *Steiermark* in Kiel on 15 September 1938. The ship was converted to
the auxiliary cruiser *Kormoran* in early 1940.

Right: Theodor Anton Detmers, Captain of HSK *Kormoran*. This photo was taken after the war as Detmers is wearing the Knights Cross he won for sinking the *Sydney*.

Below: One of Sydney's crew, Able Seaman Jack Davenport, with a ceremonial life ring.

Inside this German to English dictionary Captain Detmers hid his original 'master' account of the action with *Sydney* using tiny pencil dots under the letters. I have marked the relevant letters in red on the lower image.

B

B, b, B, b; B-flat (*Mus.*); a mark depressing the note before which it is placed a semitone lower (b-flat); das Quadrat B, or B-Quadrat, a mark (♮) rendering the note to which it is prefixed natural; das Stück geht aus —moll, —bur, the piece is in the key of B-flat minor, in the key of B-flat major; for abbreviations see Index at the end of the German-English part.

B

B, b, B, b; B-flat (*Mus.*); a mark depressing the note before which it is placed a semitone lower (b-flat); das Quadrat B, or B-Quadrat, a mark (♮) rendering the note to which it is prefixed natural; das Stück geht aus —moll, —bur, the piece is in the key of B-flat minor, in the key of B-flat major; for abbreviations see Index at the end of the German-English part.

I taped the picture of *Sydney*'s crew above my plotting table to give me, and everyone who passed by, extra motivation to find the wreck. This is where I was working when the wreck appeared on our sonar display.

This shot of *Sydney*'s 'B' turret shows just how deadly effective *Kormoran*'s attack was. The single armour piercing 15cm shell through the front screen of the turret would have killed all 20+ men inside.

Virtually every surface of Sydney's upper deck was peppered with shell holes from *Kormoran's* guns.

Every watertight door on *Sydney's* wreck was found in an open position, possibly indicating that the crew may have been trying to abandon ship at the last moment.

Sydney's main director control tower lying upside down in the debris field. The heavy corrosion marks where the paint has been burnt off by fire.

Celebrating the location of *Sydney* with the directors of HMA3S, John Perryman (in white uniform) and the former premier of Western Australia, Alan Carpenter, fourth from left.

Sister Ellen Savage, the only nurse of the twelve on board *Centaur* to have survived.

Below: With Chris Milligan in his McGill University office where for three days we trawled through all his valuable research documents.

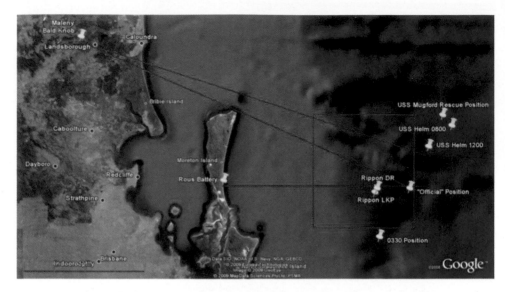

At the end of many months of research I was left with a handful clues to where the wreck of *Centaur* might have sunk. Fortunately, Gordon Rippon's navigation was spot-on.

Staring at this image of *Centaur*'s bell, my smile says it all. I couldn't believe how truly lucky we were to have found the bell with its name showing.

If the bell of *Centaur* hadn't ended up lodged between these two pipes it would have rolled off the deck and have been lost forever.

One of Captain George Murray's leather shoes at the entrance to his bedroom.

The most remarkable discovery: an Australian Army slouch hat in the debris field next to the wreck of AHS *Centaur* where it had lain for 66 years.

Above: One of the red crosses that marked *Centaur* as a hospital ship protected from attack by the Geneva Convention.

I was delighted to have received my Order of Australia Medal from the Governor-General Quentin Bryce at the Australia High Commissioner's residence in London.

e sua caravella q̃ se armou
e mocambiq̃ da madeyra
lauxada q̃ de portugal pexa
l833 lenaxão

esmexalda

bxas sodxée
pexdido e cuxia muxia
com seu jxmão v.te sodxée

Vicente sodxée
com tempoxal se pexdeo e
cuxia muxia do max. do
cabo de guaxda fuj Andado
Daxmada.

The only historical image depicting the loss of the Sodré brother's ships. Note the name *Esmeralda* above the mast of Vicente's ship.

A painting believed to be Vasco da Gama in the Museu Nacional de Arte Antiga in Lisbon.

Ahmed Al Siyabi of Oman's Ministry of Heritage and Culture overlooking the beach where the Portuguese careened one of their ships to be repaired.

This disc I found has yet to be assessed by experts, but it has features that suggest it could be associated with an early type of astrolabe, an important and rare navigation instrument.

Esmeralda's bell, after conservation and reconstruction. Dated 1498, this is almost certainly the oldest ship's bell ever recovered anywhere in the world.

The legendary *Indio*, the lost or ghost coin of Dom Manuel I. The original silver of the coin has been replaced by silver corrosion products, which is why the coin appears black.

The heavy-cruiser USS *Indianapolis*. She had a hand in changing history but it was her men who paid the price.

Charles B. McVay II, the Captain of Indianapolis, who was unfairly court martialed for the loss of his ship.

Sir Ernest Shackleton — known to his men as 'The Boss'. One hundred years on his example still inspires explorers of every type.

With Nick Lambert, former Captain of HMS *Endurance*, holding one of the ten Explorers Club flags I have carried on my expeditions.

In the Weddell Sea Antarctica, where I hope to return one day to find and film Shackleton's *Endurance*.

radio address to the nation was also broadcast to ships at sea, including *Athenia*, via coastal wireless stations. By noon, the fact that Britain and Germany were at war had spread throughout *Athenia* and a copy of Chamberlain's speech was posted in the purser's office. In *U-30*, the news of war had also reached Lemp, but specific orders were soon to follow. The first, at 12.56, ordered, 'Commence hostilities against Britain forthwith'; this was followed by a second message at 14.00: 'U-boat warfare against shipping is at present to be carried out in accordance with international rules.'

The 'international rules', or German Prize Regulations, stemmed from the London Submarine Protocol of 1936, which basically stipulated that it was illegal for a submarine to attack a merchant vessel without having first stopped and searched the ship and ensured the safety of its passengers and crew. Under the same regulations, however, it was allowable for U-boats to attack without warning troopships, vessels carrying war materiel or those being escorted by enemy ships taking part in enemy actions. There was no question that the general directive of the Kriegsmarine was for U-boat commanders to strictly observe the Prize Regulations, a written copy of which was kept in every submarine, and the later message reminding the boats about the international rules was sent by Admiral Karl Dönitz himself, head of the Submarine Service for the Kriegsmarine. Georg Högel, *U-30*'s radio operator, remembered receiving this message and handing it to Lemp.

As the rest of the world reeled from the day's declarations of war (France had also declared war on Germany), the *Athenia* was moving closer and closer to *U-30*'s position. The ship was now well beyond the coastal waters of Ireland, possibly giving some on board a false sense of security. As passengers dined in the saloons, with the war with Germany the topic of discussion at every table, they hardly noticed the periodic changes of course as the ship zigzagged one way, then the other. Twilight had already fallen and the ship was operating under blacked-out conditions. Unbeknownst to Captain Cook, however, the *Athenia* had already been spotted hours before by Lemp, who was waiting patiently for the opportunity to intercept its course.

Lemp had submerged his boat and was slowly closing on *Athenia*'s port side, still unsure exactly what type of vessel he had in his sights. Through the periscope he studied the vessel's profile and noted it was a large blacked-out ship zigzagging on a defensive course. With the war less than nine hours old, what ship other than a British merchant cruiser would be operating this way so far north of the normal shipping lanes? Under the Prize Regulations no warning was necessary to an armed cruiser, so Lemp attacked, firing a spread of two torpedoes from his bow tubes at the mid-ships of the enemy vessel. As the fish left his boat, he was unable to hide the feeling of excitement that coursed through his veins while waiting for the explosions to confirm his first kill. U-boat commanders in war were rated on one set of statistics only: the number and tonnage of ships they sank, and Lemp believed he was about to get his career off to a flyer.

Although German torpedoes were failing at an alarming rate early in the war because the magnetic pistols were causing premature firing, the pair that Lemp unleashed at *Athenia* showed no signs of any defect. He watched and waited, stopwatch in his hand, counting down the seventy seconds he calculated the torpedoes would run before striking their target. At that precise moment a single bright flash leapt up just aft of the ship's central superstructure on its port side. The time was 19.38, and Oberleutnant Fritz-Julius Lemp had made history by firing the first shots of World War II.

In total Lemp fired four torpedoes, the initial pair and then a second pair, although only one of the four found its target. Nevertheless, the powerful warhead of that torpedo caused enough damage to fatally wound *Athenia*. The realization that the ship was in a sinking condition was almost immediate. The General Alarm signal was sounded and the crew began readying the lifeboats to be lowered. Captain Cook, who was in the dining saloon, made his way to the bridge through the chaos caused by the explosion that followed the torpedo strike. The lights had gone out, plunging everything into total darkness, and the ship had taken on a six-degree list to port. When he reached the bridge, Cook confirmed that the watertight doors were closed and instructed the radio officer,

David Don, to send out coded and uncoded SOS messages. The clear, uncoded message, 'SSSS SSSS SSSS ATHENIA GFDM torpedoed position 56 42N 14 05W', was an international SOS that was picked up by radio stations on land and all nearby ships, including *U-30*, whose radio officer, Georg Högel, recognized *Athenia*'s name and call sign and immediately knew this was a civilian passenger liner and not a military troopship.

By now Lemp himself suspected that his identification of the ship as an armed cruiser might have been wrong. He asked for the copy of Lloyd's Register to check the details of *Athenia*. Next to her name was a capital P enclosed by a circle, which meant she was a merchant ship certified for carrying passengers. The owner was listed as the Atlantic Donaldson Line, well known for running civilian passenger liners between Great Britain and Canada. Lemp was shocked. He had made an enormous blunder that violated the express orders sent by Dönitz just hours before. Now he was faced with another critical decision: whether he should assist in rescuing the survivors at the risk of exposing his submarine to counterattack, or do nothing. He chose the latter.

<hr />

According to *Athenia*'s chief officer Barnet Copland, the seat of the explosion occurred in the space between the no. 5 main hatches and the hatches of no. 5 hold. This space was in the vicinity of the bulkhead separating the engine room from the no. 5 hold, and unfortunately was an area used for passenger accommodation. One passenger, a marine engineer, described how the explosion of gases came right up the trunk of hatch no. 5, blowing it into the air along with the people who had been leisurely reclining on it moments before. When Copland investigated this passenger space he found many bodies lying about: 'they were completely blackened – clothes, faces, everything. I made sure that they were all dead.'

The engine room and the no. 5 hold were flooded very quickly, leading to a total loss of power and light throughout the ship. Of the five crew manning the engine room, two were killed outright by the explosion.

Copland saw that the watertight bulkhead between the engine room and no. 5 hold was destroyed, and that there was a lot of water rising on the low (port) side of the ship. He felt that the bulkheads were in a dangerous condition and would not hold much longer. It was up to Captain Cook to give the order to abandon ship, but Copland's report of the damage would have left him in no doubt that *Athenia* was sinking and that rescue of the passengers was now a priority.

Athenia was equipped with 26 lifeboats, including one motorboat on each side of the ship with its own wireless set, providing for a capacity of 1,828 persons. The ship also carried 1,600 life jackets, 18 lifebuoys and 21 other buoyancy apparatus that could support an additional 462 persons. Captain Cook and his crew were not about to let the *Athenia* become another *Titanic*. In fact, the evacuation of the ship was a model of preparedness and seamanship. During the afternoon of 3 September, CO Copland made sure all the lifeboats were fully provisioned and ready for use; the falls were uncovered; the painters were stretched and the plugs installed. After donning their own lifejackets, the officers and senior crew responsible for each lifeboat went to their designated station and began loading the passengers. The upper boats were loaded first, swung out and lowered into the water, and then the falls were retrieved to repeat the process for the lower boats. Fortunately the chief engineer was able to get the emergency dynamo running to provide lighting at the lifeboat stations. By the time the last boats got away, the port list had increased to twelve degrees and the ship was settling by the stern. At about 9 p.m., Copland reported to Captain Cook that all living passengers, bar four men waiting with the officers, were off the sinking ship and in the lifeboats. *Athenia*'s crew had managed to safely evacuate approximately 1,350 people in about eighty minutes.

Cook, his officers and the four passengers (seventeen persons in total) had gathered on the officers' deck when they suddenly realized they had allowed every lifeboat to leave without holding one back for themselves. They had been so preoccupied with tending to the passengers that they'd forgotten about their own safety. Cook hailed the motorboat manned by the second wireless operator and instructed him to offload his passengers into the other boats before coming back for the four passengers still on board *Athenia*. While waiting for the motorboat to return after these four passengers were found places in the other boats, wireless messages were received from the Norwegian cargo ship *Knute Nelson*, the Swedish yacht *Southern Cross* and the American tanker *City of Flint* that they were en route to assist with the rescue. The *Knute Nelson* would arrive at about midnight, but the *City of Flint* was ten hours away from their location. At 11.30 p.m., having thrown his confidential books overboard and been assured that no one alive was left on board, Captain Cook disembarked *Athenia* for the last time.

For the occupants of the twenty-six lifeboats scattered on the ocean the reality of their situation was now starkly different. They were no longer passengers or crew on a large, comfortable ship safe from the elements of the North Atlantic. They had survived a ruthless torpedo attack but their ordeal was far from over. Although the sea conditions were not bad at the moment, the temperature had dropped as nightfall descended and many people had been unable to change into warm clothes before leaving *Athenia*. Most of the children, in particular, were dressed in only thin pyjamas or nighties. They huddled together for warmth and hid under coats and blankets passed to them by adults. Throughout the night the wind and waves increased and it started to rain, adding to everyone's general discomfort. Several boats had filled with water because the plugs had jarred loose, and it took hours to bail them out. The men took turns rowing and did their best to keep the boats riding safely with the assistance of sea anchors, but the lively motion caused many to be wretchedly seasick. They were living through a horribly traumatic experience and the time spent waiting for the rescue ships seemed an eternity. For the unlucky ones in boat no. 5A, however, the worst was still to come.

Once on the water in the motorboat, Captain Cook was able to distribute his senior officers to the lifeboats that had none. He did this for the starboard-side boats by placing 'the Chief Officer on one, the 1st Officer in another, and the 2nd Officer in another and so on'. The third officer remained with Cook and together they used the motorboat to distribute passengers more evenly between the drifting boats, but in general they were all full to capacity and thus riding low in the water. Before leaving *Athenia*, Cook had spotted the lights of the *Knute Nelson*, which was the first ship to arrive, just after midnight. His motorboat was the second boat to reach the *Knute Nelson*, its engine cutting out just before it arrived. It was the job of the men manning the oars to row their boats over to the lee of the rescue ships so that lines could be thrown down to secure them. Manoeuvring the heavily laden boats into position alongside the large ships was proving extremely difficult for the rowers, especially as some were already exhausted by their previous exertions. For the most part everyone had been able to safely board the lifeboats from *Athenia* without incident, bar a few people who fell in the water and had to be pulled out. Now they were facing a more dangerous manoeuvre: getting on board the rescue ships without anyone being injured or lost.

Because of the freshening wind, the waves had grown to ten feet in height by the time the rescue started, making the rowers' job even more difficult. Another problem for those trying to board the *Knute Nelson* was that the ship was in ballast and thus riding high out of the water. Rope ladders and cargo nets were hung over the side for those strong enough to climb the fifteen feet or so to the safety of the deck. For women and the youngest children a boatswain's chair was used, which Captain Cook thought a quicker and safer method. The yacht *Southern Cross* arrived not long after the *Knute Nelson* and started taking people out of the boats at 2.30 a.m. It was a miracle of sorts that two civilian ships were able to reach the scene so quickly. Given the large number of boats and vulnerable passengers involved, and the worsening conditions, it would be a greater miracle if no further lives were lost.

At one point there were six boats precariously lined up along the side of the *Knute Nelson* jockeying for the best position. The calamity everyone had feared happened so suddenly there was nothing those watching could do to stop it. The unlucky boat was no. 5A, carrying about seventy people. Able Seaman H. Dillon was in charge and already had his boat secured by a line to the deck of the *Knute Nelson*. At the worst possible time, however, the captain kicked his engine into gear, moving the ship forward and causing the line to Dillon's boat to be carried away. The boat then drifted down the side of the ship towards the stern and the spinning propeller that was half out of the water due to the ship's light condition. People screamed up to the deck for the engines to be stopped, but it was too late. The huge propeller blades were already smashing down on the wooden boat, causing it to shatter and settle rapidly in the water. Those that were not thrown into the water jumped before they too were struck by the heavy blades. Except for those strong enough to swim to the rope ladders, or the few who clung to the wreckage until they were rescued by other boats, none of the passengers from this boat could be saved, which significantly added to the total number killed.

A second deadly accident occurred at about 5 a.m. when a boat attempting to cross over to the leeward side of the *Southern Cross* got caught underneath its counter stern. As the boat rose on a wave it smashed into the overhanging stern, which was rolling in the opposite direction, and was flipped over, plunging the fifty or sixty people on board into the frigid ocean. Fast action by the other boats and sailors from the *Southern Cross* prevented the death toll from being much worse; nevertheless, it was estimated that eight more lives were cruelly lost when they were so close to being saved.

At 4.45 a.m., the British destroyers HMS *Electra* and *Escort* arrived to join the rescue effort, and together with the *Knute Nelson* and *Southern Cross* they continued to unload boats until all twenty-six were emptied. CO Copland finally got his boat alongside HMS *Electra* at 10 a.m. and later learned that he had been carrying 105 people. Boatswain William Harvey, in control of lifeboat 12A, had made it to *Electra* earlier in the morning at 6 a.m. His boat also had a full complement, which included thirty to forty children aged between ten and fourteen, and eight mothers with babes in their arms. The make-up of these two boats alone underscores the scale of the potential disaster that was averted by the quick arrival of the rescue ships, and the professionalism of Captain Cook and *Athenia*'s crew. When the final list of survivors was tallied, the total was 1,306. Of the 112 who died, 69 were women and only 16 children. The saddest reality, however, is that more died during the rescue effort than from the initial torpedo explosion.

Before he was able to get settled on board HMS *Electra*, CO Copland was faced with another crisis. From *Athenia*'s doctor he learned that a female passenger who was being cared for in the ship's hospital might not have been evacuated, although his earlier understanding had been that she was. He wasted no time, and with two crewmen got back in one of the lifeboats and rowed over to the ship, which by now was very low in the water. They found the woman lying unconscious in the hospital and brought her out. While the crewmen got her into the boat, Copland had one last look at the after bulkhead of the deep tank to see whether there was any chance the ship could be saved, as she had stayed afloat longer than anyone had expected. The amount of water flooding this space, however, convinced him that she wasn't going to last much longer. During this final examination 'he saw about 50 bodies, and they were all black'. As he reported his observations to the commander at 11 a.m., he saw *Athenia* go down: 'She heeled over to port on her beam ends, and sank stern first, the bow coming right out of the water.'

Once she had digested my feasibility report, the BBC producer Jane Merkin asked to meet me. I took this as a good sign, as too often in the world of television the first flush of excitement and enthusiasm for a project evaporates before a second round of discussions can occur. At this meeting, Jane explained that not only were they still excited, but that their vision for the project had expanded. Instead of presenting the *Athenia* search within the existing *Coast* format, they believed the story was strong enough to be filmed for a two-part series of special hour-long documentaries, with the first hour covering the search for the wreck and the second hour constituting a live dive to film it.

The live aspect was what would make the films special, but also challenging in terms of managing the filming and broadcast off a ship operating in the North Atlantic. Although we had proved on the *Hood* project that it was technically possible to broadcast live from a ship, we'd only managed to pull off a couple of short interviews, and one attempt had to be aborted at the last second when the satellite link dropped out.

What Jane wanted – relatively long choreographed dive sequences of an ROV or manned submersible filming the wreck – was a much bigger ask. To fill a one-hour documentary, billed as a live event, we'd need to provide at least twenty minutes of truly live – not taped – underwater footage. Considering all the things that could go wrong, including equipment failure, sudden rough weather or Murphy's Law, the live programme would carry a high degree of risk that would have to be very carefully managed.

Jane had the approval to develop this idea further, which she would have to do in my absence while I was in Antarctica. For the rest of the meeting I took her and her research assistant through a long list of documents they'd need to find and information that needed checking. Among other things I wanted them to locate the logbooks for the *Southern Cross* and *Knute Nelson* and to order a full set of plans and drawings for *Athenia*. We also had to find out some basic information about the ship, such as who was the legal owner of the wreck based on the hull and machinery insurance, and what cover was in place for any loss due to war risks. As this was going to be a high-profile and highly publicized project, I felt we should seek out the owner and make sure they weren't going to object to us filming their property.

I returned from Antarctica at the end of January with some potentially exciting news for Jane. A surveyor from the UK Hydrographic Office who was on the trip had told me that the Geological Survey of Ireland (GSI) had been conducting seabed surveys in the general area where *Athenia* was reported to have sunk, specifically the Rockall Bank and Trough. The surveys had been conducted using a rapidly improving technology called Multi-Beam Echo Sounding, MBES for short, which at depths below 500 metres had a reasonable chance of detecting a target the size of *Athenia*'s wreck. As these surveys were publicly funded, the data could be made available to researchers like me on application. From the GSI website I could see that *Athenia*'s reported sinking position was literally right on the border between the waters of the UK and Ireland. If the wreck was within the surveyed area it might actually show up in the MBES imagery and save us the trouble and cost of finding it ourselves.

That was a very big *if*, however, so I made no promises to Jane other than that I would look into it further.

The people I needed to speak with at the GSI were all at a meeting together, so I dashed off a quick email to the one contact listed on the website, the geologist Dr Eibhlin Doyle. The following day I was able to speak to her on the phone after sending her a more detailed email explaining what we were looking for and where it might show up in their MBES dataset. Eibhlin was very accommodating and willing to help, which was great as I've often been met by a stonewall of inaction when making similar requests to government departments in the past. I think it helped that as fellow geologists we spoke the same language, but also that I had a pretty good reputation for finding significant shipwrecks, so she didn't think I was a crackpot wasting her time. The first bit of good news, she told me, was that their dataset did indeed cover all the areas where I thought the wreck might be located. The second bit was that she had already scanned the data and found three possible shipwrecks in the area I'd outlined for her. Wow, could we really be lucky enough to discover *Athenia* without actually having to go to sea? The only way to be sure was to compare the MBES images for each wreck against what I expected *Athenia*'s wreck to look like. I asked Eibhlin to send them to me so I could make my own comparison.

I decided not to tell Jane this latest news until I'd had a look at the three possible wreck images myself. There was no sense in getting her hopes up until there was more hard information that one of them could be *Athenia*. Because there are many thousands of wrecks around the British Isles and Ireland, there were bound to be at least a couple in the area I'd designated for Eibhlin, so it was too early to get excited that one of them might be ours. I had been fooled once before during a search for a wreck on Porcupine Bank off south-western Ireland and didn't want a repeat performance. On that project we found eleven wrecks in the search area before finally locating the one we were after.

True to her word, Eibhlin emailed me on Monday afternoon to say that of the three wrecks, two looked like interesting candidates and that a colleague was preparing to send me the MBES images and associated data. When the files finally arrived just before the end of the day, however, I could see right away that the first one wasn't *Athenia*. The wreck was too small – only 120 metres long – and there wasn't much mass to it; certainly not enough to constitute a 13,465-ton passenger liner. By the way it was broken down, I also suspected it was considerably older. The second image, of wreck #111, on the other hand, was immediately exciting. To begin with, it measured exactly 160 metres end to end, just like *Athenia*. Clearly it was also a big ship, with parts of it sitting as much as twenty-one metres above the seabed. There was no mistaking this wreck for a smaller freighter or fishing vessel. It had to be a large ship very close to the size of *Athenia*, and it was less than ten miles away from the radio officer's SOS position!

The more closely I studied the image, the more excited I got. Image interpretation can be a bit of an art form, and if you're not careful, you can visualize whatever positive signs you wish to see. In my eyes, however, there was no doubt that this wreck was broken about two thirds along its length, where it was reasonable to expect *Athenia* to be broken as well. My enquiry, which had started out as a long shot, was quickly gaining real promise. Maybe this was *Athenia* and I had discovered her wreck sitting at my desk in West Sussex, nearly 700 miles away. If that was the case, it would be a first in my career, but I needed to be absolutely certain, and that meant flying to Dublin to look at the imagery in the highest resolution possible, using special image analysis software. The GSI could only send me low-resolution TIF files, which were heavily pixelated and nowhere near the quality of image I needed to be sure.

The following morning I called Jane with the news and told her I thought we needed to fly to Dublin as soon as possible. I had first taken the precaution of making sure that the GSI could make the raw MBES data and their computer systems available for the detailed analysis I wanted to conduct. Eibhlin was equally excited to learn the true identity of wreck #111 and was pulling out all the stops to help my detective work. I could tell from her enthusiasm that I was turning her into a wreck hunter. After some initial reservations that things were moving too fast, Jane agreed to join me. Part of the excitement – and fun – of wreck hunting is the chase for the next clue and the urgency to find it as fast as possible. It may seem ridiculous that there could be any urgency about a shipwreck that had been at the bottom of the ocean for sixty-seven years. It wasn't like *Athenia* was going anywhere. Wreck hunting, however, is also about being first, and if the BBC wanted to document that instantaneous moment of discovery, they needed to be present. To Jane's great credit, she realized this and against the odds convinced them to send a small film team with us to Dublin.

Eibhlin was all ready the next day when Jane and I, together with a cameraman and a sound recordist, arrived at GSI's offices in the centre of Dublin. She had two computer systems running, with large high-resolution monitors for both. One computer was running Caris software, which displayed the processed MBES data as a bathymetry map, while the second was running Fledermaus, which is used for sophisticated three-dimensional image analysis. Eibhlin and I wasted no time and sat together to start our review while the cameraman filmed over our shoulders. The first thing I wanted to do was to rule out all other wrecks or wreck-like targets so we could focus on wreck #111. We used Caris for this, and the bathymetry map quickly confirmed that the only wreck to match the dimensions and size of *Athenia* was #111, which was also the closest to the SOS position. While Caris gave us a top-down bird's-eye view of the overall seabed, Fledermaus allowed us to visualize and measure the wreck itself from any orientation in all three dimensions.

For the next hour I examined the wreck from every conceivable angle and made a series of measurements to compare against the profile drawing of *Athenia* I had found in the archives, which included notes about the flooding and damage she had suffered. This wreck was definitely broken, possibly in two locations, and there was a clear dislocation of approximately seventeen degrees between the two main parts. My feeling was that such a large dislocation was likely to have happened at the surface rather than on hitting the seabed. Generally, when ships hit the seabed, their hulls will bend or crumple but they don't fracture and dislocate to such a large degree. In *Athenia*'s case, the ship was sinking, but much more slowly than the crew had originally feared. All the damage and flooding was in the after part of the ship, with probably no appreciable flooding of the bow due to the quick closing of the watertight bulkheads. My suspicion was that the difference in loading of the hull, with one part of the ship fully flooded and the other not, would eventually cause it to fail, especially when exposed to the constant motion of ten-foot waves. It seemed likely to me that what had caused *Athenia* to finally sink was her back breaking, and the remains of that fracture and dislocation were what we were seeing in the sonar images.

My calm and methodical analysis of the MBES images belied the excitement I was feeling inside as it became increasingly clear, in my mind at least, that we were looking at the wreck of *Athenia*. I knew that without visual proof I couldn't be 100 per cent certain, but I was completely satisfied that every indication I could derive from the imagery was positive. While the cameraman filmed my exchanges with Eibhlin, Jane listened intently and then began asking questions. At first she wanted me to summarize the analytical reasons why I felt the imagery matched *Athenia*, but eventually she got to the all-important question from the BBC's perspective. If I couldn't be 100 per cent positive that the wreck was *Athenia*, what percentage *would* I apply to my level of confidence? It was a tough question, but I didn't shy away from answering, telling her that even on a conservative basis, I felt that the chances that this was the wreck of *Athenia* were 90 per cent or better.

She also asked me what this discovery meant for the project. I told her that the risk was now significantly reduced. Instead of having to conduct an uncertain and costly search, with no guarantee of success, we could plan a shorter, less expensive expedition to visually confirm the wreck as *Athenia* well in advance of the live dive. During that expedition we could assess the condition of the wreck and determine the best features to film. We'd also have back-up film in the can in case we ran into serious problems during the dive. Given that the BBC had initially been prepared to conduct a full search for *Athenia*, I thought they'd be over the moon that we had essentially discovered the wreck for the cost of some plane tickets to Dublin.

Jane herself couldn't have been happier or more positive about the outcome of the Dublin trip. This was a big opportunity for her to produce a series of high-profile documentaries to be broadcast on the BBC's flagship channel in prime time. Like me, she felt that my discovery had made the decision an easy one for Peter Fincham and Richard Klein (BBC1 controller and commissioning editor respectively), the executives in charge of commissioning the programme. As they had funded the development to determine whether it was feasible to find *Athenia*, surely the commissioning decision was a no-brainer on the basis that the wreck was now located. We agreed to take no chances, however, and spent the next three weeks conducting more research and getting a better idea of the availability and costs of potential vessels and ROVs for the expedition. We had until the end of February, when the final commissioning decision would be made, to make the strongest possible case for the project to go ahead.

Before I left for Antarctica, I initiated several lines of research that started to yield interesting results. In order to tell the full story of the ramifications of the attack on *Athenia*, we also needed to know what had happened to Lemp, Dönitz and Raeder afterwards. If Lemp had had any doubts about the seriousness of the mistake he had made in attacking a civilian passenger liner, he certainly knew he was in trouble the next day when *U-30* picked up British radio reports confirming that *Athenia* had been sunk and that the casualties included many women and children. He was shocked by his error, telling his crew 'what a mess' he had made, and asking with respect to *Athenia*, 'Why was she blacked out?' In Germany, Dönitz and the naval staff heard the same reports but thought they were false. They simply did not think it possible that a submarine commander would consciously disregard their explicit orders to operate in accordance with the Prize Regulations. Nevertheless, to be sure there was no repeat occurrence, a signal was sent to all U-boats at 23.53 on 4 September: 'The Führer has forbidden attacks on passenger liners sailing independently or in convoy.'

Lemp's ill-judged attack was a propaganda gift to the British government. Because of the large number of Americans involved, the attack was covered by all the major US newspapers, thus publicly exposing Germany as a serious threat to the security of the United States. It also allowed Winston Churchill, who had been brought back into the Cabinet as First Lord of the Admiralty, to open direct communications with President Roosevelt, who decided that the US should amend its existing Neutrality Acts in order to allow the sale of munitions to France and Great Britain.

For their part, Germany needed some form of plausible deniability of the fact that it was one of their submarines that had sunk *Athenia*. The operations division of the naval war staff decided on a policy of 'atrocity propaganda': 'In view of the method of waging war which is to be expected from the British – ruthless, and employing every conceivable means – it might perhaps be maintained that the British staged the sinking by her own mines or by her own U-boats.' And so it was ultimately left to propaganda minister Joseph Goebbels to issue a claim that it was a British torpedo that had sunk the ship, and that Churchill – in an echo of German claims in World War I that he had been responsible for the sinking of the *Lusitania* – had planned the stunt in a desperate attempt to draw America into the war.

U-30 continued its war patrol, sinking two more ships before returning to Wilhelmshaven on 27 September 1939 in a damaged state. Before their arrival, Georg Högel remembers, 'Lemp made us swear to keep the *Athenia* sinking a secret.' The submarine's commander knew that the fallout from any public disclosure of his blunder was potentially explosive, so he was determined to keep it under wraps. He met Dönitz as soon as his boat was docked and admitted to having sunk the *Athenia* in error. After questioning him about the incident, Dönitz decided that he had acted in good faith, and because there was no negligence involved, no disciplinary action was taken against him. But he was still made to report directly to Admiral Raeder, who ordered that the matter was to be kept totally secret. Presumably acting on these orders, Dönitz then had Lemp falsify *U-30*'s *Kriegstagebuch* (war diary) to eliminate all references to the attack on *Athenia*. The dirty deed was actually done by Högel, who was called into Lemp's private office and told, 'I want you to change my war diary.' Lemp then ordered all his men to swear a second oath of secrecy, binding until the end of the war.

Despite this inauspicious start, Kapitänleutnant Lemp (he was promoted on 1 October 1939) actually went on to be a very successful U-boat commander, winning both the Iron Cross and the coveted Knight's Cross for sinking twenty ships and damaging four others. However, his final patrol in command of his next submarine, *U-110*, ended on an equally disastrous note. On 9 May 1941, Lemp sank three more ships but in the process was detected by a Royal Navy escort ship, supposedly because he left his periscope up too long. *U-110* was depth-charged and ultimately forced to the surface, where she was about to be rammed by the destroyer HMS *Bulldog*. Facing a barrage of gunfire from the attacking destroyers, Lemp ordered his crew to abandon ship, which he assumed was about to be sunk along with all its confidential material. Seeing that the submarine was being abandoned, however, *Bulldog*'s commander changed his mind and decided to capture it.

Once on board the empty boat, the party from *Bulldog* found a treasure trove of intelligence material, including an Enigma machine, a short signal codebook and a short weather report codebook. Lemp tried to swim back to his stricken sub to prevent its capture but was never seen again, and *U-110* eventually sank before *Bulldog* could tow it into port. After realizing the importance of the intelligence coup, the Admiralty were happy for it to be known that *U-110* had sunk to make the Kriegsmarine believe that no vital information had been lost. All other aspects of the capture, code-named Operation Primrose, were treated with the greatest secrecy. It took some time before the documents captured from *U-110* started paying dividends, but eventually they played an important role in breaking the German naval Enigma code.

Although it was always suspected that a U-boat was involved in the sinking of *Athenia*, the true story wasn't revealed until the end of the war, when the Russian army seized the Kriegsmarine records from the naval staff HQ near Berlin. That was the first time *U-30* was identified as the offending submarine and Lemp as the commander in charge. Lemp did not live to see his secret revealed, but other members of his crew did, and one of them, who was a prisoner in a Canadian camp, spilled the beans when investigators connected with the International Military Tribunal (IMT) in Nuremburg came calling. Both Dönitz and Raeder were charged with war crimes and tried by the IMT for waging unrestricted submarine warfare contrary to the London Submarine Protocol of 1936. The charge against Raeder specifically mentions the sinking of the unarmed *Athenia*, although Raeder tried to pin the blame on Hitler for ordering the cover-up. They were both found guilty, with Dönitz serving ten years in prison and Raeder receiving a life sentence, although he was released early because of ill health.

During his interrogation at Nuremburg, Dönitz admitted that *U-30*'s war diary 'was not faked, but there was a clear order that the case of the *Athenia* should be kept a total secret for political reasons, and, as a result the log had to be changed'. When I found and inspected a copy of the war diary in the UK archives, it was immediately obvious that it had been doctored. In my eyes there were three dead giveaways. The first, and most important, was that the opening page covered too long a period and included just a fraction of the information normally recorded in these documents. This was completely out of character for a submarine conducting an operational patrol during the opening days of war and told me that a whole series of pages, including the attack on *Athenia*, had been removed and replaced by this single sheet. The second giveaway was that the first page was single-spaced, while all subsequent pages were double-spaced. And finally, the format of the dates on the first page was different from how they were typed elsewhere. Whatever the reason for such carelessness, Lemp and Georg Högel did an incredibly poor job in falsifying the war diary that in the end fooled no one.

By the time of Jane's big meeting with the BBC, where she expected to get the green light for the project, we had found a lot more background information on the broader historical significance of the sinking of *Athenia*. In addition to the ramifications for Lemp, Dönitz and Raeder, a researcher helping Jane found a fantastic-looking builder's model of *Athenia*, and some American survivors were also located who were happy to speak on camera about living through the ordeal as young children. Being able to tell the stories of these survivors was extremely important, especially as they were American, which Jane thought might help the BBC get National Geographic on board in a co-production deal. The other exciting American connection was that Joseph P. Kennedy, who had been the US ambassador to the United Kingdom at the time, had sent his twenty-two-year old son John F. Kennedy to be present as his official representative when the survivors began arriving on the rescue ships in Glasgow. In addition to meeting the survivors in hotels and in hospital – which was captured on film – the young JFK also carried the promise of money, in the form of a fund organized by his father, to help pay for clothing and temporary accommodation. Jane thought there was nothing like a bit of that Kennedy stardust to stimulate the interest of American television.

The mood leading up to the all-important commissioning decision couldn't have been more positive. We had a strong pitch based on two sixty-minute films, with all the technical details of the search and live dive phases worked out. Further research on other shipwrecks lost in the general area of *Athenia* had also raised my confidence level from 90 to 95 per cent. I hadn't even contemplated anything other than a 'yes', which was why I was shocked when Jane called me to say that the decision was a hard 'no'. It made absolutely no sense to either of us that after funding such an extensive development to determine whether it was feasible to find *Athenia*, the project would be terminated by the BBC executives after we'd told them we had actually found the wreck. What more did they expect from a development that had delivered on everything they'd asked for, and more?

Jane wrote to me the following day saying how shocked and disappointed she was. She explained that the decision to cancel the project was due to a lack of finance. I had a hard time accepting this given that during the course of the development my estimates of the total costs actually went down by a huge amount because the wreck's position had been discovered and because I had found the ideal expedition vessel and manned submersibles willing to conduct the project at below market rates. A few days later I was told by another source that the real reason was 'audience viewing terms'; in other words, they thought the programmes would rate badly. Apparently this concern had crept into the executives' minds because a documentary broadcast just the week before on the shipwreck of HMS *Ark Royal* – the aircraft carrier involved in the sinking of *Bismarck* – hadn't pulled in anywhere near the audience numbers expected. While I was obviously biased, I couldn't help but think back to the bold decision made by Tim Gardam at Channel 4 when he commissioned the riskier and more expensive *Hood–Bismarck* project, and feel that the BBC had bottled it.

Despite the BBC's decision, which they were perfectly entitled to make, I haven't given up on the project to film *Athenia* and to tell the story of how her sinking kicked off the Battle of the Atlantic. I have got close to making it happen once or twice since then and I refuse to give up. Others may look at this situation and say that I've already failed, but in my mind you only fail once you stop trying. I am currently making another push to fund the dives to *Athenia*, so watch this space.

V

HMAS *Sydney* (II) and HSK *Kormoran*

SOLVING AUSTRALIA'S GREATEST
MARITIME MYSTERY

HMAS *SYDNEY* (II)

HMAS *Sydney* (II)   HSK *Kormoran*

SUNK 19 NOVEMBER 1941   SUNK 19 NOVEMBER 1941

645 died   *81 died*
0 survived   *318 survived*

The common perception about shipwreck searches is that they are akin to looking for a needle in a haystack. And to a certain extent that is true. The oceans can be an infinitely large place to be looking for a relatively small object if you don't have a decent starting point. As a minimum you have to be reasonably confident that whatever you are looking for is within a defined area. Uncertainty is what determines the size of the haystack. If the essential facts about a lost ship, like the time and approximate position it sank, are either unknown or unknowable, then the haystack is going to be infinitely large. In the case of Australia's most famous and controversial naval loss HMAS *Sydney*, even the experts conceded that the location was impossible to predict with any certainty.

The *Sydney* was the one ship everyone in Australia hoped would be found one day – preferably in their lifetimes. Ever since the ground-breaking discovery of the *Titanic* in 1985 had highlighted the possibility of locating long-lost shipwrecks in the deepest of waters, the prospect of finding the *Sydney*, and finally solving the many mysteries surrounding her loss, had become a national debate – and, dare I say, obsession – conducted in books, academic papers, television documentaries, newspaper articles and ultimately an unparalleled parliamentary inquiry, but most importantly by relatives of the 645 men whose lives were seemingly lost without trace that tragic day in November 1941. For a nation increasingly proud of its ANZAC heritage, it became imperative to find the *Sydney* in order to commemorate the lives of her crew and to ensure their service and sacrifice was remembered.

Why the *Sydney*? Of all Australia's wartime shipwrecks still undiscovered – a list including HMAS *Parramatta*, HMAS *Yarra*, HMAS *Canberra*, HMAS *Nestor* and Australia's first ever submarine *AE1* – what made the *Sydney* so special that the Commonwealth and state governments would risk over $5 million of taxpayers' money on a high-tech gamble to find a wreck that many considered to be unfindable?

HMAS *Sydney* (II) was a modified Leander-class light cruiser that had begun life in a Tyneside shipyard. She was originally slated to serve as HMS *Phaeton* in the Royal Navy. A year after construction started, however, the Australian government purchased the partially completed hull in an effort to bolster the country's paltry sea defences, which at one point in 1933 consisted of just four ships. Prior to its launching on 22 September 1934, the *Phaeton* became the *Sydney*, the second Royal Australian Navy (RAN) ship to proudly carry the name of the country's largest city.

Like her World War I namesake, *Sydney*'s place in the hearts of the Australian public was earned primarily in battle. For *Sydney* (I), it was her destruction of the dangerous German cruiser *Emden* in the Battle of Cocos on 9 November 1914, when Captain John Glossop, RN, used his ship's superior speed and firepower to inflict such severe damage on *Emden* that his opposite number was forced to run his ship hard aground on North Keeling Island to prevent it from being sunk and causing further casualties among his men. Remarkably, *Sydney*'s battle with *Emden* was the first ship-to-ship engagement in Australia's history, and having ended in a decisive victory over such an illustrious adversary, it led to wild celebrations at home and deep respect abroad. Such was the admiration for *Sydney*'s service to the nation that after she was decommissioned in 1928, her tripod mast was erected as a memorial on Bradleys Head in Sydney Harbour.

For *Sydney* (II), the chance to demonstrate the same fighting spirit as her predecessor came in May 1940, when she was ordered to join the Royal Navy Mediterranean Fleet in Alexandria. Following the fall of France and the aligning of Italy with Germany, the Mediterranean became a key naval battleground. The Admiralty was determined to control it, and had enlisted *Sydney*'s help in that cause. Captain John Collins, one of the first cadets to enter the Royal Australian Naval College, was in command of *Sydney*, and it wasn't long before ship and crew were distinguishing themselves in action as part of the 7th Cruiser Squadron. In addition to numerous patrols, *Sydney* took part in the bombing of the Italian fortification in Bardia, Libya, and was called on to sink the disabled Italian destroyer *Espero* and rescue any survivors. Collins' humane treatment of the survivors resulted in many of them requesting to remain on board rather than being transferred to a prisoner-of-war camp.

Sydney's next battle honour came in a major action with the Italian fleet, which became known as the Battle of Calabria. Despite some fifty-seven ships being involved, the action, fought mostly at long range, was effectively a draw, with relatively minor damage sustained by both sides and just one Italian destroyer, the *Zeffiro*, sunk. Nevertheless, *Sydney* was again at the heart of the action, as witnessed by the fact that she expended more than 400 rounds of her main six-inch ammunition against the Italian ships and all of her four-inch anti-aircraft ammunition against the repeated waves of Italian bombers. Having avoided any significant damage thus far, she was gaining a reputation as a lucky ship, but she was yet to come up against her greatest test under fire.

This occurred on 19 July off Cape Spada, north-west Crete. *Sydney* had been detailed to conduct two separate operations: a patrol in the

Gulf of Athens to intercept Italian ships, and cover for a squadron of destroyers conducting anti-submarine sweeps further south. Realizing he couldn't be in two places at once, Captain Collins took it upon himself to position his ship, in company with HMS *Havock*, closer to the destroyers, roughly a hundred miles south of where he should have been patrolling. When one of the four destroyers reported a sighting of two fast Italian light cruisers closing on them menacingly, Collins immediately turned *Sydney* and *Havock* south and raced to intercept the enemy, all the while maintaining radio silence to keep their position unknown.

His cunning tactics ensured that when *Sydney* opened fire on one of the Italian cruisers, the *Giovanni delle Bande Nere*, at 20,000 yards, the faster cruisers were caught completely unawares. Now under attack from both *Sydney* and *Havock* and the four destroyers, who were close enough to launch their torpedoes, the *Bande Nere* and the other cruiser, *Bartolomeo Colleoni*, decided to beat a quick retreat, but not before the *Colleoni* was damaged and finally stopped by accurate shell fire from *Sydney*. Leaving the stricken *Colleoni* to be finished off by the destroyers, Collins set *Sydney* after the fleeing *Bande Nere*, hitting the cruiser once and taking a relatively harmless shell through her own funnel in return. Although the *Bande Nere* ultimately escaped, this was a resounding victory for *Sydney* and a personal triumph for Captain Collins, who was rewarded by being made a Companion of the Order of the Bath.

Sydney returned to Alexandria to a tumultuous welcome, but not before the commander-in-chief of the Mediterranean Fleet, Admiral Sir Andrew Cunningham, made a special trip some forty miles out to sea to personally congratulate Collins and the entire ship's company. As *Sydney* entered the harbour, the deck of every ship was lined with cheering men. Collins and the crew were feted as heroes, and over the next few days the ship received literally hundreds of congratulatory messages and cables from around the world.

To HMAS *Sydney* from HMS *Hyperion*: 'Up jumped the *Sydney*, and gripped her with glee, and we sang as we watched the digger gunners shooting, you come a waltzing Matilda with me.'

HMAS Sydney *(II) and HSK* Kormoran

To HMAS *Sydney* from HMAS *Penguin*: 'From all old *Sydney* (I) serving in the *Penguin*. Well done. You are a chip off the old block.'

To Captain J. A. Collins, HMAS *Sydney*, from Sir Stanley Crick (Lord Mayor of Sydney): 'Sydney delighted and thrilled by gallant exploit of HMAS *Sydney* in sinking Italian cruiser *Bartolomeo Colleoni*. We are proud that you have emulated your predecessor's destruction of German cruiser *Emden* in 1914. Lord Mayor heartily congratulates you, your officers and men on behalf of citizens. Sydney eagerly looks forward to showing appreciation when you return.'

Although *Sydney*'s men had seen enough action to feel they deserved a trip home, they spent a further six months serving with the Mediterranean Fleet, conducting numerous patrols and escorting convoys and bombing missions against the enemy. By mid January 1941 it was time for them to return to Australia, as the government was becoming increasingly concerned about the depleted strength of naval units defending its own waters. Whilst *Sydney* was away fighting in the Mediterranean, German raiders had penetrated Australia's home defences and started scoring some successes of their own. From August to December 1940, the raiders *Orion*, *Komet* and *Pinguin* were collectively responsible for sinking a number of merchant vessels and for the destruction of the phosphate mining and ship-loading facilities on the Micronesian island of Nauru, which Australia relied upon for supplies of phosphate for the production of munitions and fertilizer.

Sydney arrived in Fremantle on 5 February for a brief port call for fresh provisions. When she finally arrived at her berth in Circular Quay in Sydney five days later, the reception she received was overwhelming. Thousands of people and numerous VIPs, including the governor general, the minister for the navy and Sydney's lord mayor, who had indeed delivered on his promise of a heartfelt and appreciative welcome, had turned out to greet the ship and her gallant crew. Throngs of people lined the parade route to applaud the men, resplendent in their dress whites, as they marched from the quay to the town hall, where a formal civic reception awaited them. One estimate put the crowds at a quarter of a million people. At the reception, all 645 crew, from Captain Collins down to the most junior rating, were given a handsome personalized medal commemorating their action in defeating the *Colleoni*.

Although the focus of celebrations for *Sydney* took place in the city whose name she proudly bore, this should not detract from the widespread feeling of intense pride and appreciation shared by the public across Australia. Her homecoming was indeed a great national celebration, which was repeated on a more personal level as every community welcomed back their returning sons. Her remarkable victory had touched the nation and raised its spirits in a way that would never be forgotten. She had gained an almost legendary status that guaranteed *Sydney* would forever be the most famous name carried by an Australian warship. It is for these reasons that the country was thrust into a period of deep sadness and despair when the shocking news was announced later that same year that *Sydney* had been sunk, and that not a single man from the 645 on board had survived.

The nineteenth of November 1941 had all the makings of a completely routine day at sea for HMAS *Sydney*. Having escorted the Australian troopship *Zealandia* north to the Sunda Strait, she was on her return transit south to Fremantle and expected to arrive the following afternoon. Under the command of Captain Joseph Burnett, who had taken over from John Collins in May, the ship was observing strict radio silence, so absolutely no one had heard from or seen her since she handed over the escort of *Zealandia* on 17 November to HMS *Durban* for onward passage to Singapore. This was *Sydney*'s fourth voyage to the Sunda Strait under Burnett's command, but unlike the previous three, it was not to end peacefully.

Approximately eighty-five miles west of Dirk Hartog Island, a completely chance encounter was about to take place that would have grave consequences for Burnett, *Sydney* and all her men. The German raider *Kormoran*, cleverly disguised as a harmless neutral merchant ship (the Dutch

Straat Malakka), was steaming in the opposite direction on an almost perfect intercepting course. *Kormoran* had been on the hunt for Allied merchant ships, and the last ship she wanted to tangle with was a better-armed and much faster warship; in fact this was exactly the type of encounter that Theodor Detmers, her cagey and experienced captain, had been ordered to avoid at all costs. Without hesitation, Detmers turned his ship sharply left and began to flee, taking full advantage of the late-afternoon sun, which made sighting *Kormoran* difficult for *Sydney*'s lookouts.

The events of the next two and a half hours are without question the most controversial and intensely debated in Australian naval history. According to Captain Detmers, and those of his men (316 out of 398) who survived to tell their stories, *Kormoran* was able to defeat *Sydney* after drawing the cruiser extremely close alongside before revealing her Nazi war colours and launching a lethally effective surprise attack. Detmers had pulled off a classic *ruse de guerre* that was entirely legal and permissible under international law. When *Kormoran*'s full arsenal of 15 cm main guns, 3.7 cm anti-tank guns, and 2 cm anti-aircraft machine guns was finally ordered by Detmers to 'decamouflage' and 'fire free', *Sydney* was steaming on a parallel course a mere 900 metres away, with Burnett apparently unaware he was literally staring down the barrels of a dangerous adversary.

By the time the shooting stopped, *Sydney* had been allegedly hit hundreds of times by every calibre of gun *Kormoran* fired. She had also suffered a torpedo strike and explosion on her bow, and when last seen, she was drifting away with fires blazing from stem to stern. *Kormoran*, on the other hand, was hit just three times, with most of her casualties caused by an engine room fire that eventually led Detmers to scuttle his disabled ship later that night. Despite the fact that both ships were lost in the end, this was a completely one-sided battle and a clear but improbable victory for the Germans. Never before in the annals of naval warfare had a converted freighter/raider defeated a light cruiser.

Sydney's loss of men was total. All 645 were killed, either as a direct result of the shelling and fires, or when the ship finally sank, or when the few still alive were cast into the unforgiving sea. With no survivors, the Australian side of the story has never been heard, and all the mysteries, controversy and questions that have swirled around this sad day in Australian history stem from this single fact.

While the upper echelons of the navy and government were quietly, perhaps secretly, prepared to accept the German accounts at face value, the public and the families of *Sydney*'s men were not. Because what those stories indicated was that Captain Burnett and his officers had made fatal mistakes that allowed a seemingly inferior raider to comprehensively defeat the glory ship of the Australian navy. Surely Burnett would know better than to approach an unidentified ship during wartime without determining exactly what it was? Hadn't he written to his superiors about the possibility of German raiders in Australian waters and cautioned his crew to be alert to their presence barely six weeks before? Why then wouldn't he use the full capabilities of his ship, the Walrus aircraft on board or indeed his superior long-range gunnery to positively identify and control the *Kormoran* before coming so dangerously close alongside? Most disturbingly, even if Detmers had caught Burnett unawares, why were there no Australian survivors – not a single one – when more than 80 per cent of the Germans lived to breathe another day?

With the passage of time and no new information coming to light, these questions – or more precisely, the lack of acceptable answers – began to fester in people's minds. The Australian government didn't help matters by the slow and incomplete way they informed the relatives and released information. Nor did the RAN or the Admiralty by not conducting a formal board of inquiry into the sinking of *Sydney* as was to be expected for the loss of a capital ship. The void left by the lack of a definitive and authoritative explanation for *Sydney*'s loss was inevitably filled by speculation and unfounded conspiracy theories that took root and persisted for the next sixty years. Rather than dying down, the controversy over the loss of *Sydney* and her men simply intensified.

<hr />

I credit the well-known and highly respected Australian maritime archaeologist Graeme Henderson with first making me aware of HMAS

Sydney. We were at a conference together at the International Maritime Organization in London in early 1996 when Graeme first told me how *Sydney* had been taken by surprise by a slower and seemingly less powerful German raider and in the ensuing battle had been thoroughly defeated. I vaguely knew about German raiders: how they were essentially warships disguised as merchant vessels let loose by the Kriegsmarine in World War II to patrol the oceans and either destroy or, better yet, take as prizes Allied supply ships. But I had never heard of the *Sydney.*

Two things stuck in my mind about the story Graeme told me. The first was that not a single soul from *Sydney* survived. I knew of other ships where everyone had died in battle, but this was a relatively rare occurrence and it was even more surprising in light of the large number of Germans who did survive. The other curious thing was that no one seemed to have a clue where *Sydney* had sunk. Graeme explained that because there was such a cloud of uncertainty about where the battle had taken place, there was essentially no starting point at which to begin a search and therefore the two ships were bound to remain lost forever.

This discussion with Graeme had started with me saying how I believed that virtually any shipwreck could now be found, regardless of how deep it was, if the latest deep-water search equipment was being used. I gave some recent examples of wrecks I had found with Blue Water Recoveries to make my case, but I could see Graeme wasn't swayed by my confidence. As a sixteen-year-old, he had found the first seventeenth-century shipwreck, the Dutch *Vergulde Draeck*, discovered in Australian waters, and he was one of his country's top maritime archaeologists. So he knew the simplest truth about shipwreck hunting as well as I did: that to find a wreck you needed to be looking in the right place. I had to accept that he knew what he was talking about: that finding *Sydney* was impossible because no one knew where to look. However, I couldn't hide my doubts about there being no documents or information of any sort to point to the wreck location. Graeme must have sensed my scepticism, or perhaps he was simply annoyed by my questioning attitude, because at this point he issued a surprising challenge: 'If you're so good, why don't you come to Australia and find the *Sydney?*'

People had asked me before which shipwrecks I wanted to find, but no one had ever put such a challenge to me so bluntly. It was a remarkable meeting and one that I never forgot. Although the opportunity for me to get seriously interested in a search for the *Sydney* was still years away, I did have this feeling afterwards that my conversation with Graeme was a fateful one, and that one day I would take up his bold challenge.

By the time I started to think seriously about searching for the *Sydney*, some two dozen books had already been published about her fatal battle with *Kormoran*. Of these, none was more provocative than *Who Sank the Sydney?* by Michael Montgomery, whose father was the ship's navigating officer. Michael's 1981 book made several sensational claims, including that Detmers signalled an SOS to lure Burnett into lowering a boat to assist the stricken *Kormoran*; that he fired on *Sydney* using an underwater torpedo tube while flying a neutral (Norwegian) flag in violation of international law; that a Japanese submarine assisted *Kormoran* in the defeat of *Sydney* by firing the final torpedo that put her down; and most egregious of all, that the Germans brutally machine-gunned *Sydney*'s survivors in the water before setting fire to the oil slick enveloping them to destroy any evidence of their evil deeds.

Along with the release of newly declassified documents that had been kept closed for thirty years after the war, *Who Sank the Sydney?* effectively opened the floodgates for a prolonged national debate that became increasingly angry and divisive. Barbara Winter was the first historian to tackle Michael's claims head-on. Relying on exhaustive research and the first use of German naval documents, her book *HMAS Sydney – Fact, Fantasy and Fraud* dissected and debunked more than forty rumours put forth by various conspiracy theorists. Despite her excellent work, in some circles her book polarized people's opinions to an even greater extent. Wildly opposing positions were being staked out by an increasing number of researchers and amateur historians, with the occasional relative of a *Sydney* sailor being sucked innocently into the fray. People either believed the German accounts of the action or they didn't, instead

210

coming up with their own scenarios that to a greater or lesser extent depicted Detmers and his men as liars.

The lack of consensus about the battle and the conduct of *Kormoran*'s crew fuelled the debate that ran on unchecked into the 1990s. Its nature was about to change, however, as the recent discoveries of the *Titanic* and *Bismarck* shipwrecks raised the possibility for the first time that the wrecks of *Sydney* and *Kormoran* could also be found and investigated to shed new light on what had actually happened. These deep-water discoveries stimulated a symposium organized to coincide with the fiftieth anniversary of the battle and held at the Western Australian Maritime Museum (WAMM) in Fremantle, a key institutional player in deliberations about *Sydney* owing to the fact that it had responsibility for the wrecks under the Historic Shipwrecks Act of 1976.

The symposium's main objective was to determine whether the spread of floating wreckage and German lifeboats recovered after the battle could be used to pinpoint where the ships sank, and whether that position agreed with the testimony of Captain Detmers. While the conclusion of the assembled experts was that a search for the wrecks wasn't feasible because the area to be searched was far too large, the focus of the debate had clearly changed in a positive way. Everyone's attention had shifted to determining where the wrecks might be located and what physical evidence could be obtained from them to help uncover the truth about what had happened.

Ten years passed before a second symposium was held, in 2001. This Wreck Location Symposium was also convened at the WAMM, but it was a mark of the progress being made that this time it was sponsored by the RAN Sea Power Centre. The symposium, and the RAN's involvement, stemmed from one of the recommendations of a 1997–1999 joint parliamentary committee that was asked to conduct a fresh inquiry into the loss of *Sydney*. The inquiry was remarkable in a number of ways. First, that it was taking place at all: it was truly unique that a government committee was openly considering the possibility of mounting a search for a sixty-year old shipwreck in order to answer historical questions. Second, the breadth of the inquiry and the public's response to it was astonishing. The committee held public hearings across the country and considered submissions from 201 persons and organizations, which were subsequently published in nineteen volumes totalling 5,281 pages. This was an unprecedented level of interest for such an inquiry. It seemed that nearly everyone in the country had either an opinion or some evidence to offer about the loss of *Sydney*.

As its name implies, the aim of the 2001 Wreck Location Symposium was to more accurately define the potential search area for the wrecks. The participants were all volunteers well known for their research on *Sydney* and/or expertise covering the key aspects involved in potentially defining a search area. The RAN, represented by the deputy chief of the navy at the time and the military historian Professor Peter Dennis, fully expected the presentations to be passionate, but what they obviously hoped for in the end was that there would be some consensus about where to look for the wrecks. Without such a consensus it would be impossible for the RAN to support a search or recommend to the government the outlay of the millions of dollars required. Professor Dennis summed up his closing remarks to the symposium on this discouraging note:

> I have to say, as a concluding comment, that I am disappointed that several years on from the parliamentary report and given all the work various groups here today have been doing, a greater measure of agreement and precision does not yet seem to be emerging. Until it does, talk of mounting searches at this stage is still premature.

When the Chief of Navy, Vice Admiral David Shackleton, announced his decision in June 2002 following an evaluation of the Wreck Location Symposium report, it was predictably negative. In deciding that a suitable basis did not exist for a government-sponsored search for *Sydney*, Shackleton specifically pointed to a 'lack of consensus among historians and researchers as to where the wrecks might be, and hence the huge size of any potential search area'.

Reading the symposium report myself when it arrived on my desk months later, I could see exactly why Shackleton had decided against supporting a search. The two main workshops in the symposium dealing with the question of location couldn't have been more of a contrast, in both their methodologies and their conclusions. One workshop, chaired by Glenys McDonald and based on the oral testimony of people living along the coast who reported witnessing signs of the battle itself, pointed to a location in about fifty metres of water less than thirty miles off the coast north-west of Port Gregory, whereas the other workshop, headed by Wes Olson and based on archival records, recommended a location 185 nautical miles away in water depths of 2,700 metres. Faced with a stark choice between the two locations, which was bound to be highly controversial whatever he decided, Shackleton chose neither.

Looking back now, I can hardly believe that this was the moment I decided to commit myself to becoming involved in a search for the *Sydney*. Shackleton's decision was unequivocal. It dashed any hopes that the government would risk public funds in mounting a search for the wrecks, which was arguably the key recommendation of the parliamentary inquiry. On top of this, the Wreck Location Symposium had revealed that the field of *Sydney* researchers was already a very crowded, competitive and acrimonious space that was probably not going to be very amenable to an outsider elbowing his way in. Professor Dennis's colleagues warned him as much, describing his assignment as 'a poisoned chalice, a minefield, and a vipers' nest'.

Perhaps a more realistic or rational person would have steered clear of the *Sydney* at this time and waited for a change of opinion or a shift in the government's position. I didn't see it this way, though. Despite all the research that had been done previously by many others, I had a definite feeling that there were gaps in the archival record and that some key documents hadn't been found. I kept thinking about the quotation in the parliamentary inquiry report at the top of the chapter on documentary evidence: 'There has to be a large box somewhere holding all these missing documents.' If I could find that box, or any new documents, it might just help to reverse the navy's position. I realized it would probably take something dramatic to get people's attention, like the discovery of a previously unseen primary source document, or one whose significance had been overlooked by other researchers. Most importantly, the information would have to have a direct bearing on the precise location of the battle. Nothing less would do. I knew these were big expectations, but I felt it was worth trying, given that Shackleton himself had left the door slightly open by saying: 'Notwithstanding this conclusion, the navy remains interested in hearing from anyone who believes they may have information that might help in locating the final resting place of so many Australian sailors.'

I've had some great days in archives making important discoveries, but I don't think anything will compare with 16 January 2003, the first day I began serious research into the loss of *Sydney*. I was in the UK's Naval Historical Branch (NHB), which at the time was still based in London, to review some documents I had pre-ordered from the archivist, Kate Tildesley, over the phone. I had given Kate a reference number, PG/11875/NID, which I knew to be a British registration number assigned to documents captured from the German navy after the war had ended, but was wondering why this particular number was on a document from the National Archives of Australia shown to me by Peter Hore, a historian and retired RN captain. Peter had been part of the archive record workshop with Wes Olson, and was now, along with Wes, helping me in my search for fresh information.

When I arrived at the NHB, I was shown into the office of the head of the branch, Captain Chris Page, where Kate was waiting for me with an old-style archive box resting on her lap. Chris began explaining to me the origin of the box and how Kate had found it lying uncatalogued in their basement after her investigation of the PG reference number I had given her. I found it almost impossible to take in what he was telling me, however, as my attention was drawn to the spine of the box, on which was written in very large block letters the single word KORMORAN.

Outwardly I remained calm, but my pulse quickened as my mind raced with excitement about what kind of new or secret information might be inside the box. From the way Chris and Kate were acting, taking the time to carefully explain how it had become effectively lost in their archives, I sensed that whatever documents it contained had to be very important. Because the box was uncatalogued, it didn't appear in any of the normal indices or finding aids that archivists and independent researchers use when searching for records. This meant it was unlikely that anyone had seen the documents it contained since 1947, the year the box was created. Even Peter Hore, who had supplied the document that had led me to the box and who had spent a full year on behalf of the RAN Sea Power Centre searching all the UK archives, including the NHB, for new material on the disappearance of *Sydney*, was unaware of its existence. It seemed too good to be true, but on my very first day in the archives I appeared to have stumbled across the 'large box' of missing documents mentioned in the 1998 parliamentary inquiry.

The mindset that drove the way I conducted this research was the same one I adopted for every shipwreck I attempted to find. I was single-minded in my determination and persistence to find the right documents, and I focused on two things. The first was a strict reliance on original primary source documents as the starting point or backbone for my research. Because information can change or degrade with the passage of time, or as it is passed from one person to another, my aim was always to use the earliest documents created by the key participants or well-placed witnesses to an event. The principle is similar to that used by a court of law in determining what testimony the jury is allowed to hear. While the testimony of credible participants or eyewitnesses is allowed, the testimony from second- or third-hand witnesses is considered hearsay and is not. In researching *Sydney*'s loss, this meant I started with historical documents in archives before reading other people's published books. Peter Hore called my approach getting back to the 'factual ground zero'.

My second focus was to devote all my study and analysis to the navigational clues about where the battle between *Sydney* and *Kormoran* took place. These clues – courses, speeds, ranges, bearings, positions in latitude/ longitude, wind and sea conditions – were the only things that mattered to me because ultimately they were the pieces of information that would lead me to the wrecks. Whereas other researchers were concerned with how the battle was fought, for example, I wouldn't allow myself to get bogged down with questions that didn't directly relate to where the ships sank, which was my sole focus. I knew that the truth about the battle could only be revealed by a damage assessment of the wrecks. So from my perspective, finding the wrecks was all-important and for this reason my research was 100 per cent focused on location, location, location.

While waiting for the NHB to catalogue and scan the documents in the *Kormoran* box so that they could be released to me, I switched my attention to an account that Wes Olson told me was the master copy of an encoded report made by the German Captain Detmers. A coded account found on Detmers when he briefly escaped from a POW camp in January 1945 was already known to researchers, but referring to a document as a 'master' implied it was the first ever created and therefore one I needed to see. Unfortunately, finding this account looked like it could be a real problem. First off, it was contained somewhere within Detmers' personal dictionary, so it was privately owned and thus not publicly available. Secondly, the dictionary had last been seen thirteen years earlier by Barbara Winter, but Barbara had moved on from the *Sydney* saga after her book was published and had given most of her research materials away.

If I wanted to get my hands on the dictionary, therefore, my only choice was to retrace Barbara's steps and hope the current owner would allow me to see it. The only information Barbara could give me was that Detmers' widow Ursula had died in 1997, and as they had no children, the dictionary was most likely with a nephew, Hans-Günther Janzten, who probably lived in Hamburg where the Detmers had also lived. With the help of a German naval contact I'd met during my research on *Bismarck*, I was able to quickly turn those two probabilities into a reality, and within days I was speaking with Hans-Günther, who did indeed have his uncle's dictionary containing the all-important master account. Most importantly, he readily agreed to show me the book and answer any questions I had about his uncle.

Looking back now on these breakthroughs, which occurred less than two weeks apart, I can safely say that it is highly unlikely I would have remained so committed to solving the problem of where the ships had sunk without having so much good fortune so early in my involvement. I knew there was an element of luck in how I stumbled on the box of *Kormoran* documents in the NHB, but I feel that fate also played its part. Like my conversation with Graeme Henderson in 1996 that first piqued my interest in *Sydney*, or my work on *Bismarck* that gave me the contacts to find Hans-Günther Janzten and gain access to Detmers' dictionary, I believe I was destined to find the missing documents that started the process for reversing the RAN's negative stance about a search for the *Sydney*. Everything in my career up to this point, both the successes and the failures, had equipped me with the skills and experience to tackle this enormous challenge. I wasn't just ready to find the wrecks; I was confident I could do it.

My research into the loss of *Sydney* wasn't being conducted as an idle pursuit or for the sake of yet another book. For me, it was strictly a means to an end. I wasn't discouraged by how difficult the challenge might be or by other people declaring that the wreck could never be found. I had set my sights firmly on leading a search for the *Sydney*, which meant it was also up to me to produce the information to make the chief of the navy sit up and take notice. The RAN chief was key, because without his backing the prospect of getting any Australian government funding for a search was zero. Because the cost of such a complex deep-water search would run into the millions, I personally saw no alternative other than it being a largely government-funded project. The one positive was that a search for the *Sydney* had already been elevated to the highest levels, and the 1998 parliamentary inquiry had recommended that the government provide up to $2 million in matched funds.

At the time, I had no knowledge of how the Australian navy worked or whether I could get the attention of the RAN chief, but I decided to write to him in early September 2002 about my interest in *Sydney*. A couple of weeks earlier I had given an interview to a journalist from *The Australian* and told him that contrary to all the pessimistic pronouncements, I believed the wrecks could be found as long as the right technology was used and the search was conducted where the Germans said the battle had taken place. As my name was now publicly linked with a search for *Sydney*, I wanted the navy to hear my thoughts directly rather than reading snippets in newspapers. A changing of the guard had recently taken place, with Vice Admiral Chris Ritchie taking over from David Shackleton as the new chief, which I took as a good omen. I had no issues at all with Shackleton or his decisions regarding *Sydney*, but I thought a new man might have a fresh outlook and not feel shackled by the polarizing Wreck Location Symposium.

I didn't try to get Ritchie to immediately reverse Shackleton's decision, but I did make him aware of a couple of technical misconceptions that might have been a factor in his predecessor's thinking. The first was that a search area had to be no larger than 500 square kilometres in order for it to be technically feasible or viable for a search to be made. This was a throwback to the size of Bob Ballard's searches for *Titanic* in 1985 and *Bismarck* in 1989, which somehow became thought of in Australia as the standard. Using my search for HMS *Hood* as an example, and also to establish my bona fides, I explained to Ritchie how we were prepared to search an area of 2,200 square kilometres, which would take us only eight days to cover with the superior technology that was now available compared to the forty-five days it took to locate *Titanic*.

The second misconception was that there was a credibility problem with the information from German sources as to where the battle had taken place. I agreed that there was indeed a problem, but in my view it was with the people who were analysing the information rather than the information itself. Most people who discounted the German positions did so on the basis of their own personal, and unsubstantiated, opinion that Detmers and his crew were purposely lying about the location of the battle to hide misdeeds or illegal actions. These same people pointed to a lack of consistency in the reported positions as proof, but I told Ritchie this was not unusual in my experience researching World War II ship losses. In fact, when comparing the actual position of wrecks found on the seabed against where they were reported sunk, I had found that in most cases the most accurate positions were from German sources.

I was pleased to receive a written reply from Ritchie the following month, even though he didn't take up my offer to initiate a more extensive search of the German archives. His advice, from official and academic sources in Germany, was that all relevant documents were already in the public domain. I knew this was incorrect because it was well documented that earlier searches had been incomplete and that many sources previously inaccessible were now open, while additional records had been returned from the US and UK. Despite my protestations in a subsequent email, however, Ritchie held firm, refusing to fund any further archival searches as he wasn't convinced that 'new information would help to narrow down the search area for the wrecks'.

I had been knocked back twice by Ritchie, but at least I had opened a direct line of communication with him and he could see that my interest in finding *Sydney* was a serious one. Ultimately this contact did have the desired effect on his attitude towards a search, but nearly three years would pass before that moment came. As I was beginning to learn, changing opinions in Australia about virtually anything connected with *Sydney* was an uphill struggle, calling for extreme patience and persistence on my part.

After my initial research breakthroughs, the next year was spent making the type of slow, steady, undramatic progress that is more typical of the work that needs to be done before an at-sea search can be mounted. At my suggestion, Wes, Peter and I had been working together as a team since late 2002 to try and answer a number of questions I believed were important in pinpointing the probable location of the battle. If we could nail down where the two ships fought each other, we would have a good chance of finding *Kormoran*'s wreck, because the ship wouldn't have drifted very far after being disabled by *Sydney*. And once *Kormoran* was found, we'd then have a fair chance of finding the *Sydney*, since her last sighting had been from the position of *Kormoran*. So that was my strategy in a nutshell: find the wreck of *Kormoran* first, and then use it as a pointer to find the *Sydney*.

From my perspective as a shipwreck hunter, this was a logical and straightforward plan, but I knew it was also very risky and would be difficult to sell to potential sponsors and to the relatives, who understandably wanted the focus of any search to be on *Sydney* alone. Finding one lost deep-water shipwreck is hard enough, but for this project to be judged a success, two would have to be found in a single expedition, because we all knew we'd only get one shot at a funded search. I could easily imagine the distress and anger that would follow in Australia if we found *Kormoran* but not *Sydney*, especially if public funds were used. However you tried to dress up finding *Kormoran* as a good thing, it would never be appreciated unless *Sydney* was also located.

In any search, whether it is deep or shallow, you need to have a starting point. Generally that is a geographic position in latitude and longitude coordinates or some other navigational information of reasonable accuracy and precision upon which the parameters of a search box can be based. As no survivors from *Sydney* lived to tell their side of the story, the only information that made it into the historical record was provided by the German crew from *Kormoran*. The first and most important question in my mind, therefore, was whether the Germans in general, and Detmers in particular, had been truthful in their accounts. If Detmers had lied, as a fair number of Australians believed, the wrecks would probably never be found. In fact no one – and certainly not the navy, given their original negative stance – would even contemplate backing a search if they believed that the only survivors of the battle were lying about where it had happened.

With so many Germans ultimately making it to safety in lifeboats and rafts after abandoning *Kormoran* (316 out of a total 395 on board survived), the number of positions they gave when interrogated by their Australian captors was plentiful. By far the most common position was 26°S, 111°E, which was roughly 120 nautical miles SSW of Shark Bay, where Detmers had intended to lay the 340 mines he had on board before running into *Sydney*. It was obviously a shorthand position, truncated or rounded to the nearest degree of latitude and longitude, but this didn't surprise me, as I had seen the same thing countless times before in records from both world wars. Whether it was abbreviated because the person had forgotten or didn't know the full position in the first place, or had decided for whatever reason against giving any more details when questioned, it didn't change the reality that 26°S, 111°E was a very rough approximation that rendered it unreliable as the basis of mounting a search.

Although 26°S, 111°E wasn't a strong enough clue by itself to justify a search for the wrecks, it did demonstrate to me and others that the German accounts were valid. The number of *Kormoran* survivors who gave this position was simply too large for it to have been a fabrication or a conspiracy to hide something from the Australian interrogation teams. It also came from some of the ship's more senior crew: officers or people in a position to know the movements of *Kormoran*, like Hans Meyer, the navigating officer; Wilhelm Bunjes, the prize officer; and the wireless operators Hans Linke and Ernst Pachmann. For me, the clincher was that the men who gave these positions had been spread throughout the ocean on seven lifeboats and rafts, with precious little time together beforehand if they were supposed to be concocting a fictitious story.

Having left all their logbooks and charts on board to be destroyed when *Kormoran* blew up and sank, Detmers decided very early on during his incarceration as a POW in Australia to secretly re-create a logbook-type report of his ship's action against *Sydney*. This was a confidential account known only to him and was proof of his utter determination to inform the Kriegsmarine about the action and the loss of his ship. Detmers might have been stuck in POW camps from his capture until early 1947, but that didn't stop him from fulfilling his professional duty as a raider captain. He was also understandably proud of his famous victory against a superior warship and obviously wanted it recorded for posterity. Using a standard Cassell & Co. German-to-English dictionary purchased with the regular pay that officers received, he ingeniously recorded his first and thus master account of the battle by a series of barely visible pencil dots placed under individual letters. The two accounts spelled out by the letters – one for *Kormoran*'s bridge and the other for the engine room were fortunately in plain German and not encoded, as Wes Olson had initially thought.

Detmers' dictionary was the stuff of World War II movies like *The Great Escape*, which I had seen many times as a kid, so to hold such a remarkable piece of history when I relocated it with Hans-Günther Jantzen in Hamburg was truly thrilling. The dots were so small and faint, it was easy to see how Detmers had kept it hidden for all those years, even though he was regularly searched. The only way I could make out some of the indentations was by holding the page at just the right angle in daylight. The dictionary spoke volumes about Detmers' time as a prisoner. I imagined how in the privacy of his bedroom he might have silently read the account, perhaps by candlelight, and relived his fight against the *Sydney*. Was this one of the ways he stayed mentally strong as the years, and the war, passed him by?

When we got home with magnified copies of the dictionary, Peter Hore took on the job of turning the dots into letters – we called this 'dedotting' – and then translating the plain German into English. Detmers had a system for dotting the letters, which Barbara Winter had first worked out, making Peter's job a bit easier, although no part of the job could really be called easy. The account contained numerous substitutions, abbreviations and misspellings, and no punctuation or spacing to indicate when one word ended and the next began. The damned dots drove Peter mad and nearly ruined his eyesight, but by the first week in July he had a draft of the dictionary account in both German and English ready for me to read.

The first thing that impressed me about the account was the amount of detail it contained. It was in the format of a standard *Kriegstagebuch* but contained a lot more information than I typically saw in other such war diaries. The entry for 8 November noted the completion of the change of disguise into the Dutch merchant ship *Straat Malakka*, and Detmers' intention to survey the coast off Shark Bay. The remainder of the account dealt with the deadly actions of 19 November and was essentially a blow-by-blow description of the raider's surprise attack against *Sydney* recorded in five- to 10-minute intervals. The key entry occurs at 17.30 'G' zone time (GMT minus seven hours). Having been unable to provide the secret signal requested by *Sydney* to verify their identity as the *Straat Malakka*, Detmers gave a general order that was chilling in its simplicity and lethal intent: '*Entarnen! Feuer Frei!*' ('Decamouflage! Fire free!')

1730 – Order, off disguise. Battle flag flies clear at the mainmast. Time taken to reveal identity 6 sec. Enemy drops astern. Two torpedoes on inclination 90, enemy speed 14, at the same time alter to 260 [degrees]. Single shot from No. 1 gun short, adjust range 1,300 metres. 2nd salvo [guns] 3, 4 and 5 up 400 metres, fire about 4 sec later gave hits on bridge and gun direction tower. Immediately after full salvo [from] the enemy falls wide no hits. Then at 5 sec salvo intervals about 8 salvoes fired. Hits amidships on bridge and aircraft and with bearing correction left between forward turrets. A/A guns and starboard 3.7 [cm gun] fire on [torpedo] tubes, A/A guns and bridge. Up to our 5th salvo no reply, then No. 3 turret [*Sydney*'s X turret] good and quick. No. 4 turret [*Sydney*'s Y turret] shoots only 2 to 3 salvoes all wide. A and B turrets shoot no more. About our 8th to 10th salvo torpedo hits at front of A turret, 2nd torpedo just misses the bow. Enemy turns towards. B turret roof flies overboard.

Although the lopsided battle continued for another fifty minutes, with *Sydney* taking all the subsequent hits, in that opening five-minute exchange both ships had already suffered the fatal damage that would ultimately lead to their loss. At 18.25, with *Sydney* beyond *Kormoran*'s gun range at 10,400 metres, Detmers ceased fire and began to take stock of his own situation. The decision to scuttle *Kormoran* was quickly taken and the next six hours were spent by the crew getting the lifeboats and rafts into the water and abandoning ship. The last Detmers saw of *Sydney* was at about 22.00 hours, when the glow from the fires that had engulfed the stricken cruiser could no longer be seen. At thirty-five minutes past midnight, the fires set off in *Kormoran* by the scuttling charges finally reached the aft cargo holds, where all the mines were stored, causing the ship to blow up and rapidly sink by the stern.

Detmers' account of the action made for fascinating reading, and in time would be verified by the damage we observed in both the wrecks, but for my purposes the most important entry was the following couple of lines at the start of the next day, in which he recorded a more precise position than his crew did, at least in terms of the latitude component.

Wednesday 19.11 – 111 East, 26 34 South, [Wind] SSE 3 to 4, sea [state] 3, medium swell from SW, [visibility] very clear, course 025, [speed] 11 kt.

I recognized the format of the entry straight away as the string of weather and sea conditions recorded in standard *Kriegstagebuch* logs at the start of each day. It confirmed my suspicion that Detmers intended the dictionary account to be an ad hoc replacement for the war diary lost in *Kormoran*. The entry was slightly different however, in that it also included navigation information – namely a latitude/longitude position, course and speed – that isn't normally recorded at the same time as the daily weather conditions. The conflation of weather and navigation information was slightly confusing, but would only become a problem if I couldn't figure out the time of the day, or event, the position referred to. Because no time was given for this entry, I had to basically guess the time of day *Kormoran* was actually in the position 26° 34' south, 111° east. As the ships were moving at considerable speed when they sighted each other (*Kormoran* at eleven knots, and *Sydney* at probably seventeen), an incorrect guess about the time could equate to a large offset of the search box in the wrong direction.

The most logical time the position would have been taken was 12.00, when all ships in the days of celestial navigation would determine their position by taking a 'sun-sight' using a sextant. Wes Olson favoured noon, as did I, as it placed the position of the battle very close to 26° south, 111° east, where the other Germans said it happened. The problem with this assumption however, was that when Detmers' lifeboat was picked up by the passenger liner *Centaur* and he was brought in to Carnarvon to be interrogated, he reportedly said that the action had taken place at a different position: 26° 32' south, 111° east. When this position was relayed to the Navy Board the same day, it was changed ever so slightly to 26° 31' south, 111° east. The one-minute change in latitude is so small it is almost immaterial, but it demonstrates how errors – in either the transmission or receipt of such information – creep into the historical record and cause confusion. In fact, it appears that both positions were the result of errors, as every single written account created by Detmers himself, starting with the dictionary, recorded this position as 26° 34' south, 111° east.

That still left a major question mark over the time of the position. Was 26° 34' south, 111° east the position of the battle at 17.30 or the noon position of *Kormoran*? The problem got even more confusing when Detmers published a book account in 1959, which included a line in the Appendix that listed 26° 34' south, 111° east as the position where *Kormoran* actually sank, which we know didn't happen until 00.35 the following day. There were numerous other questions and unknowns that would occupy my mind right up to the day the search began in early March of 2008, but this conundrum over the time of Detmers' position was by far the most critical, as it would have the biggest impact on the size and placement of the search box for *Kormoran*. Most of my research and thinking over the next five years was spent trying to come up with other ways to independently corroborate Detmers' position. However, as 2003 ended, I was sure that the German accounts could be trusted and that the additional precision of Detmers' position made a search for the wrecks feasible.

Peter and I staked our belief on the veracity of Detmers following a methodical and painstaking analysis of all the extant versions of his account. In addition to the dictionary account he made soon after arriving at HM Dhurringile Prison outside Melbourne in early 1942, there was his 1959 book *The Raider Kormoran* and three other accounts directly connected to him. Two of those, surfacing between the years 1943 and 1945, were already known before I found the fifth and final account, which Detmers called the *Gefechtsbericht*. This was a neatly typed two-page version of his battle report, confiscated from him when he was repatriated back to Germany at the end of the war. The Director of Naval Intelligence (DNI) in Australia had had reason to believe that *Kormoran* survivors had produced a 'Secret Official Report' regarding the action and were going to turn it over to whatever authority was still in charge of the Kriegsmarine. Acting on this tip, DNI Admiralty searched the survivors when they landed in Cuxhaven on 21 February 1947 and found relevant papers on nine of the men, including Captain Detmers.

These confiscated documents were sent back to Australia for analysis by the naval war historian, and then finally returned to DNI London in late 1947, where they were placed in the box I found in the NHB archives marked *Kormoran*. Until I rediscovered this box in 2003, its full contents had not seen the light of day and previous researchers were unaware of the critically important information it contained. This included two new sinking positions for *Kormoran* that had never been revealed before – 26° south, 111° 21' east; and 26° south, 111° 40' east – and a four-panel diagram of the battle hand-drawn by Lieutenant Wilhelm Bunjes, the prize officer. Shortly after *Sydney* was sighted, Bunjes had climbed up into the crow's-nest to act as lookout and thus had as good an overall view of the battle as anyone. This was exactly the type of primary source information, with a direct bearing on how the battle was fought and its location, that the Australian authorities had been seeking since the 1998 parliamentary inquiry.

As exciting as these documents were, the trophy amongst them from my point of view was Detmers' *Gefechtsbericht* account. It was tucked away inside the NHB box in a manila file folder with his name, incorrectly spelled, on its cover. What we found when comparing it to the master dictionary account and the coded account that was found on him when he briefly escaped from the POW camp in January 1945, was that other than some small differences, the three accounts were basically the same. In fact, from the very first time he committed the details of his fight with *Sydney* to paper in the pages of the Cassell's dictionary, right up to the book he published seventeen years later, Detmers' story of what happened on 19 November 1941 never varied. I felt this spoke volumes for his credibility.

As I was to learn, not everyone in Australia shared the same belief in the German accounts as Peter, Wes and me. On the all-important issue of position, there was still a raging debate between those who advocated a more southerly location for the wrecks and those, like us, who accepted the German accounts as the truth. This disagreement had derailed the prospect of navy support for a search after the Wreck Location Symposium, and unless a consensus of opinion emerged, the controversy would ensure that there would be no action from the government. I decided it was worth one more go trying to convince Vice Admiral Ritchie that I was on the right track with my analysis into the most probable sinking positions for both *Kormoran* and *Sydney*. Fortunately, my strategy of staying in contact with the RAN chief finally paid off. Ritchie's reply to a lengthy letter I sent him summarizing my research started by congratulating me for my persistence and went on to agree that I had found some interesting information. Most importantly, he indicated a willingness to reconsider the RAN's position on supporting a search for *Sydney*, with the caveat that he wanted to see more cooperation between interested parties. He didn't mention the other party by name, but I knew exactly to whom he was referring.

As word of my interest in leading a search for the wrecks spread, the number of people contacting me to offer opinions, criticism and the odd tip about where to find new information ballooned. Mainly these were people who had conducted their own research into the battle and wanted to share their thoughts with me. Their attitude was that if someone like me, with runs on the board as a shipwreck hunter, was interested in searching for *Sydney*, then that could only be a good thing. But there were also a handful of 'researchers' who vigorously disagreed with my faith in Captain Detmers and seemed to take great pleasure in telling me that if I planned on searching in the general vicinity of the northern 26°S, 111°E position, then all I should expect to find was barren seabed and no shipwrecks.

There is no doubt that the issue of *Sydney*'s loss and whereabouts generated very strong and passionate views from both sides. I heard from people who accepted Detmers' version of the battle at face value, while others warned me to distrust anything he said, with one person describing him as a 'trained and professional liar of great skill and cunning'. The question of position was just as polarizing, with the northern and southern location camps lining up to defend their ideas while lambasting the other side for their ignorance or sheer stupidity in not being able to grasp what they believed was the true situation. The debate had trundled on for decades before I got involved, so I was under no illusions that my ideas would be treated any differently. I expected to be scrutinized and criticized, and I was.

Fortunately, however, I wasn't alone in being confident about the northern 26°S, 111°E position, and found an ally in a group of volunteers from Western Australia led by Ted Graham. Ted wasn't a researcher, but he had a background in marine survey, which gave him the requisite technical understanding about how to conduct a deep-water shipwreck search. He also had a gritty determination to mount an Australian-led search for the *Sydney* and was involved in several non-profit companies formed for the express purpose of doing just that. Although the first two companies collapsed, he soldiered on and finally found a more stable partnership with two other professionals who, like him, had a long-term interest in finding the *Sydney*. With Dr Don Pridmore (a geophysicist and director of an airborne magnetic survey company) and Dr Kim Kirsner (a cognitive psychologist and naval history buff), he co-founded the HMAS *Sydney* Search Pty Ltd (HMA3S), whose sole purpose was to find the wreck of *Sydney* and commemorate its crew.

Ted was the first person who contacted me after I was interviewed for a popular ABC radio programme hosted by Liam Bartlett on the morning of 16 September 2002. This was typical of him, reaching out to anyone he thought might help realize his dream of finding the *Sydney*. Having discussed my interview with Kim, who was also listening, he wanted to start a conversation about us possibly joining forces, as they liked how I believed in the careful review of primary source information before deciding where to conduct the search. A few days later, I heard from Kim, who was essentially responsible for providing the research ideas within HMA3S. Although his approach was different from mine, he believed there was 'clear and compelling evidence that HSK *Kormoran* lies in the vicinity of 26°S 111°E'. As our mutual belief in the northern position seemed to be good common ground for a possible partnership, I responded to their overtures and started to regularly correspond and share information with HMA3S.

By late 2004, when I finally bit the bullet and decided to travel to Australia for a series of meetings with other researchers and the RAN, the make-up of volunteer directors at HMA3S had changed quite a bit. Joining Ted and Don, whom I had had a chance to meet during their travels to the UK, was businessman Keith Rowe, project manager Bob King, barrister Ron Birmingham QC, and retired RAN commodore Bob Trotter. The day before I arrived in Australia, Kim Kirsner had resigned over an internal dispute about the ownership of some research he and a colleague (John Dunn) had done on the search area. Kirsner's absence during my first full meeting with the HMA3S directors had no effect on me, but it did point to underlying tensions within the group. From my perspective I was simply pleased to be meeting them, as our most recent correspondence hadn't been smooth and I was getting the distinct impression they viewed me more as a competitor than a potential partner.

I was sensitive to the fact that many people saw me as an outsider – I was even labelled a Pom despite being born in New Jersey – who hadn't put in the years of effort they had. But I had made a real contribution with the discovery of the box of missing *Kormoran* documents and tracking down Detmers' dictionary, so I wasn't about to be fazed by mindless personal attacks. I also knew that I was the only one with the bona fides the navy and the government would require of someone leading a search costing millions of taxpayer dollars. Indeed, although I didn't know it at the time, Vice Admiral Ritchie had already met with HMA3S and basically told them that he'd support a search provided they appointed me as the search director.

Looking back now, it is easy to see the dilemma Ritchie faced. On one hand, his predecessor Shackleton had put paid to a search because of the Wreck Location Symposium, which aimed for a consensus but proved to be completely counterproductive. However, that didn't end the public interest in finding *Sydney*, which if anything just got stronger.

The chasm between those advocating differing positions for the wrecks wasn't narrowing. In fact the voices were just getting louder, making it even harder for Ritchie to act. What he needed, and what he got with HMA3S and myself, was two credible groups agreeing about where to search for the wrecks. I imagine he probably saw this as a prerequisite before taking the bold step of reversing Shackleton's decision and throwing his support behind a search, which in turn was the key to unlocking government funds. At the same time, he needed a shipwreck hunter he knew could deliver, and that was me.

Ritchie did eventually reverse the RAN's position, which as expected was the signal the government needed to back the search for *Sydney*. Prime Minister John Howard used the sixtieth anniversary of the end of World War II (14 August 2005) to announce that $1.3 million in federal funds was being granted to HMA3S to bankroll the search. In his statement, Howard pointed to my partnership with HMA3S and my experience in finding the wreck of HMS *Hood* as key reasons for the award. Shortly afterwards, the Western Australian and New South Wales governments added to the total with their own grants of $500,000 and $250,000 respectively. The $2.05 million raised was an impressive sum of money, and a serious vote of confidence in my partnership with HMA3S, but it still fell short of the total amount I felt was needed to safely search a large area of seabed in depths that might exceed 3,500 metres in a worst-case scenario.

The worst-case scenario was a point I tried to convey to Ted Graham and Bob King, my main contacts in HMA3S, who didn't have the experience I had about the margin of safety you needed with such complicated deep-water operations. In addition to all the things that could go wrong during the search – rough sea conditions, equipment failure, breakdown of the search ship and so on – we were about to search for not one, but two wrecks that many people thought could never be found. In my mind this called for extreme caution with our search plans and budgeting. Everyone knew we'd have only one chance with funding, so the biggest mistake would be to go out underfunded and run out of money before both wrecks were found and filmed.

In the end, it took another two years of political lobbying before HMA3S were able to secure additional funds from the Commonwealth government to bring the total funding available for the search up to $5.3 million. In that time there were more personnel changes in HMA3S, with Ron Birmingham and Bob King resigning and Glenys McDonald joining as the fifth director heading into the search. The loss of Ron and Bob was unfortunate, as I liked both men and they had made important contributions right up to the moment they decided to stand down for personal reasons. Glenys's joining was an interesting addition, because her very strong belief in the oral testimonies that placed the battle some 200 nautical miles south-east of the 26°S, 111°E position was one of the reasons the Wreck Location Symposium was deemed a failure. To her great credit, however, she kept an open mind about the northern position and agreed that this was the location that should be searched first.

The new-look HMA3S still relied on the core directors to pull off what would be considered a remarkable feat in anyone's eyes. Bob Trotter was the main conduit to the navy; Keith Rowe possessed the key political contacts to secure federal and state funding; Don Pridmore provided his substantial business acumen and management skills at critical stages; and Ted drove everything forward in his typically forceful manner. They were all volunteers with busy lives, but committed themselves to finding *Sydney* so that every living relative of the 645 men who died in her would know where their bodies rested. I too had volunteered my time since 2002, but in late 2007 my role changed and I was contracted to HMA3S as the offshore search director responsible for the operation. Ted had asked me a couple of times to join as a director, but I believed our partnership would have greater credibility in the eyes of the government if I remained independent, and this judgement ultimately proved correct.

The final player of note was someone who worked quietly behind the scenes within the navy, but who turned out to be absolutely instrumental in guiding the new Chief of Navy, Vice Admiral Russ Shalders, and the Minister for Veteran Affairs, Bruce Billson MP, in their push for the navy to recover the administrative lead after it had languished with another government department for two years. John Perryman was the senior naval historical officer at the Navy's Sea Power Centre, and in this role he was the one tasked to deal with any letters that arrived arguing for more action or claiming some special knowledge about the location of the wrecks. He therefore had a grasp of all the historical information and issues connected with the search, making him the best-informed person in the government. In the end, it was Billson who convinced Prime Minister Howard to approve the additional funding and transfer the administrative lead, but without John providing him with the ammunition for his submission, Billson's request wouldn't have packed the persuasive punch that it did.

I was pleased to hear from Ted, who rang me, on my birthday no less, with the fantastic news that the government was increasing its funding for the search, but an hour later I received another call that had the potential to destroy everything we had worked so hard to achieve. This second call was from Peter Meakin, the director of news at Channel Seven in Sydney and one of Australia's most experienced news journalists. I had met Peter before and stayed in touch, and immediately thought that the government's good news had leaked out and he was calling for a comment. Instead, he had unbelievable news that could only be described as bad: that a group of amateur divers were claiming to have found *Sydney*, and had underwater video of the wreck to prove their claim!

The discovery hit the front page of the *West Australian* newspaper the following day, with the banner headline: 'FOUND. Local amateur historians say they have discovered HMAS *Sydney*, solving Australia's greatest military mystery.' The timing of the news couldn't have been worse for us, as the prime minister was planning to announce the increase in funding for HMA3S in a press conference the very next day from the deck of the current HMAS *Sydney* (IV), accompanied by Minister Billson, RAN chief Shalders, Ted and Glenys. Now, with all of Australia fixated on the possibility that *Sydney* was already found, the government had no choice but to postpone the funding announcement until the purported discovery could be confirmed one way or the other. Until such time, we had to wait patiently, even though I immediately knew from the photographs Peter Meakin sent me of the wreck that it wasn't the *Sydney*.

A fortnight later, a navy hydrographic survey vessel, HMAS *Leeuwin*, confirmed that the wreck wasn't the *Sydney* or the *Kormoran*. It was just a small wreck, 30 metres long compared with *Sydney*'s overall length of 169 metres, that the locals in Shark Bay believed was a barge sunk there after the war. The whole episode was a fiasco that unnecessarily upset the relatives and caused us to lose valuable time in starting the real search effort. The amateur divers walked away from the experience feeling bruised and chastised, but it was the newspaper that suffered most of the damage for its poor judgement and irresponsibility in running such a questionable story without proper verification. It was a lesson to me about how careful we'd need to be in managing any news generated during our search, which after six long years of trying was now cleared to get under way in earnest.

∞∞∞∞∞∞∞∞∞∞∞∞∞∞∞∞

The economics of a shipwreck search are pretty simple: money buys time on the water. It follows, therefore, that the more money you have, the more days you can search and the greater your chances of finding what you're looking for. For that reason, HMA3S's $5.3 million pool of money was a figure I thought about only in terms of the number of days of search time it would buy, and whether that would be enough to find the *Sydney*. Excluding all the fixed expenditure, like the company's overheads and the substantial mobilization costs for the vessel and search equipment, I reckoned that the budget could fund an absolute maximum of forty-five days of in-water search time. The number of days I would get for planning the search would be fewer, however, as HMA3S wanted to reserve some for contingencies. Normally these details are agreed in advance, but as Ted was unable to give me a clear answer on the number of days I'd be allowed, I based my search plans on having at least thirty, which was two to three times the number I was normally given to find a single World War II shipwreck. As we all knew, this was not a normal search.

If I were to summarize my strategy for finding *Sydney*, it would be: 'plan for the worst, but hope for the best'. As it was imperative the wrecks

233

were found whatever the cost, I wanted to make sure my plan covered every possible location, even the unlikely ones. In the end, I opted for a search box that was thirty-four nautical miles wide by fifty-two long. This was a huge area covering all potential locations derived from the German positions, and also incorporating large margins for error. It was a conservative plan, but one I could afford in view of the time I had at my disposal. This was the 'planning for the worst' part of my strategy.

Hoping for the best meant that within the 1,768 square nautical mile area, I would first search the parts where my analysis indicated the wrecks had most probably sunk. If my analysis was right, *Kormoran* would be found first, early enough to leave sufficient time to conduct the second search and locate the *Sydney*. However, if *Kormoran* wasn't located where I initially thought, I would move on to the next highest probability area based on the principles of Bayesian search theory. In practice, this meant choosing the sequence of track-lines to search in order of the highest to lowest probability. While I favoured certain parts of the box based on the relative weight I assigned to each of the German positions and the other loss clues, I didn't go as far as placing an 'X' on the chart where I thought *Kormoran* had sunk, as that would be an arbitrary decision and contrary to the way I conduct such searches.

The length of the search box was born out of the uncertainty about whether Detmers' 26° 34' south, 111° east position was the noon position or the battle position. Although my box would cover both scenarios, I had one final clue to help me determine the more likely of the two. One of *Kormoran*'s life rafts, heavily loaded with twenty-five young Germans, was the earliest to be picked up at sea by a passing troop transport, the converted Cunard liner *Aquitania*. As the time the raft left *Kormoran* and the time and position it was recovered eighty-two hours later were all known, I'd be able to calculate where it started if I could just determine the speed and direction it had drifted over that period. Since the sinking position of *Kormoran* and the starting point of the raft were essentially one and the same, I saw this reverse-drift analysis as a way to corroborate the German positions using independent information, and to help me resolve the uncertainty about Detmers' 26° 34' south, 111° east position.

An object floating on the ocean drifts at a speed equal to the sum of the velocity of water (the current) and the velocity of the object, with respect to the water, as a function of the wind speed (leeway). Every object reacts differently to the force of the wind, depending on how much of its surface area is exposed. For example, objects floating high with a lot of area above the sea surface will drift quickly, whereas objects that lie low in the water will drift more slowly. An empirically derived number called leeway factor can be used to calculate leeway speed as a percentage of wind speed. Fortunately, leeway factors for different types of life rafts, including World War II vintage rubber rafts, are known from field studies conducted by the search and recovery (SAR) community. In the case of *Kormoran*'s raft, I'd be able to get a reasonable estimate of its total drift as long as I could somehow determine what the current and wind speeds and directions would have been in late November 1941.

To get the most scientifically robust data, I decided to commission two brand-new studies, one on historical wind conditions by Len van Burgel, a professional meteorologist, and the other on currents by Dr David Griffin, a physical oceanographer from the CSIRO Marine and Atmospheric Research Division based in Hobart. Other researchers had tried this for the 1991 symposium, but in early 2008 we had the advantage of computerized databases that gave us access to much more precise data over long periods, 1968–2007 for winds and 1992–2006 for currents, to pick out dates that most closely matched the conditions in late November 1941.

Dr Griffin, who ran BLUElink, an ocean-forecasting system based on satellite-derived data and advanced supercomputer processing, produced the final chart of trajectories that showed possible locations where the rafts could have been launched. In addition to the raft picked up by *Aquitania*, I also had analysis run for a second raft picked up 115 hours later by the SS *Trocas*. The picture was complicated by the fact that the dominant current in this part of the world, the Leeuwin current, is highly variable because of random instabilities leading to the formation of transient eddy currents. Before the BLUElink project was initiated in 2003, physical oceanographers were more or less unaware that these large, swirling and unpredictable currents even existed. For this reason any previous attempt to pinpoint the sinking position of *Kormoran* based upon drift analysis was pretty much an exercise in futility. There was simply no way of knowing what the current field looked like in late November 1941 or how eddy currents might have affected the rafts' drift.

Instead of a single trajectory based on the average current conditions, which would have been entirely misleading, I explained to Griffin that what I wanted to see was the full range of trajectories from as many representative scenarios as he could extract from the BLUElink dataset. He produced a total of 300 for each raft, each representing daily surface currents from the latter half of November for a consecutive fifteen-year period up to 2006 and for years when the southward-flowing Leeuwin current was weaker than normal. The final results reached me the day before our survey vessel *Geosounder* was scheduled to leave Geraldton for the long voyage north to the search area, and they had an immediate impact on my search plans.

Griffin's chart of back-calculated launch points didn't tell me where *Kormoran* had sunk. It wasn't intended to. What it did do was show me all the possible positions where the two rafts could have departed from the ship. Because some of the launch points did start at or around Detmers' 26° 34' south, 111° east position, it meant I couldn't automatically rule this out as the battle position and still needed the search box to be large enough to cover this area. However, as most of the launch points were all well north of Detmers' position, the clear indication was that 26° 34' south, 111° east was more likely to be the noon position of *Kormoran* on the day of the battle as Wes Olson first postulated and I had been trying to prove ever since.

At a press conference in Geraldton's town hall the next morning, I gave the media and the public a short presentation about my research and search plans. I explained the science behind the reverse-drift analysis and the confidence it gave me in the belief that *Kormoran*'s wreck was most probably located in the north-eastern quadrant of my overall search box. This was the new high-probability area and I would arrange the sequence of track-lines to ensure it was searched first. Although I still expected to find *Kormoran* before *Sydney*, I prepared everyone for the possibility that the sequence of discovery could actually be reversed. This was one benefit of maintaining the extended length of the search box: that it also covered areas to the south where there was a good chance *Sydney*'s wreck was located. The long period of research had come to an end. It was finally time to see whether the Germans had actually been telling the truth about where they fought the *Sydney*.

⋅∞∞∞∞∞∞∞∞∞∞∞∞∞⋅

Searching the deep ocean for wrecks is a complex and risky operation that rarely goes perfectly to plan. All sorts of problems crop up that in extreme cases can mean the difference between success and failure. The main culprits are rough weather and equipment failure. But over the course of my career I've had projects stopped dead in their tracks for a host of other unexpected reasons, including ships breaking down or catching fire, injuries to personnel requiring helicopter medevac, loss of sonar towfishes and ROVs, and in one instance literally running out of food, prompting an urgent return to port. This is why we always have contingencies in the form of extra days built into our schedule. The other form of protection is to carry plenty of spare parts, especially for mission-critical pieces of equipment. So for a lengthy search like the one we were about to start, I had to expect some problems and losses of time. What I didn't expect was that the problems would start before we even left Geraldton harbour.

The company that HMA3S had hired to conduct the search operations was the Seattle-based geophysical survey contractor Williamson & Associates. Despite their small size – maybe twenty or thirty full-time employees and offshore contractors – they were one of the world leaders in the field of deep-water searching and had the type of wide-swath side-scan sonar systems I felt were best suited to this particular search. I knew the company very well, having worked with them in the past, and was happy to be going to sea with such an experienced team. I also knew, however, that because the pool of deep-water search contractors to choose from was so small, HMA3S had had little choice beyond

Williamson. In fact they were virtually handed the job when the only other company in the running had to withdraw from the selection process because of a scheduling conflict.

Having worked all around the world, Art Wright, Williamson's party chief, understood the challenge of running such a high-profile job in a remote part of the world far from their Seattle base. For that reason he made sure he brought lots of spare equipment; most importantly, a back-up deep-tow winch fitted with 10,000 metres of cable and a second side-scan sonar system. These were essential pieces of equipment, because while without a sonar you obviously can't conduct a search, equally, without a winch or cable you can't operate a sonar. The comfort I felt from seeing the two large, heavy winches welded to the deck of the *Geosounder* turned to dread the morning we were scheduled to get under way when Art broke the bad news that the high-voltage electrical core of their primary tow cable had shorted, rendering this winch-and-cable combo unusable. That meant that the spare winch was now the primary, and the broken primary served no further purpose for the entire duration of the search other than being twenty-two tons of dead weight whose position on deck left the ship with a nauseating tendency to roll in moderate seas.

I was still bemoaning the loss of the winch and cable a few hours later as the *Geosounder* pulled away from the quayside and began to make her way towards the protective breakwaters at the entrance of Geraldton harbour. It was a big moment for the town, and for Western Australia. People gathered to wave us off while television news crews filmed us from land and a helicopter circled overhead. There was a palpable sense of excitement and anticipation for what was to come over the next several weeks, and the cloud of anxiety that had hung over the ship throughout the day had lifted in the fresh southerly breeze. I had come up to the boat deck to take in the scene and was joined by Glenys and Carter Huynh Le, Williamson's electrical engineer. We watched as the pilot boat came alongside to retrieve the local port pilot, signalling that command had been handed back to *Geosounder*'s captain, Blair Cliffe. Not a minute later, the ship's loudspeaker shattered the mood by requesting my urgent presence on the bridge. The last thing a captain wants when entering or leaving port is extra bodies on the bridge, so I knew something was drastically wrong.

When I got to the bridge, the concern etched on Blair's face confirmed my worst fears. A fuel leak in the engine room had caused it to fill with great clouds of smoke. No fire was reported, but the risk of one breaking out meant we needed to urgently find a safe anchorage to stop the ship and assess the severity of the situation. There were some very tense moments on the bridge trying to get *Geosounder* over to the anchorage just north of the harbour without running aground on an area of shoals and without getting the props fouled on the floats and lines of crayfish pots. While John Perryman acted as an additional lookout, I was steadily calling the depth under the keel from the echosounder to help Blair and his second mate get the ship anchored without further incident.

The leak wasn't big, but it was in the worst possible place, with the fuel streaming down from a hairline crack in an overhead tank on to a red-hot engine exhaust covered with lagging that started to smoulder and create smoke. Fortunately the engines hadn't been running long and the leak was caught early. Had we been at sea for longer, the combination of fuel oil and heat could easily have ignited a real fire. The ship's engineers worked feverishly to patch the crack, but nothing they tried could stem the persistent stream of diesel pissing down on them. As much as I hated the idea, and the embarrassment, of limping back into port with our tail between our legs, we had to get the tank bottom properly repaired by welders, which also meant pumping out the forty tons of diesel fuel it contained. To make matters worse, it was a holiday weekend, so I could see us losing two to three days, or worse, to make the repairs.

At times like these you rely on your teammates, and I had one of the best in Patrick Flynn, the project manager HMA3S had hired to handle all the shore-side logistics and media enquiries. As soon as I called Patrick from the ship, he was immediately on the case. The next morning when we returned to port he had two road tankers waiting for us to remove all the fuel and vent the tank of dangerous gases so the welders could get to work. With the ship patched up and refuelled, we were ready first thing

Monday morning for the pilot to guide *Geosounder* back out through the harbour entrance, this time without a repeat of the dramas of Friday afternoon. The leak did cost two and a half days of my schedule, but without Patrick driving the repair, it could have been twice as bad.

My original plan to cover the 1,768 square nautical mile search box was based on searching a grid of twelve north-to-south track-lines using Williamson's wide-swath SM-30 sonar. This low-frequency sonar (30 kHz) had enormous capability for covering a swath of up to six kilometres while still being able to detect wrecks the size of *Kormoran* and *Sydney*. If any wreck-like targets were detected, we'd then switch to the high-frequency AMS-60 sonar (60 kHz) to collect better-quality images that would allow us to identify the wrecks. As my entire search plan relied on being able to cover the seabed at a high rate with the SM-30, I was pinning all my hopes on how it would actually perform. Sonar performance was so important that before steaming up to the search area, we conducted a series of commissioning tests west of the Abrolhos Islands using both sonars just to make sure they were producing the quality of images I expected.

I chose a central track-line (number 7) just east of the 111° longitude to get an indication of what the seabed through the heart of my search box looked like. We purposely began outside the box to give the Williamson technicians plenty of time to tune in the SM-30 and for the ship's drivers to find the best heading and speed to run the track-line. With 7,000 metres of tow cable deployed and *Geosounder* crawling along at 2.8 knots, the sonar towfish was more than three nautical miles behind the ship at all times. I considered this to be a practice/reconnaissance line that, if need be, could be rerun if the sonar imagery wasn't up to scratch or we weren't getting 100 per cent coverage of the seabed. At one ping per four seconds it takes a fair amount of time before a coherent image of the seabed can be generated.

An hour later the verdict was in and it wasn't good. Despite performing well during the commissioning test, a technical fault with the SM-30 cropped up that left it producing poor-quality images. So poor, in fact, that I asked Art Wright to recover the towfish after it reached the end of the track-line to fix whatever problem had developed. The imagery was also being blighted by acoustic noise at the sea surface. One quirk of side-scan sonars is that they can 'see' above as well as below, and in situations where the range to the surface is less than the operating range of the sonar they will pick up acoustic noise from surface waves. As we were towing the SM-30 at a depth of 2,200 metres, but searching at a range of 3,000 metres (half the six-kilometre swath), the sea surface returns were effectively masking the outer 800 metres of seabed imagery on both the port and starboard channels. The net result was that we were only searching 70 per cent of the seabed I planned for each track-line to cover.

I drew a diagram to show John Perryman how the geometry of the situation was working against us, and explained that it would be better if we were searching in deeper water. He was surprised by this, but understood the dilemma I faced. I could no longer cover the search box with twelve track-lines, because of the reduced range, and had to increase the number to seventeen, costing us an additional four days. Together with the fuel leak and an extra day needed during the mobilization, the total hit from these losses to my planning was seven and a half days. Suddenly the original healthy budget of thirty at-sea days was beginning to look a little anaemic, and the safe margin of extra time I thought we would have was evaporating. We were off to a bad start that put everyone under pressure, and that pressure was about to mount in the coming days.

The man in the hot seat to fix the ailing SM-30 sonar was Carter Le, a hard-working engineer who had emigrated from Vietnam to study electrical engineering at the University of Washington. At first the problem appeared to be minor. A damaged cable was found and replaced, eliminating the electrical ground fault. However, further tests revealed more serious faults of the core electronics within the towfish. For the next forty hours, Carter slaved over the twin five-foot-long racks full of electronic circuit boards and power supplies to fix what seemed to be an endless array of broken components. It was tedious, complicated work made incredibly difficult by lack of sleep, pressure from the search being stopped, and the pitching and rolling of the *Geosounder*. The small white patch behind Carter's ear marked him as someone who suffered from seasickness, and I was concerned how he was holding up in the less than ideal conditions.

The culprit behind the rough weather was a tropical cyclone that had been harmlessly spinning her way into the Indian Ocean well north of our position before gradually veering south towards us. Ophelia was a Category 2 cyclone with maximum sustained winds of 100 kilometres per hour that changed course and headed menacingly our way as soon as we arrived on site to begin the search. I was receiving updates from Australia's Bureau of Meteorology (BOM) every twelve hours, and each one showed Ophelia getting closer. Only the gods knew how hard we were going to be hit, but our current misfortune was certainly about to be compounded by a spell of weather downtime. If there was a silver lining to the forecast it was that Ophelia was bound to weaken as she reached the colder southern waters where we were searching.

The SM-30 towfish was in the water for yet another test when the latest BOM forecast showed that Ophelia had changed course again and was now heading directly towards our position. I was getting seriously hacked off by this storm, which possessed an unerring ability to home in on us like a guided missile. The wind was blowing 27 knots and the weather conditions were already marginal for attempting a towfish recovery, but with expected 45–50-knot winds on their way, we had to make a decision. Art Wright wanted to leave the towfish in the water and hunker down till it passed, but I thought that was tempting fate with such an unpredictable storm. We'd have more options with the towfish out of the water and safely secured, so I overruled Art and had his team prepare for a night-time recovery. Once the SM-30 was on deck, I asked Blair to begin steaming to the north-east, which would keep us ahead of Ophelia and give us the most comfortable ride into the three-and-a-half-metre swells.

Once Ophelia had finished toying with us and was on her way to give Carnarvon a good soaking, we headed back to the search box to resume operations. I chose to search track-line 6 next as I wanted to complete the centre of the box before shifting over to the eastern side, where I thought we had a better chance of finding the wrecks. We also needed to see whether Carter's efforts to fix the sonar had been successful. I was prepared to have to rerun the two centre track-lines because the image quality had been so poor and the coverage suspect. Even if these two lines were written off, it was more important in the long run to give Carter the time he needed to get the SM-30 working at optimum performance. There were thirty-one people on board the *Geosounder*, but he was the only one with the ability to get the sonar working, and thus the full weight of everyone's expectations was riding on his shoulders. All eyes would be on the sonar imagery to see if he managed to improve it.

By the time the SM-30 had reached the start of track-line 6, we already knew that the imagery hadn't improved. In fact, it was now worse than before. No data was being created at the outer ranges of both channels and the imagery elsewhere was badly obscured by acoustic noise and dropouts. These were symptoms of yet more electronic faults inside the towfish that Carter was unable to find and repair. I believed we had taken a step backwards, but now the pressure was on me to make the right decision. I could get Art to recover the towfish immediately and put Carter back to work to find the source of these new problems. However, I also needed to take into account that morale on the ship was spiralling downwards and that Carter had had very little sleep over the past few days. The sonar imagery was far from perfect, but we were covering some ground and I needed him to be rested before he returned to the job of finding these elusive faults. I decided to keep the search going at least through the end of track-line 6.

Twenty-one hours later, my next decision was an easy one. Towards the end of track-line 6, the sonar imagery deteriorated further, so I made the call to bring the towfish back on deck and the troubleshooting regime kicked into gear once more. Carter brought the two electronic racks back into the ship's survey lab while the other Williamson technicians remade the electrical termination for the steel tow cable, which they suspected was the source of the ground fault. There is only so much testing you can do with a side-scan sonar on the surface to see if it is working properly. The truest test is what the imagery of the seabed actually looks like, which means you have to deploy the sonar all the way to its working depth without knowing whether it is fixed or not. It can be a frustrating process if repeated once. Repeated multiple times, it is downright maddening.

I came down to the survey lab early next morning knowing that the SM-30 would have been searching along track-line 8 for more than an hour with enough imagery produced to judge whether it was fixed. One look was all it took to see that it wasn't. Although the port channel looked OK, the starboard channel was so badly obscured by a combination of extreme noise and dropouts that to continue in this way would be like searching for the wrecks with one eye closed. Art didn't like what I had to say, but I found the situation totally unacceptable. He was worried about Carter after another marathon day and night session, as was I, but I explained it would be six hours before the towfish was on deck again and he'd be needed, and in that time he could continue to sleep.

The situation was now becoming critical and I felt the project was in real jeopardy. I had been keeping Patrick Flynn up to date with periodic phone calls, and he in turn was briefing Ted, who was understandably very nervous. So nervous, in fact, that over the past few days he had started another fund-raising campaign, aimed at finding $1 million to pay for an additional ten days of search time. His case to the federal and state governments was that if *Kormoran* was located late in the search, we'd need the extra funds to find the *Sydney*. He was also completely bewildered by the fact that the success or failure of the project seemed to be riding on just one person, and that Williamson didn't have a backup engineer on board to work opposite Carter. Throughout the day, as Carter and the other technicians made their fourth attempt to produce a working sonar, pressure was also put on Williamson to urgently fly out another more experienced engineer to help.

My last discussion with Patrick very late in the evening of 10 March centred on whether we should suspend the search and bring *Geosounder* back to Geraldton to pick up a relief engineer. Apparently Ted had lost confidence in Art and Carter and wanted to see some form of action from us to rectify the current position. I recognized this as a typical reaction from someone sitting in an office, on land, with no feel for what was actually happening at sea. At any rate, it was my call and I told Patrick I still had confidence in Carter. While lesser individuals would have wilted from the prolonged pressure he was under, he never gave up. And when he got wind that a relief engineer was being lined up to fly out, he made it clear to me that he didn't need the help and was determined to solve the problems himself.

Sometimes, even in situations as seemingly dire as ours, the best thing to do is not to act. I believed this was one of those situations and told Patrick I wanted to let Carter, and the Williamson team, continue to work on the problem. The improvements I saw in the port channel of the SM-30 imagery indicated that Carter was closer to getting us a fully functional sonar. Besides, I knew that returning to port now would be seen by the press, and others, as the drastic step of a failing mission, which would only serve to wind up the pressure on the team one more notch. As I turned in that night, I suspected that the next launch could either make or break the project.

It was 12 March, twelve full days after we'd left Geraldton to look for *Sydney*, and I had very little to show for our time at sea. Our search had been plagued by one problem after another, and the entire project was in serious jeopardy of ending in abject failure and humiliation for everyone involved. We all knew this was the only opportunity we'd ever get to find the wreck, but I feared our chance was slipping away from us.

At least the SM-30 sonar was back in the water. The imagery wasn't up to my very high standards, but it was much improved thanks to Carter's tireless efforts. With a functional sonar I was satisfied the search could continue, and any suggestions about returning to port were rejected. We had completed track-line 8 and even detected a possible target at the southern start of the line. It was unlikely to be a wreck, but at least our eyes were now focused on possible targets rather than poor, noisy imagery. Still, $2.3 million, which represented a big chunk of the expedition's funds, had been spent with only a fraction of the search box covered. We could suffer no further downtime, of any type, if we were to complete the box before running out of money.

As if this wasn't already a precarious situation, I was constantly reminded of the enormity of our challenge by the photo of *Sydney*'s crew I'd taped above the chart table where I worked. We expected to find *Kormoran* first, and in light of our very poor start, that would be a significant achievement in itself. But *Kormoran* was only a means to an end; a pointer to where I needed to search for the *Sydney*. If we found *Kormoran* but ran out of time and money to find *Sydney*, I knew the relatives would view our expedition as a failure. *Sydney* was the prize, pure and simple.

Track-line 9 was next on my list. My plan from this point forward was to search the entire eastern side of the search box in geographic order from west to east. Having run track-line 8 from south to north, we were now starting no. 9 from the north. This meant we were finally in a position to search the favoured north-eastern quadrant with a working sonar and reasonable weather conditions. It had taken twelve, mostly painful days, but we had reached a point where I could start to feel encouraged. How long the weather would hold was a worry, however, as Captain Blair had told us that gales were forecast for the following day. As the only luck we had had so far was bad, I prayed for a reversal of our fortunes.

And then, in an instant measured literally in the seconds it took to recognize the distinctive colours and patterns on a computer screen, our gloom was transformed into the type of elation reserved for Olympic gold medal winners. Nearly 2,600 metres beneath the deck plates of the *Geosounder*, the transducers of Williamson's troublesome SM-30 sonar began to reverberate from the returning energy of an acoustic pressure wave that could only have been produced by a large steel-hulled shipwreck. John Perryman was nearest to the sonar display and saw the large target first. He called over to me by the chart table four metres away. 'Hey, David, what's that?' I knew immediately it was ship wreckage and shouted, 'That's it!' At the same time, I could hear one of Williamson's technicians commenting from the next room about the target, albeit at a far lower decibel level.

John and I were now seated side by side as I began to explain the scene of utter destruction below us. I recognized the first target as outermost debris by the simple fact that it was by itself with no other targets nearby. A few minutes passed and then smaller targets began to appear across the screen. I believed this was the start of the main debris field, and as the density of objects increased I was proved correct. Someone had found Glenys and she was now sitting behind me and John. I could sense the adrenalin coursing through her as she patted my back and said, 'Well done, David, well done.' The main debris field was contained within a large oval typical of a ship that had experienced a catastrophic explosion. I already suspected that this was the remains of *Kormoran*, but needed to see more. I told John that we hadn't seen a large section of hull yet, but it was surely coming. He couldn't believe I was able to predict these things, but I had seen it before with *Lucona*, HMS *Hood* and plenty of other wrecks destroyed by explosions.

The whole ship was buzzing with excitement and people poured into the survey lab to witness history in the making. The documentary film team had missed that instantaneous moment of discovery, but their cameras were now in position, pointed directly at me and John. There was a brief period when the targets petered out as the sonar glided past the main debris field, but I was sure we hadn't seen all of this wreck. And then, as I'd predicted, a large rectangular shape slowly scrolled down the screen. It was the remains of the hull and appeared to be upright. I hoped to see one more sign that proved we were looking at a shipwreck, and for all the anguish it had given us the SM-30 sonar didn't disappoint. 'A shadow!' I yelled as a classic angular acoustic shadow appeared behind the section of hull. It was one of the most dramatic and satisfying shipwreck discoveries I had ever experienced. Glenys, who like Ted and the other HMA3S directors had dedicated a major portion of her life to this quest, started to feel the magnitude of the moment and began to sob.

We had found the *Kormoran* in the location and condition expected, and despite the intense frustration of numerous days lost to Ophelia and myriad equipment failures, we had done so on just the fourth track-line after sixty-four hours of active searching. Against any standard this was a remarkably fast result, but for a shipwreck that many had predicted would never be found it was stunning to the point of disbelief. Most importantly, it meant that the testimony of Captain Detmers, upon which I had based my entire search plan, had proved to be truthful and accurate, leaving us with a proper chance to find the *Sydney*.

In my career as a shipwreck hunter I have experienced just about every emotion imaginable for a person in charge of such costly and technically complex adventures conducted on the high seas. Searching for shipwrecks is basically an all-or-nothing proposition, where you either find what you are looking for or go home empty-handed. The risks and rewards are great and so are the emotions that you go through as a result. I have experienced the feeling of searing disappointment when a search ends in failure and there is nothing to show for your efforts. If I am honest, this fear of failure is a driving force in the meticulous approach I adopt for every search, although I'd rather think it's all about being professionally motivated. Fortunately, I have also known the feeling of unbridled joy that success brings.

So finding *Kormoran* was simply brilliant: another chance to experience that elation and watch a dedicated team celebrate a great result. There was huge relief that we had overcome all the problems thrown at us, but for me personally, there was a greater sense of satisfaction that I had been right to trust the German accounts in deciding where to search for the wrecks. The best feeling was yet to come, however, for I now knew that we would find *Sydney*; that it was just a matter of time. What I didn't know was the extent to which the incredible highs and lows of the first two weeks would be repeated several times over the next month until the project was complete, nor the extent to which many people's lives – including my own – would be changed forever by what we were about to achieve.

The mood on board the *Geosounder* was instantly transformed by the finding of *Kormoran*. There were beaming smiles and high-fives all around, especially for Carter, who was most pleased that the wreck had been picked up on the troublesome starboard channel that he had worked so hard to fix. Word spread like wildfire around the ship, and crew who would not normally take an interest in the survey functions of the vessel flooded down to the lab to share in the excitement and take pictures of the magnificent sonar images. For a brief period we were the only people in the world who were aware of what all Australia had been waiting to hear for so long. It was time to share the news. However, the timing of the public announcement had to be carefully choreographed between the navy and HMA3S. In the meantime, I switched my focus to finding the *Sydney*.

The first and most important conclusion I was able to draw from the position in which we found *Kormoran* was that it proved that Detmers' account was credible and he was telling the truth about where the action had taken place. *Kormoran's* wreck was located in the north-eastern quadrant of my search box, less than seven nautical miles from the nominal 26° south, 111° east position. The vexing question over whether Detmers' 26° 34' south, 111° east was the noon position or the action position was solved. It was indeed the noon position, but that was now a matter for the history books and no longer a concern for me in the search for *Sydney*. The important thing was that I could use Detmers' last sighting of the burning *Sydney* to create a new search box using the navigation clues in his written account of ordering a halt to the shelling of *Sydney*:

1825 – Cease fire! Last range 9,000 metres. Last shot range 10,400 metres, last ship's bearing 225. Ammunition expenditure 500 Bdz 50 Cz. Decision: prepare to scuttle ship. All officers to the bridge. Order to XO: turn out all boats and life-saving equipment. Lensch and Noll inform impossible to get through to the engine-room. Check this myself. No. 2 generator is still ready but useless. Mine-deck continuously under watch. Outline of the enemy lost in twilight. At around 16,000 metres out of sight. Enemy course about 150 true. Large fire seen until about 22 hours.

The simplest navigation calculation taught to children in schools is how to determine a position from a fixed point, knowing the distance and course from that point. Using the position of *Kormoran's* wreck as the fixed point, it follows therefore, that *Sydney* should be found very close to the distance and course last seen by Detmers. The rationale where to look for *Sydney* was as simple as that, and it is why our chance of finding her was predicated on finding the *Kormoran* first.

Of course, it can't really be that simple when looking for a ship that no one saw sink. As with the hunt for *Kormoran*, I would be searching an area that conservatively covered all possibilities, not a single point on a chart. There were risks and uncertainties I had to factor in to the calculation of my search box. For example, where was *Sydney* heading when she broke off the action with *Kormoran* and was seen heading to the southeast? Detmers estimated that the last course was about 150° true, whereas the final gun bearing at 18.25 was 225° relative to *Kormoran*'s heading. This would have made *Sydney*'s course 120° true. Was her crew steering the ship towards Fremantle (160° true) or Geraldton (130° true)? It seemed that both were a possibility and would need to be factored into my calculations.

The biggest unknown, however, the one that could truly make finding *Sydney* difficult, even impossible, was the question of when the ship actually sank. Did *Sydney* go down after Detmers stated he saw a large fire, 'about 22 hours', but before midnight, when he abandoned *Kormoran* and, having 'looked for *Sydney*', could see nothing but blackness, and that 'she was gone'? Or – and this was the really worrying possibility – did she limp away, over the horizon and out of Detmers' sight, and sink at some time and place that was impossible to determine? The problem always came back to the fact that no one actually saw her go down, and there were no other physical clues to her whereabouts.

With everyone still celebrating the finding of *Kormoran*, I stole away to the chart table, where I studied the Quarter Million Plotting Sheet I had used to define the final search box for *Kormoran*. A repeat performance was called for, and as I looked at that photo of *Sydney* I'd taped to the wall a fortnight before, I took inspiration from the smiling faces of the 645 men in gleaming white uniform gazing down at me. Taking every documented clue, navigational analysis and my gut instinct into account, I outlined an area in red pencil where I thought, hoped and prayed their resting place would be found. At 360 square nautical miles, the box was five times smaller than that for *Kormoran*, which was a direct reflection of my greater confidence. The key question was how far *Sydney* would be found from *Kormoran*'s position. It depended in part on the speed she was still making after she disengaged from *Kormoran*. Having plotted various German observations, it was apparent that she slowed drastically, to the point where she was probably making little headway at the very end. I believed the distance between the two wrecks would be ten to fifteen nautical miles, but I set the southern boundary of my search box at twenty-five nautical miles just to be sure.

After the long transit south to complete track-line 9 and to have a second look at the target we'd detected at the start of track-line 8, which turned out to be geology, it was the early hours of Friday 14 March before we were in a position to start searching for *Sydney*. The SM-30 sonar was three quarters of the way up track-line 10 when I got a call from Patrick that caused me to immediately change my plans and break off the search. The word from Patrick was that the public announcement about the discovery of *Kormoran*'s wreck was scheduled for Sunday, and that it would be announced by Prime Minister Kevin Rudd himself, flanked by the Chief of the Defence Force, the Chief of Navy and a defence minister.

Seeing the importance the government was placing on this announcement, I decided to play it safe and get some high-resolution images of the wreck to confirm my original interpretation. I personally had no doubt the wreck was *Kormoran* and wrote this in my live log of the operations. However, if my identification of the wreck, based on imagery alone, was the basis for breaking this extremely important news to the country, I wanted to have the best possible evidence to support my interpretation. I wasn't in the line of communication with the navy or the prime minister's office. But I imagined that if I was, the first question I would be asked was: 'David, are you sure you've got this right?'

Running the two high-resolution lines past *Kormoran*'s wreck consumed twenty-one hours, but the time spent was well worth it. We came back with two terrific images, at 1,500-metre and 750-metre swaths, which were as good as any I've ever seen. Like a bird's-eye view of an accident scene, they revealed the complete destruction of *Kormoran*'s back half, caused by simultaneous detonation of the 340 mines stowed on the mine deck. The final image was a stunning shot of her intact bow, which was only made possible by flying the SM-30 towfish incredibly close to the wreck and about fifty metres off the seabed. Afterwards, someone asked me to put this manoeuvre into perspective, and I could only say it was like threading a needle at a depth of 2,560 metres using a thread that was five miles long.

Flying the towfish so close and so low accentuated the acoustic shadow emanating from the bow, similar to the way a light shining on an object from an angle will make the shadow behind the object stand out. My aim was to reveal the bow's overall shape and I was delighted to see that it perfectly matched *Kormoran*'s flared design and raised forecastle deck. From the same image I was also able to accurately measure the wreck's beam, which was important in distinguishing one ship from the other. I had the original builders' plans of both ships and knew that *Kormoran*'s beam was 20.2 metres, while *Sydney*'s, by comparison, was just 17.3. I carefully took the measurements from a number of different points to be sure, but they were all the same: 20 metres breadth, 13 metres height and a length of 106 metres. The dimensions fitted *Kormoran* like a glove.

Having prepared the high-resolution images and accompanying explanations for the prime minister's press conference, we were able to get back to searching for *Sydney*. After completing track-line no. 10, we made a slow turn to starboard to run line 11, heading from north to south. The track-line progression was taking us to the east, into shallower water. This meant less time spent in turns, and together with the much shorter track-line length, the pace of the search had picked up, although it could never be described as fast. There was plenty of time to work on other tasks and to generally shoot the breeze. When Sunday morning, 16 March, rolled around, John Perryman and I were talking about the PM's press conference, wondering what the public reaction would be to the news we had known for four days. I don't know if it was a premonition, or because we were in a good rhythm with the SM-30 working really well, but I turned to John and said: 'Watch, I bet we find *Sydney* while the PM is having his press conference.' It was such an outrageous suggestion, John let it pass without comment.

That same morning, at 07.05, the towfish had entered the search area on track-line 12, the third line in the search box I had designated for *Sydney*. This line was being run from south to north, and 7,400 metres of cable was deployed off the back of *Geosounder*. I was over at the chart table looking at a plot I had made on transparent Mylar of possible courses *Sydney* might have taken after the action. The sonar was currently passing through the zone I considered to be the highest probability, roughly midway between ten and fifteen nautical miles from where we'd found *Kormoran*, when John shouted out: 'Oi! What's that!' I could hear in his voice, and see in the way he leaped out of his seat, that he was really excited by something. He had every right to be, because at that very instant the sonar was passing over the upright stern of *Sydney*.

The sonar image was so perfectly and beautifully vivid, there was no need for interpretation. This wasn't the geology that had fooled me earlier; it was a shipwreck, and as no other shipwrecks were missing in this area, it had to be the *Sydney*. The distinctive image of a ship-shaped target with a well-defined acoustic shadow was only seconds old, but with one look at the monitor I knew it myself and screamed: 'That's it, we've found her!' The moment of discovery was as instantaneous as that. At 11.03 Australian Western Standard Time, the sixty-six-year search for HMAS *Sydney* was over.

Like little kids on Christmas morning, John and I started jumping around the room. As excited as we'd been at finding *Kormoran* just four days before, this was so much more. After the years of struggle to convince the navy, the government and countless naysayers, and the difficult start to the search, perhaps I should have felt relief, or even vindication. I felt neither of those things, although I'm sure they added to the intensity of my emotions. I was just so incredibly happy that *Sydney* was found, and looking back I remember it as the single most exhilarating moment of my professional life.

I looked at my watch and couldn't believe we were so close to finding *Sydney* at the exact moment the prime minister was addressing the press, as I had mischievously suggested to John. Imagine everyone's reaction had Kevin Rudd started his statement by saying he was there to announce the finding of *Kormoran*, but as I stand in front of you, I've just heard from David Mearns on the *Geosounder* that he has also found the *Sydney*. It was a delicious thought that was still going through my head when I called Ted Graham. He was walking through the streets of Canberra having just left the prime minister's office when I rang him on his mobile. I asked him if it was possible to get the PM and the press back to reconvene the press conference. When he asked me why I said, 'We just found the *Sydney*. She's sitting upright in a small debris field. The Prime Minister or anybody else can announce it. We have found HMAS *Sydney*.' Ted was so choked with emotion that all he could say was 'Thank you, David.'

<hr />

Once the good news was relayed to the prime minister's office, it was agreed that the announcement about *Sydney* would be made the following day. That left me with just enough time to collect a number of high-resolution images of the wreck for confirmation, as I'd done with *Kormoran*. Fuelled by the adrenalin of our amazing success, everyone on board the *Geosounder* worked round the clock to get the imagery I wanted to accompany the PM's official statement. Three sonar passes were made at increasingly narrow swaths, with the SM-30 set at 750 metres swath for the final pass, which produced one of the best sonar images of a deep-water shipwreck I had ever seen. The detail in the acoustic shadow was remarkable, and it confirmed beyond doubt that the wreck was a heavily damaged warship with a nasty fracture where its bow had once been. Within the debris field there was a large square sonar target, which I tentatively identified as the remains of *Sydney*'s bow.

And so on Monday morning, the press reassembled in the prime minister's courtyard to hear Kevin Rudd say: 'This is an historic day for all Australians and it's a sad day for all Australians, as we confirm the discovery of HMAS *Sydney*. This is a day which begins a process of closure for many families of the crew of *Sydney*. It's also a time for the nation to reflect on the bravery of all those who gave their lives in defence of their country, in this particularly bloody and brutal naval engagement.'

When the floor was opened to journalists, the first question put to the PM was: 'Considering there hasn't been any photography taken yet of the wreckage, how has it been confirmed that it is the *Sydney*?' Rudd's reply was that the government had sought confirmation from the navy before proceeding with this morning's announcement. He added: 'It's very important that these things are got right.' He then asked navy chief Vice Admiral Russ Shalders to provide a fuller explanation. Shalders described how we'd used the high-resolution sonar images to compare the dimensions of the wrecks to the drawings, and how the distinctive profile of *Kormoran*'s bow was an important factor. He finished by telling the reporters: 'David Mearns has indicated that there is no doubt that this contact is the *Sydney*.'

Because the prime minister's announcement about the *Sydney* had occurred the day after he'd made the exact same announcement about *Kormoran*, it left everyone with the impression that we had located the two wrecks on consecutive days. While that wasn't true – in fact it took sixty-seven hours of searching before *Sydney* was found – people could still scarcely believe the speed with which the two wrecks had been discovered and positively identified. Australians, after all, knew *Sydney* as this great, unresolved mystery, in which one of the central question marks was that nobody knew where the wreck was located. Wasn't *Sydney* supposed to be the needle in a thousand haystacks; the wreck that was truly unfindable?

The significance of what we had achieved, and the overwhelming public reaction to our success, was apparent as soon as the *Geosounder* arrived in Geraldton after the end of the search phase of the project. Because of its proximity to where the battle had been fought, and the fact that it was home to a prominent memorial to *Sydney* and her 645 men, Geraldton was as closely intertwined with the cultural history of *Sydney* as any other community in Australia, and its residents turned out in droves to celebrate our success.

A media conference was held the following morning, Good Friday, at the Geraldton-Greenough Chambers for us to present our findings from the search phase, and an outline of our plans for the visual investigation of the wrecks by ROV. Although I had been posting regular technical updates on a search diary for the HMA3S website, which racked up an amazing 12.5 million page views from 3.2 million unique viewers, this was our first opportunity to discuss the search in person and to show the packed audience how the wrecks had been identified from the sonar images alone. One point I especially wanted to make, in order to prepare people for the type of photographs they would soon be seeing from the ROV survey, was that I expected to find both wrecks in a badly damaged condition. My primary concern was for the relatives and friends of the *Sydney* crewmen, who would be understandably upset by photos that showed the torpedo and shellfire damage inflicted on the ship.

Ten days earlier, I hadn't thought we'd be in this position. But now, with both wrecks found and identified well ahead of schedule, we actually had a reasonable amount of time left in the budget to visually investigate them. The purpose and scope of that investigation was still an open question, however. While some people were calling for a full-blown forensic examination to conclusively determine what caused *Sydney* to sink, HMA3S only had a mandate to visually identify the wrecks, which if taken literally could be achieved by a single ROV dive on each, lasting an hour or two at most. The navy and HMA3S clearly needed some guidance about striking the right balance. All the good achieved by finding the wrecks would be quickly undone if the ROV investigation became politicized.

The decision in the end was to follow my recommendation to comprehensively document both wrecks and their debris fields, but to stop short of trying to answer specific questions about cause or blame. As this might be the only time a survey vessel equipped with an ROV capable of reaching both wrecks was available along this coast of Australia, I believed we did have a responsibility to use that asset to its full capability. I also suggested that certain parts of the wrecks should be filmed and photographed in greater detail to allow direct comparison against the

German accounts. However, we did not have the equipment or personnel, or the mandate, for a true forensic examination in keeping with the legal standard expected by a court. Our job was to be evidence gatherers, not judge or jury.

The changeover for *Geosounder*, involving off-loading the Williamson equipment and mobilizing a 3,000-metre-rated ROV named *Comanche*, was meant to take two to three days. However, any hopes I had of a smooth ROV mobilization were dashed the evening before we were scheduled to leave port when I got a call saying that the special HMI lighting system I had brought to be installed on *Comanche* had suffered a catastrophic electrical failure. It turned out that a simple mistake by a junior technician had led to the lights being wired incorrectly, which in turn had caused a high-voltage short circuit that had completely destroyed the twin 400-watt light heads. Losing this expensive and vital piece of equipment was painful enough, but what made it worse was that this disaster triggered a soul-destroying stream of related electrical failures that kept us in port for six more days and once again put the project in serious jeopardy. Day after day the ROV crew fought a losing battle to trace the source of the problem as component after component failed and either had to be replaced or eliminated. These included a camera pan-and-tilt mechanism, countless thruster motors, the starboard seven-function manipulator, the scanning sonar, the sound velocity profiler, the bathymetric survey system, a bulkhead connector, a gyro power supply board, a drive motor for the tether management system, and both the primary and spare rotary slip ring units without which the ROV would not work.

To the credit of everyone involved, they never gave up, and the *Comanche* was eventually repaired, allowing us to get under way on 29 March. Nevertheless, the episode left us with strained relationships on the vessel, a severely diminished ROV and a very anxious ROV crew. In fact, the only reason we left the day we did was because we were forced out of port by the second tropical cyclone to plague our project. This one was named Pancho, and like Ophelia, it steered its way south directly towards the wreck site, arriving there as a Category 2 storm on the morning of 29 March. Like a bully standing in a narrow hallway,

Pancho stopped us from steaming north while it moved menacingly to block our path. Its long reach meant we were no longer safe tied up in Geraldton, so we were advised to head out to sea. Unfortunately, the weather was also too rough to make the necessary test dive of the ROV in deep water west of the Abrolhos. Pancho appeared to have us snookered.

By now my frustration level was off the scale. We had lost a big chunk of time with the malfunctioning ROV and were looking at another extended period of downtime because of Pancho. I had bought some fishing equipment in Geraldton, which provided the crew with a few hours of fun trying to hook and haul in some of the dolphin fish that were always lurking on the port side of *Geosounder*. The fishing was a little diversion to lift everyone's spirits and to help restore lost morale. We did our best to try and complete the ROV test dive during a lull in the weather, but when that failed, I asked *Geosounder*'s new master, Deland Van Wieringen, to get under way to the wreck site at reduced speed. I was taking a calculated gamble that we could deal with the worst of Pancho as we moved north while it slid by us to the south, and that upon reaching the wreck site the *Comanche* would pass its required 500-metre test dive and be allowed to continue down to *Sydney*.

After a slow and rough transit north, the *Geosounder* arrived at the wreck site in the early hours of 1 April with surprisingly good conditions for a launch. The aim was to deploy the ROV to a depth of 500 metres for an hour or so of testing before descending deeper. Sixteen days had passed since finding *Sydney*; as the delay got longer, more and more people were asking when we would produce photographs of the wreck. Was this the day we would finally set our eyes on the 'Grey Gladiator' and put the minds of the relatives to rest? At only forty metres depth we got the answer: not a chance.

When I heard the catalogue of problems with the ROV, I felt like I'd been kicked in the stomach. It sounded as if every item that was supposed to have been fixed before leaving Geraldton had somehow managed to break again. Thrusters, scanning sonar, bathymetric sonar, gyro power supply: all malfunctioning. We had gone through some really low moments during the search and ROV mobilization, but the next two days were the absolute worst. To add to our misery, the brief window of decent weather was gone, replaced by 30–35-knot winds and three-metre swells. I wasn't the superstitious type, but it was hard not to feel that there was some greater force preventing us from completing the job.

The morning of 3 April broke with promise that a dive might be possible. The weather had improved and the ROV, while not fully functional, was in a reasonable state for testing. Normally it was the decision of the ROV supervisor whether a dive could continue, or whether it would have to be aborted in case of a serious problem. However, because of all the delays we had experienced, it was agreed that this critical call would be mine to make. It was also agreed that if the 500-metre depth test was successful, we would continue straight to the wreck at 2,500 metres without having to surface again.

It didn't take long for me to have to wield my client's prerogative, as during the deck checks – before the vehicle had even touched the water – the main camera pan-and-tilt unit failed. Although this meant we'd be diving with a fixed camera, one that couldn't be mechanically pointed at an object, I decided to press on and gave the okay to proceed. The vehicle was in the water at last, but at 100 metres depth we suffered another failure, causing my heart to sink. This one was a potential showstopper, and I couldn't see how we'd ever reach 2,500 metres at this rate. The chain that drove the level-winding mechanism on the main lift umbilical winch had snapped, and with no spare chain or links on board, I thought we were stuffed. Fortunately, a fix was found by jury-rigging a chain hoist that would manually shift the level-wind into position as cable was paid out or hauled in. Crisis averted, at least temporarily.

The ROV needed to pass, or at least survive, one more series of tests when we reached 500 metres before proceeding to *Sydney*. However, when the pilot pushed his joystick forward to fly the *Comanche* ROV out of its protective garage, nothing happened. The vehicle was stuck, frozen in place, because the subsea winch on the garage wouldn't release its floating tether. Part of me wanted to scream in frustration at the extent of our rotten luck. Without the freedom to fly around the wreck site, the ROV was nothing more than a video camera at the end of a very long cable. We'd have no ability to control its position relative to the wreck, other than moving the *Geosounder* around on the surface 2,500 metres away. To top it off, we couldn't even point the camera because of the broken pan-and-tilt unit. In the offshore world, we had a dismissive name for ROVs with such limited functionality. We called them 'dopes on a rope'.

A decision was needed, and because this one was so important, I called a meeting on the bridge with all the ROV supervisors and senior pilots, along with Captain Deland and John Perryman. In my mind it came down to a simple choice. We could either abort the dive and hope the ROV could be mended before the next wave of bad weather hit us, or we could just go for it. I knew the decision was mine to make, but I wanted to hear everyone's opinions first. I could see that the ROV crew felt defeated and had lost confidence in their ability to fix the system. To be fair to them, the sheer number of equipment failures they had faced was unprecedented. It was a huge long-shot, but if we were able to drop the garage close enough to the wreck, we might get one or two pictures that would confirm it was *Sydney*. If the ROV was truly unrepairable and we were forced to terminate the whole project, at least we would have achieved HMA3S's minimum requirement of identifying the wreck. 'OK, let's go for it,' I said.

Once the decision to continue the dive was made, I wanted to make sure we avoided all risk of entanglement of the garage with the wreck. I explained to the navigator, Nigel Meikle, that I wanted to drop down on *Sydney*'s port quarter in order to stay clear of the fractured bow, where there would be a lot of jagged steel and fouling hazards. When the vehicle reached 2,300 metres, John and I joined the ROV pilots in their control van on the boat deck so I could speak to them directly. At this point everything depended on the quality of Nigel's calculations. If he was wrong, we could search for hours without getting anywhere near the wreck. As the vehicle descended the last fifty metres very slowly, I expected to see the lights beginning to reflect off the featureless seabed. Instead, a faint shape appeared in front of us and I was the first to recognize what it was: 'We're on it, it's a gun!'

In a remarkable display of his skills, Nigel had put us exactly where I had asked him to. The excitement level in the control van exploded with the realization that we were staring directly at the twin barrels of 'X' turret, pointed to port in the position they had last fired on *Kormoran*. John perfectly summed up the significance of this wonderful picture: 'My God, there is no doubt about that at all!' My gamble had paid off. We had the pictures Australia had waited decades to see, and the project could have ended right at that moment if it had had to. Fortunately, it didn't. The ROV crew persevered in patching up the *Comanche*, and over the next four days we comprehensively documented the damage to both *Sydney* and *Kormoran*, resulting in a collection of stunning photographs that cleared up once and for all any historical questions about their fierce battle.

∞∞∞∞∞∞∞∞∞∞∞∞∞∞∞

When we returned to film the wreck of *Sydney*, this time with an ROV set free from its garage, the evidence of the damage it had suffered at the hands of *Kormoran* was plain to see. The German account of the action, which was scarcely believable at the time, was that they had fired 500 15 cm shells of mainly the armour-piercing type, and of those, 150 or more had scored direct hits on the cruiser. We didn't attempt to count the shell hits during our investigation – that analysis was performed afterwards by another organization, using the footage we shot – but the number of gaping holes and impressive indentations we could see suggested that the German accounts were not exaggerations. In fact, when taking into account the numerous hits from *Kormoran*'s smaller-calibre guns, it was fair to say that *Sydney*'s hull was riddled by gunfire. The evidence certainly backed up Detmers' account that the fighting started at close range, and that *Sydney*'s bridge and forward guns were targeted early on in the action. In particular, one 15 cm shell hit on the port-side base of the director control tower, a strike that Detmers claimed took place at the outset of the battle, would have had a devastating effect on all bridge personnel, including Captain Burnett.

The large sonar target I had spotted on the northern edge of the debris field, some 480 metres away from the hull, was indeed the remains of *Sydney*'s bow. We found it upside down on the seabed at a slight angle that perfectly exposed its port side, where *Kormoran*'s G7a torpedo had hit. The break in the bow was very irregular, with large sections of hull plating bent and torn in a multitude of directions. The port-side plating also showed signs of being indented, which was consistent with *Kormoran*'s torpedo exploding on contact. Finding the bow such a large distance from the rest of the hull indicated to me that it had detached at the surface before plummeting to the seabed. As the hull and all the other wreckage forming the debris field had been found to the south-east of the bow in the direction that *Sydney* was last seen travelling, I took this as a clear sign that the bow snapped off first, and that this was the likely trigger causing *Sydney* to sink. The final orientation of the hull on the seabed, of 140° true, also supports the likelihood that the crew were trying to reach Geraldton, and that the ship was under way right up to the moment the bow was suddenly and catastrophically lost.

Like *Sydney*'s hull and bow, the parts of the ship found in the debris field tell their own story about the demise of the ship and the last moments of her men. By far the biggest surprise of our investigation was finding five of *Sydney*'s nine boats at various locations in the debris field. It never occurred to me that the wooden boats would have survived on the seabed for sixty-six years, or that they wouldn't have been ripped off the ship as she sank and carried away to disintegrate in a different location. Each time another boat was found, it brought home the helplessness of *Sydney*'s men. For it was clear that these boats had never been launched or used.

On 31 March, a few days before we'd started documenting the wrecks, an announcement had been made in Canberra that caught us completely off guard. Deputy Prime Minister Julia Gillard, standing in for Kevin Rudd, announced that a commission of inquiry (COI) was going to be held into the loss of *Sydney*. It was to be headed by Terence Cole, one of Australia's most eminent judges, who had been selected in part because of his expertise in maritime law. I had mixed feelings about the inquiry when it was announced. My main concern was that it would upset the relatives and open old wounds, just when we felt our expedition was close to providing the closure everyone desired. On the other hand, it might put an end to the debate and controversy surrounding the loss of *Sydney* once and for all. During the government's announcement, the point was made that 'no board of inquiry was conducted during World War II after the loss of *Sydney*'. Had the navy held an inquiry in 1941, as they should have done, it is likely that the controversy would never have started in the first place.

While on balance I agreed with the need for an inquiry, I couldn't but help wonder what effect it would have on the ongoing operations. Would I get a call from Justice Cole instructing me about the type of evidence he wanted collected? Or would the navy send me a plan for forensic examinations in light of the fact that the COI was to be a legal inquiry in front of a sitting judge? Neither of these things happened. In fact, we received no instructions or advice about how we should conduct the photo/video documentation of the wrecks. I found it a bit odd that the collection of vital visual evidence upon which the outcome of the inquiry would likely be based was being left to me without input from those actually responsible for the inquiry.

Looking back today, in light of the extent of the scientific analysis of the photographic evidence we gathered, and the conclusions that the inquiry was able to derive from that evidence, I can only say that I am very proud of the work the expedition teams did in helping to solve the mystery surrounding the loss of *Sydney*. I was also pleased that the evidence we provided to the COI spoke for itself, and that there was no need for me to give any written or oral testimony. I did briefly meet with Justice Cole, at the commission's offices in Sydney, before a longer meeting to help his legal and technical teams with some specific questions. He greeted me with the words: 'I guess I owe my job to you.' Although we both laughed at his attempt to break the ice, there was some truth in his throwaway line. His next words were in the form of a question I'd been asked many times before: 'Tell me, how did you become a shipwreck hunter?'

Cole's inquiry, as it became known, turned out to be an extraordinarily comprehensive and wide-ranging investigation that grew from the broad nature of its single term of reference: 'To inquire and report upon circumstances associated with the loss of HMAS *Sydney* II in November 1941 and consequent loss of life and related events subsequent hereto.' The raw statistics of the inquiry highlight what a complicated and monumental effort it was. Over the course of sixteen months, Cole's team of solicitors reviewed nearly 31,000 archival documents and examined 77 witnesses, producing 2,563 pages of transcript. Including Cole himself, forty-two personnel worked on the inquiry. Cole separately appointed the Australian Defence Science and Technology Organization (DSTO) and the Australian division of the Royal Institution of Naval Architects (RINA) to give expert advice on the damage both ships suffered based on the video footage and photographs we provided. Their technical report alone was 392 pages long, while Cole's three-volume report covering the evidence, conclusions and various conspiracies ran to over 1,350 pages. The cost of the COI to the Australian taxpayer was $6.66 million dollars.

As for the battle between *Sydney* and *Kormoran*, Cole's conclusions pretty much agreed with what the underwater footage showed and what I wrote in my own book, *The Search for the Sydney*, which came out at virtually the same time as his report in August 2009. The empirical evidence produced by our expedition verified that the German accounts were truthful with respect to where the action took place; that the initial action was fought at close range of 1,000 to 1,500 metres; that *Sydney* suffered a torpedo strike on her port side; that she was heavily damaged on both her port and starboard side by *Kormoran*'s main armament of 15 cm guns with a minimum of eighty-seven hits being recorded; that *Sydney*'s bridge and director control tower suffered severe damage early in the battle, resulting in the deaths of many officers, probably including Captain Burnett; and that she suffered from extensive fires, making it unlikely that the ship's boats would have been usable by survivors. Based on the DSTO analysis, Cole assessed that the casualties on board at the end of the battle were in the order of 70 per cent of the ship's total company. Those still alive were unable to escape because of the damage to the boats and the Carley floats, and were killed when *Sydney* sank.

Cole was critical of Captain Burnett, and pretty much laid the blame for the loss of *Sydney* and his men at his feet, although he stopped short of saying his actions were negligent. He concluded that Burnett knew of the possibility that a raider was off the Western Australian coast at the time, and that the unknown vessel they sighted on 19 November might be German. Burnett's fundamental error was that he assessed *Kormoran* to be an innocent ship, and in approaching her so close abeam he lost all tactical and armaments advantage, thus placing his ship in grave danger that was otherwise avoidable. In other words, he fell for the trap set by Detmers, and by the time he realized his error it was too late. *Sydney* was not at action stations when she drew up alongside *Kormoran*, and the first-strike advantage that Burnett handed to Detmers resulted in such critical damage that she was unable to recover. Cole's final, damning conclusion was that 'Capt. Burnett made a serious error that had terrible consequences.'

Each year, on the anniversary of the battle and the loss of *Sydney*, a sunset service is held at the beautiful Dome of Souls memorial situated at the summit of Mount Scott in Geraldton. Since its dedication on the sixtieth anniversary in 2001, the memorial has become so significant in commemorating the loss of *Sydney* and the sacrifice of her 645 men that after the wreck was located the site was declared a national memorial in 2009. Our discovery of the wreck site also allowed the fifth and final element of the memorial to be constructed: a pool of remembrance to symbolize the finding of *Sydney*. The granite floor of the circular pool forms a map that shows the exact location of the wreck, indicated by the wingtip of one of the 645 gulls representing each of the lost souls, along with the geographic coordinates that radiate out from the gull's wing.

I had attended the commemoration service in November 2008, which was an amazingly emotional and well-attended event, with an estimated 2,000 people coming from all parts of Australia. Visiting Geraldton and attending the service is the one chance I get to meet the relatives, learn the stories of their families and hear what our discovery of the wreck means to them. So when Shane Van Styn, the mayor of Greater Geraldton, invited me to be present at the seventy-fifth anniversary service in November 2016, and to speak at a civic reception for the relatives afterwards, I wasted no time in accepting.

With it being a major anniversary, the service was again well attended, and I was able to reconnect with many of the relatives I had first met in 2008. Whereas the 2008 service was almost raw in its emotional intensity, eight years later I sensed a different, more accepting mood from the families. Many relatives told me that this service was easier for them and that knowing the wreck had been found helped with relieving the pain and sadness they still felt when remembering their loved ones. There were still tears, but these were tears of remembrance rather than mourning. The families of *Sydney*'s men suffered a terrible loss, and many lives were irrevocably changed as a result. However, it was plain to see how proud they were of the service their father, uncle, brother or son gave in defence of their country, as they filed one by one under the Dome of Souls to place a wreath in remembrance.

For all the professional fulfilment and recognition I got from leading the successful search for the wrecks, it doesn't compare to the deep feelings that come from someone thanking you for finding their relative. As I stood looking at the huge array of colourful wreaths at the end of the service, a woman came alongside me and gently took my hand. As she looked at me, I could see that her eyes were moist with tears. But her voice was calm and steady as she said simply: 'Thank you for finding my dad.' It took me a second or two before I could reply, as my throat immediately choked with emotion. Her words had hit me in a way I hadn't expected. I wanted to hear her story, so I asked about her father's service and what the discovery of the wreck meant to her.

Helen Miller was a month away from her third birthday when her father, Chief Petty Officer Louis Nicholas Sampson, went down with the *Sydney*. At the age of thirty-four, CPO Sampson was one of the oldest and most experienced crew aboard *Sydney*, having served two long tours on the ship, which included taking delivery from England and her famous service in the Mediterranean. He had joined the Australian navy as a Boy Sailor 2nd Class at the tender age of fourteen, and as Helen describes 'was navy through and through'. His two brothers also joined up when they were young, and all three served with distinction and rose to become chief petty officers. With three young children at home (Helen's brother John was six and her sister Fran was four), his death had a devastating effect on his wife Mary and the family. As you would expect, their lives changed forever.

In some ways, the loss of a family member at sea, where there is no known location and no body to be buried, can be a double tragedy. For many of the *Sydney* relatives, this was certainly true, as the mysterious circumstances of her loss compounded the void they were left trying to fill. As Helen explained, finding the *Sydney* had finally provided the answers they so desperately needed:

It was lovely to meet you and say in person 'Thank you for finding my dad.' You asked did it really mean so much to us and I guess unless you had lived with that mystery your whole life you would find it hard to understand. I suppose lots of people lose a parent when they are young but to have the uncertainty, rumours and innuendo attached to it makes it all such a big mystery and we really just needed to know what had happened to him and all his shipmates.

My mother married again and did not speak often of our dad. I think it was too painful for her. It was her sisters and brothers who told us little snippets of what he was like. His brother Fred also talked about him and was desperate for his resting place to be found. Unfortunately he died before she was found.

When we heard *Kormoran* was found first we thought no God could be so cruel and virtually did not breathe until *Sydney* was also found.

VI

AHS *Centaur*

SUNK ON A MISSION OF MERCY

AHS *Centaur*

SUNK 14 MAY 1943

268 died
64 survived

The incredible discovery of HMAS *Sydney* generated so much public interest in Australia's naval history that for once I didn't need to think about the next wreck I wanted to find when the inevitable question was asked. The first enquiries about the AHS *Centaur* had already started trickling in while I was still on board the *Geosounder* taking photos of *Sydney* and *Kormoran*. A few enterprising people had found my email address, and in writing to congratulate me about *Sydney* also asked if I would publicly support their calls for a government-funded search for the *Centaur*. The fact that *Centaur*, in an earlier and different wartime role, had played a part in the aftermath of the action between *Sydney* and *Kormoran* added an intriguing twist. The *Centaur* Association, a small group of families who had lost loved ones when the *Centaur* was callously and illegally attacked by a Japanese submarine in May of 1943, were hoping to use our discovery of *Sydney* to draw attention to what they considered was Australia's second most famous wartime shipwreck.

I was not surprised, therefore, when someone in the crowd of reporters that surrounded me after the final HMA3S press conference in Fremantle asked if I thought I could find the *Centaur*. The little I knew about where the ship was sunk already told me it would be an easier puzzle to solve than *Sydney*. That didn't mean easy, I explained, just easier than *Sydney*. In my opinion the *Centaur* was definitely findable. I wasn't worried about the technical aspects; the main challenge was whether the government would have the political will to commit public funds for yet another deep-water shipwreck search. As I told the reporters: 'How it's organized and who funds it, those are the big questions.'

When the *Centaur* set out from Darling Harbour, Sydney, on 8 May 1943 for what turned out to be her final voyage, she seemed oblivious to the possibility that she might become a casualty of the war. Whereas all other ships transiting at night were totally blacked out to avoid attracting the attention of enemy submarines, the *Centaur* was fitted with oversize floodlights and a row of green lights encircling the ship at deck level that blazed brilliantly. In contrast to all other merchant ships, the *Centaur* was purposely designed and liveried to be easily seen at day or night. As one officer later commented, she was 'lit up like glory'. The passengers on board during this voyage were equally carefree about making excessive noise that might be heard by passing ships. Hours before the attack, one group were having what can only be described as a loud and raucous deck party, accompanied by musical instruments, not the type of behaviour a merchant captain at the time would normally tolerate. But if the *Centaur* was acting as if she was immune from attack, that was precisely because she *was* immune from attack. Or at least she should have been.

AHS *Centaur* was an Australian hospital ship. Properly designated and marked hospital ships, whose function was purely to provide medical assistance to wounded, sick or shipwrecked personnel, were afforded protection under the terms of the Geneva Convention of 1907. Under the convention, to which Japan was a party, an attack on a hospital ship constituted a war crime. Despite this protection, hospital ships were nevertheless attacked and sunk during the war by both the Allied and Axis forces. Most of these casualties were caused by aircraft bombings, where it could be argued that the ships' distinctive markings couldn't be clearly seen by a plane at high altitude or whose vision was obscured by cloud cover. Attacks on hospital ships by submarine, which by their very nature had to take place at relatively close range, were very rare. Although it was never admitted during the war, the Japanese submarine *I-177*, commanded by Hajime Nakagawa, was identified as most likely responsible for the attack on *Centaur*.

The attack was also one of the most deadly in history, with 268 of the 332 persons on board dying as a result. This callous and cold-blooded killing of doctors, nurses, field ambulance personnel and ship's crew – all innocent non-combatants on a mission of mercy – caused outrage in Australia. It left an open wound in the hearts and minds of countless relatives that would never totally heal until *Centaur*'s wreck was found and those who died within her were properly remembered. It also left the relatives, and the sixty-four survivors, with numerous questions. Why did the *I-177* attack such a brightly lit and clearly marked hospital ship? Why did the Japanese navy not admit to the attack, even if it was an unintentional mistake? Why did Nakagawa never face a war crimes trial? And finally, for the families who organized themselves as the *Centaur* Association, and for the last living survivor of the ship's crew, Martin Pash: where was the wreck of the *Centaur*?

◇◇◇◇◇◇◇◇◇◇◇◇◇◇◇◇◇◇◇◇◇◇

Compared with the inordinately long time it took campaigners to raise awareness and the funding for the *Sydney* search, the government's backing for an expedition to find the *Centaur* was organized so quickly it felt to me like the decision was made virtually overnight. That wasn't literally the case, of course, but when a joint press release came out later in 2008 from Prime Minister Kevin Rudd and Queensland premier Anna Bligh announcing up to $4 million to fund a search, it revealed that my initial feeling wasn't far from the truth.

Within a fortnight of our discovery of *Sydney*, prominent newspapers in Australia had already started running articles suggesting that the next ship on the list should be the *Centaur*. These early articles featured interviews with members of the 2/3 AHS *Centaur* Association, an organization with about 300 members that included *Centaur* survivors, descendants, relatives and friends. This was how I learned about Ted Leask, whose father lost three brothers when the *Centaur* went down. In echoes of the movie *Saving Private Ryan*, Ted's father Malcolm would also have been on the ship serving with his brothers but for a decision by the authorities that he stay home to look after his pregnant wife, and because they thought that three brothers serving on one ship in wartime was enough. Ted was the first to contact me while I was still in Australia to ask me to support the association's calls for a search, which I gladly agreed to do. The newspaper articles began a groundswell of public support that was quickly taken up by local politicians. Peter Slipper, the federal MP for Fisher, Queensland, was one of the first to add his voice to the campaign, and went as far as having one of his aides contact me to get my opinion on the likelihood that the *Centaur* could be found.

On the sixty-fifth anniversary of the sinking, the stage was set for intervention from a higher source. Anna Bligh, the first woman elected as a state premier in Australia – and incidentally a descendant of Captain William Bligh of the infamous HMS *Bounty* mutiny – wrote to Kevin Rudd requesting federal assistance with a search for *Centaur*'s wreck. Bligh made a ministerial statement in Queensland's parliament the same day, recounting the loss of the *Centaur* and those who perished in her, among whom were forty-five Queenslanders. Seven months were needed to work out the details between the two governments, but in early December they announced plans to jointly conduct the search and to equally share the $4 million budgeted costs. Premier Bligh summed up her reasons for taking this extraordinary action:

> If we can find the *Centaur*, it will be of considerable emotional significance for many of the relatives of those lost on her. This site is part of our state and nation's history and we should try and find, preserve and protect it. For them and their families it is right that we try to find and mark its exact

location. The 268 Aussies and their lost ship are reminders of what we as families, a state and a nation have lost in war. It is right that in our remembering them we can say exactly where they lie.

With the necessary funds allocated, it was agreed that the Queensland state government would take the lead in managing the project, and that it would be a fully contracted affair with all contractors chosen via a competitive procurement process. A joint steering committee (JSC) made up of representatives from the army, navy, Queensland Museum, Maritime Safety Queensland, and Queensland's Department of the Premier and Cabinet (DPC) was formed to oversee the project and to take major decisions regarding the selected contractors. The first order of business for the DPC, who would be chairing the committee, was to select an experienced project manager to assist in planning and conducting the search operations. With the formal structures being put in place, this was certainly shaping up to be run very differently from how the volunteers of HMA3S had operated. The experience was undoubtedly a new one for the DPC too; in the history of Queensland's procurement practices, I have no doubt this was their first time hiring a shipwreck hunter.

With the *Kormoran* and *Sydney* I had earned my place to lead the search expedition based on my own pro bono research, my years working alongside HMA3S in the same voluntary capacity and my prior track record of success. I naturally wished to be involved with the search for *Centaur*, but I would be given no special advantages. The interest and competition for such an unusual and high-profile project would be intense. The DPC's advertisement for the provision of project management services issued in early January of 2009 attracted corporations from around the world, and in the end more than ten national and international firms tendered offers. I had no idea who the other bidders were, but I eventually learned that even Ted Graham himself had submitted an offer. Despite my good work in leading the search for *Sydney* on behalf of HMA3S, Ted continued to promote the idea that such iconic Australian shipwrecks should be found by Australians. Fortunately, his parochial attitude wasn't shared by the navy in the case of *Sydney*, and

I could only hope the DPC were similarly open-minded regarding the *Centaur*.

The DPC's procurement process was an arduous undertaking, but I was pleased to make it through as one of the three short-listed tenderers invited to make a thirty-minute presentation to the JSC. In the weeks after the contract was announced I had received five unsolicited requests to join other teams planning to submit offers. Some were large professional project management companies with offices around the world and thousands of employees. I was slightly concerned how the JSC members would evaluate my offer compared to the larger firms with limitless resources. I was confident I had all the necessary skills to run the project myself and that my track record was hard to beat: by this time in my career I had led the research and discovery of more than twenty major deep-water shipwrecks, with a success rate of 85 per cent. But would the JSC take the risk of trusting such an important project to one person located over 10,000 miles away, or would they be swayed by Ted's political call for an Australian to lead the search?

I had to wait a couple of weeks following my presentation to the JSC before I was notified of the final decision. Because of the ten-hour time difference between Brisbane and the UK, I learned I had won the contract from a voice message left on my mobile by Anthony Crack, the DPC's executive director for state affairs and the chairman of the JSC, who was as keen as I was to get the project started. Despite a delay caused by the state government going into caretaker mode because of a general election, we quickly negotiated a contract, and on 19 March another joint press release from Kevin Rudd and Anna Bligh officially announced that Blue Water Recoveries had won the tender process and would be managing the search. Two days later Anna Bligh was elected premier in her own right, capping off what might be considered a winning weekend for both of us.

Opportunities to work directly for a foreign government at a very high level don't come along that often, so I was delighted to have won the contract, but more than anything I was extremely relieved to have done so in the face of all the competition and politics. In the lead-up to the decision I had developed a very good relationship with some of the key members of the *Centaur* Association, including Ted Leask, Jan Thomas and Richard Jones. My role as project manager was a responsibility I took very seriously, and I knew I could only repay the trust they had shown in me by finding the wreck for them. However, I quickly realized that the search wasn't going to be as easy as perhaps everyone was expecting. We had set the bar very high in locating *Kormoran* and *Sydney* so quickly, but as I frequently remind people, there is never a guarantee when it comes to finding long-lost shipwrecks at the bottom of the ocean.

If Michael Montgomery's provocative book *Who Sank the Sydney?* was the spark that restarted the debate and re-evaluation of the battle between *Sydney* and *Kormoran* that ultimately led to the 2008 discovery of the wrecks, then it would be fair to accord Chris Milligan and John Foley's *Australian Hospital Ship Centaur, The Myth of Immunity* the same credit. This book, which began as a 'short-term personal history project' by Milligan, a Montreal-based social sciences professor, and evolved into a fourteen-year research project and subsequent writing partnership with Foley, a Queensland Coast and Torres Strait pilot and maritime historian, is the definitive history of the *Centaur*. In fact, its contents were so important to me that meeting Chris was my first priority once I began my own research into the sinking.

McGill University in Montreal might seem an unusual place to start researching an English ship sunk off the eastern coast of Australia, but it is where Chris Milligan teaches and lives, and if I wanted a direct line to the key documents about where the *Centaur* was lost, there was no better person in the world to help me. Chris's interest in the *Centaur* stemmed from the fact that his father's brother, Able-Bodied Seaman David 'Sunny' Milligan, lost his life when the ship was attacked and sunk on 14 May 1943. Chris was investigating the life of this uncle he had never met for a teaching assignment when he got hooked on the bigger story of the *Centaur*. At the time, in the late 1970s, there were no books or articles about the ship, so it fell to Chris to begin piecing together her history, which included the unverified submarine attack resulting in her loss. In his words, 'Somehow it did not seem right to allow this tragic incident, with its cloud of rumours, to slip unnoticed into history and be forgotten.'

Other than a brief dalliance with the *Kormoran*, having rescued Captain Detmers' drifting lifeboat after the battle with *Sydney*, *Centaur* had enjoyed a fairly uneventful war. Built in 1924 for the Ocean Steam Ship Company, the main operating subsidiary for the legendary Blue Funnel Line, she was a 96-metre-long, 3,066 gross ton passenger liner and cargo ship employed on the Singapore to Australia service. Throughout their long history, Blue Funnel ships were well known for their excellent build quality, their upright smoke stacks painted powder blue with a black top, and for being named after characters from Greek mythology.

In late 1942, the Australian army were in need of a fourth hospital ship for evacuating sick and wounded soldiers from the north-east coast of New Guinea to Townsville. Because it could provide accommodation for 200 passengers and had the shallow draft needed to reach the embarkation ports, the *Centaur* was identified as the most suitable vessel available. A simple conversion costing £20,000 was envisioned, but this quickly ballooned into a major internal refit once the army officers, medical officers and seaman's union had all had their say. In the end, *Centaur* was transformed into a modern hospital ship capable of carrying 280 cot cases, enabling two major operations simultaneously, and being able to proceed on voyages of up to eighteen days equipped with special ventilation for tropical use. By 12 March 1943, she was ready to begin her new life on the eastern coast of Australia and left port for her maiden voyage as a hospital ship. The additional improvements had driven the final bill for the conversion up to £55,000.

Along with her new interior, the *Centaur* was also sporting a vivid new paint scheme that completely changed her outward appearance. Gone was the original jet-black hull, repainted a stark white against which the iconic Geneva Convention markings for a hospital ship would stand out. These consisted of a four-foot-wide green band that encircled the entire ship, with three seven-by-seven-foot red crosses painted on each side of the hull, at the bow, amidships and at the stern. The *Centaur*'s distinctive blue funnel was changed to dark yellow, as were the masts and derricks, and two four-by-four-foot red crosses were installed on either side of the funnel, up high so that they could be seen from a distance. The ship was still called *Centaur*, but in place of the name on the bows, a white '47' was painted within a black square to represent the hospital ship number registered under the Geneva Convention.

From virtually every angle, whether from a plane up high or a ship on the water, the new paint scheme and prominent red crosses, brilliantly illuminated at night by electric lights and flood-lamps, ensured that *Centaur* was easily and immediately recognisable as a hospital ship. Neither could the Japanese forces, and their submarine fleet in particular, have been unaware of the ship's change in status: on 8 February, they were notified by the Swiss legation that *Centaur* would be commissioned as a hospital ship on 1 March 1943, with details of her particulars and distinctive markings.

Whether he had murder on his mind or was simply frustrated at recording just one kill during his current war patrol, the Japanese submarine commander ignored all the visible signs that the ship framed in his periscope was an innocent non-combatant. Travelling alone, without an escort, and lit from stem to stern against the backdrop of a moonless night, *Centaur* was too inviting a target to pass up, and he brought his boat up to a surface shooting position. With no threat of a counterattack by gunfire or depth charges, he could close to within a few hundred metres for a strike that was unmissable. At that distance there was no need to fire a spread of torpedoes in case one veered off target: a single shot aimed to hit the ship dead centre was all that was needed.

The torpedo slammed into *Centaur*'s port side just forward of the bridge, with devastating effect, opening up a huge hole and instantly flooding the no. 2 hold and patient wards with seawater. A second explosion immediately followed, erupting from the belly of the ship in a tower of flames and burning fuel oil that ignited the bridge and boat deck amidships. The torpedo's detonation had also caused the main fuel oil tank to explode, shooting flames and oil in every direction. The twin explosions rocked the ship to port, and then over to starboard, throwing people violently to the deck and against bulkheads. Those nearest the seat of the explosions were killed outright by the shock wave, while many others were badly burned. As the ship began to rapidly settle by the head, it became apparent that she was sinking quickly. Anyone still alive had scarcely two minutes to escape the confines of the hull. Some dived into the water and swam clear, while others were pulled downwards by the suction of *Centaur*'s death plunge. The unlucky ones trapped inside could only count on the mercy that their deaths came quickly. The last anyone saw of the ship was the poop deck sticking vertically out of the water. It was 4.12 a.m.

The speed with which the ship sank left no time for an organized abandoning or for an emergency SOS call to be transmitted. None of the seven lifeboats on board were launched, and the two that eventually did float clear of the ship were damaged and unusable. *Centaur* was also equipped with four large drum rafts designed to free themselves automatically, but of these only one released properly and was seen to have as many as thirty men clinging to it. Everyone else was left scrambling for whatever pieces of wreckage they could find to support their weight. In these crucial first minutes survival came down to the seriousness of a person's injuries, whether they could swim or not, or whether they had enough time to grab a life jacket before reaching the deck. In the end, the fact that only 64 out of the 332 on board survived showed how much luck also played a part in the difference between life and death.

When daybreak came, it provided the survivors with the first opportunity to take stock of their situation, which in any analysis was grave. In addition to the men on the large drum raft, about half a dozen more were balancing on the roof of the navigation bridge – otherwise known as the monkey island – which had been blown off the ship by the explosion. The rest of the survivors were mainly in groups of two or three, hanging on to oil drums, a broken lifeboat, hatch covers and small rafts. There were undoubtedly some individuals floating by themselves, helped by a life jacket or other wreckage, but they were difficult to spot amongst the waves and would soon be drifting away from the main group. The second mate, Gordon Rippon, realizing that he was the senior surviving deck officer, began working to pull the survivors together into a larger group. Eventually, by paddling towards each other or through the efforts of a volunteer swimming out with a rope, they were able to get more than fifty people together into a single makeshift raft lashed together by whatever ropes or cables they could find. They would drift this way, on 'survival island', as Rippon called it, for the next thirty-five hours before being spotted and rescued.

Other than the fortunate few who had been on duty or awake at the time of the attack, and were thus fully clothed, everyone else was wearing only their bedclothes, or in some cases nothing at all. About half had life jackets or lifebelts, although these would begin losing buoyancy after twenty-four hours. Injuries varied from severe burns and broken bones to lacerations and bruises. Jack Walder's burns covered so much of his body he eventually died and was given a Christian burial at sea by the group. Another man paddled away by himself, saying he wanted to get some fresh water, and was never seen again. They had all swallowed their fair share of diesel oil, which covered everything and worked its way into people's eyes, noses and stinging open wounds. Those were the physical ailments, but each of them also had to deal with the mental shock and trauma of witnessing the horrific deaths of so many friends and colleagues.

When Sister Ellen Savage jumped from *Centaur*'s deck, she had two fellow nurses alongside to bolster her courage. She was sucked down deep into the ship's vortex, got tangled in ropes and for a moment thought she might never rise. When she did, the other two nurses were gone and she herself had been badly injured by debris swirling around as *Centaur* sank, sustaining fractured ribs, a broken nose, a fractured palate and burst eardrums due to the pressure. It might have been worse, however, as she couldn't swim and was probably saved by the lifebelt she grabbed just before jumping. Of the twelve Australian Army Nursing Service (AANS) nurses on board, Sister Savage was the only one to survive. She realized this after she was transferred to the survival island raft and was shocked to see no other women.

Undeterred, and hiding the true extent of her injuries, Sister Savage did what she could to comfort the injured and tend to their wounds. Lieutenant Colonel Leslie Outridge, the commanding officer of the 2/12 Field Ambulance Corps, was the only other medical personnel to have survived, though he had sustained severe burns. Despite their own injuries, the pair did their best to treat the injured with the limited supplies from a medical kit that had floated past Outridge as he was leaving the ship.

When it came to lifting the survivors' spirits, it was Savage, through her sheer presence and positivity, that the men remembered and responded to. If morale dipped, like when a plane flew past without spotting them, she would rally them with a singsong. When prayer was called for during Walder's burial, she led the group with her rosary beads clutched tightly in her hands. When hunger or thirst beckoned, she took on the crucial responsibility of rationing and doled out the small amount of food and water they had found in the lifeboat. Their journey on survival island was beyond miserable: constantly wet, often cold, and harassed by sharks looking for a meal. Throughout the ordeal, however, Sister Savage suffered alongside the men without complaint.

When rescue did come, on the afternoon of their second day at sea, Savage even made sure that everyone else was tended to before she was. Her remarkable character and the selfless way she helped her fellow survivors was plain to see. Quite rightly, the press and the government decided that she was a heroine, and one of the few positive stories resulting from this awful tragedy. The photos of her being interviewed in a hospital bed sporting a blackened eye spoke volumes about the callousness of the illegal Japanese attack on a hospital ship carrying nurses. Similarly, her image – or at least an image of a nurse symbolizing her plight as a shipwrecked survivor – was the basis for a famous and dramatic war poster that called for Australians to work, save and fight in order to 'Avenge the Nurses!' Savage was later awarded the George Medal for 'conspicuous service and high courage': only the second Australian woman to receive such a high honour.

After failing to attract the attention of a number of planes and ships that passed by without seeing the emergency flares they shot into the sky, the survivors finally got lucky when a plane and a warship spotted them at virtually the same time just before 14.00 on 15 May. When the plane, an Australian Avro Anson, dropped down to sixty feet for a closer look, the pilot and his navigator were stunned to see, within an oil slick and scattering of debris two miles wide, a cluster of rafts with people waving at them. By Morse lamp the plane transmitted, 'Will send help.' Sister Savage, who had kept her emotions in check throughout, finally broke down and sobbed.

The US destroyer *Mugford* was on routine duty escorting a merchant ship into open waters forty miles east of Cape Moreton when their zig-zagging track happened to intersect the path of the drifting survivors. An alert lookout, who would have been scanning the waters for enemy submarines, sighted rafts on the horizon at 13.47, causing *Mugford* to break off from the escort to investigate. After receiving a signal from the Avro Anson, *Mugford* closed on the first group and had them safely on board by 14.10. Over the next seventy-three minutes the destroyer stopped six more times to rescue survivors, including one solitary man, Owen Christensen, who had spent the entire time adrift on a hatch-board by himself. The ship's log recorded: 'Total number of survivors on board 64.' Another vitally important piece of information recorded in the log was the position where the *Centaur* had sunk: '070° true from Pt. Lookout 24 miles.'

When the newspapers were cleared by military censors to report the sinking, they pulled no punches in describing the act as a despicable and cowardly war crime. A sampling of the headlines included: 'WANTON AND DELIBERATE ACT' (*Argus*, Melbourne); 'TREACHEROUS NOCTURNAL ACT' (*Cairns Post*); 'LIMITLESS SAVAGERY' (*Liverpool Echo*). There was very little doubt that the attack had been made by a Japanese submarine. Twice during their time in the water survivors reported seeing and hearing a submarine on the surface, which in one instance came so close that men could be seen in the conning tower. But which submarine was it, who was the commander and why did he attack a clearly marked hospital ship in violation of the Geneva Convention?

Without specifically stating that the attack had been made by a Japanese submarine, Prime Minister John Curtin made it clear in his statement to Parliament that he directly blamed Japan:

The attack on the *Centaur* bore all the marks of wantonness and deliberations. The deed will shock the conscience of the whole civilized world. It will demonstrate to all who have any lingering doubts the unscrupulous and barbarous methods by which the Japanese conduct warfare. To the next-of-kin of those who are lost, the government and nation extend heartfelt sympathy, which is the deeper since those persons were non-combatants engaged on an errand of mercy, and were by all the laws of warfare immune from attack.

Chris and John's book on the *Centaur* was a wonderful introduction to the ship I would spend the next year determined to find. More importantly, it gave me a huge head start in identifying the archives and historical documents I needed to consult. A controversial incident like the sinking of *Centaur* was bound to leave a large documentary record, and one look at the extensive list of references in their 1993 book showed this to be the case. I still had to visit each of the archives they mentioned, but their book read like a virtual roadmap of the research I planned to conduct.

Two notes in the book especially caught my eye, and they were the main reason I made seeing Chris Milligan a priority as soon as the project began. The notes referred to information Chris had received from *Centaur*'s second mate, Gordon Rippon. Rippon had had the midnight to 4 a.m. watch, which meant he probably took the last navigation fix before the ship was attacked at 4.10. As the senior officer amongst the survivors, it was probably also Rippon who provided the first recorded sinking position – '070° true from Pt. Lookout 24 miles' – to the captain of the *Mugford*. Whoever had taken this position – and everything pointed to Rippon – I would be relying almost entirely on their skills as a navigator in deciding where I should conduct the search.

In some ways the challenge in finding the *Centaur* was going to be similar to the problem I had faced with *Kormoran*. Because the Japanese submarine commander never owned up to the attack, I had a navigation position from only one of the vessels involved. And the only other clue I had to work with was the time and position where *Mugford* rescued the survivors. It meant having to conduct another reverse-drift analysis using the meteorological and oceanographic methods we'd perfected with *Kormoran* to see whether the drift of the survivors' rafts agreed with Rippon's position. There was also the strange case of the 'official position' to resolve. This position (27° 17' south, 154° 05' east) appeared in official documents such as the movement card for the vessel, and in authoritative World War II reference books, but it wasn't clear what its origin was, or who was the source.

Chris was all ready for me when I pitched up to his office at McGill University on a cold but sunny day in mid-May. He had retrieved four boxes of research material from storage and was reacquainting himself with documents he hadn't looked at for over sixteen years. With the boxes at our feet we sat on a small sofa in his cramped office and for three days went through each of them page by page. It was an absolute treasure trove of unique information. Because Chris had started his research in the late 1970s, when a good number of the *Centaur* survivors were still alive, he had been able to record their personal recollections when writing his original version of the ship's history (he published a monograph in 1981, before teaming up with Foley to write their 1993 book). To my great good fortune, his most extensive correspondence was with Gordon Rippon.

The first thing Chris was able to confirm beyond any doubt was that Rippon was indeed on watch just before his ship was attacked, and it was he who worked out *Centaur*'s last position and provided it to the *Mugford*. His testimony was recorded in several documents, including a remarkable questionnaire that Chris was able to get nine of the survivors to complete. With respect to his responsibility on the ship, Rippon replied that it was 'safety of the vessel both from a navigational viewpoint and the avoidance of other vessels'. In response to a question about what he did during his time on board the *Mugford* after being rescued, he wrote: 'The captain asked me to go on the bridge to advise him about the position of the ship when torpedoed. I remained there until the search was abandoned.' And if I needed any further proof that Rippon was the source of the last known position of *Centaur*, he provided it in a 1980 letter to Chris:

> Now as regards your 2nd letter on the *Centaur*'s movements. The position of the sinking is firmly based on the 3.30 a.m. position when Point Lookout light was abeam bearing 270° – 23 miles. This means she was steaming due north. Steaming at 12 knots she would be 9 miles further north by 4.15 a.m. when she was struck. By the way a knot is one nautical mile per hour.

Short of the original plot that Rippon kept on the bridge of *Centaur*, I couldn't have hoped for better information.

In a long letter he wrote to his father about the incident, Rippon made clear his sorrow 'to lose the poor old *Centaur* as I was very happy and would have liked to have had her for a year as mate'. He remembered it 'as one of the happiest times in a life-time at sea'. He was also very distressed by the deaths of the ship and hospital personnel who had become his friends during the short period *Centaur* served as a hospital ship. What especially hurt him was the loss of eleven of the twelve nurses: he called them 'marvellous lassies'. Rippon's physical injuries were minor and his stay in hospital was short, but in an indication of the effect the incident had on him, he admitted to his father that he later suffered what he described as a 'delayed shock reaction'.

Gordon Rippon made the grade to mate all right. In fact he went on to become one of the youngest captains in the merchant navy and stayed with the Blue Funnel Line right through to retirement. There could be no greater testament to his skill as a navigator than the fact that he was chosen by this first-rate company to be the senior instructor for their in-house navigation school. He died in 1996 so I never had a chance to meet him. I would have liked to have talked to him about navigating at sea, even though our careers were spent at opposite ends of the change from celestial to electronic and satellite-based navigation. At least I had Chris's papers to get a sense of the man and the skill with which he performed his duties.

The papers also helped me understand Rippon's answer in a contemporary interview he gave, a copy of which I found in the UK archives. He explained that when his watch ended, he gave a chit with his 4 a.m. position to the wireless operator on duty. This was normal practice during the war to make sure that the operator always had the most current position to transmit in case of an emergency. In this interview Rippon also testified that the final position he reported 'was based on a 4-point bearing of Pt. Lookout'. A four-point bearing is a special type of running fix that requires the navigator to take two bearings off a fixed point, in this case Point Lookout lighthouse, first at 45° and then at 90° when abeam of the light. With *Centaur* travelling at a constant speed (twelve knots) and course (north), its range from the lighthouse would be the same as the distance the ship would have run since the first bearing was taken. This was a relatively simple and accurate method of using the lighthouses along the eastern coast of Australia to help navigate a course, but it depended on the ship always being within the range at which Rippon and the other watchstanders could see the lighthouse lights.

One of the discrepancies about *Centaur*'s route up the coast was why she didn't follow the Green Route prescribed by the Naval Control Service. The Green Route was defined by a specific set of waypoints that would have put the ship about 110 nautical miles off the coast as opposed to roughly 25, and was supposedly a safer option for merchant ships as submarines generally patrolled nearer the coast for potential victims. *Centaur*'s captain, George Murray, decided to ignore this advice because it would have taken him too far offshore, and as a hospital ship he believed he was immune from attack in any case. This decision cost him his life, and in death he did come in for some criticism for choosing his own route. But was it fair for his reputation to be tarnished because of the illegal action taken by another commander?

However history was to judge Captain Murray's command of the *Centaur*, the insights I gained from Gordon Rippon about the ship's reliance on the coastal lighthouses for navigation narrowed my focus for future research. I would need to investigate these lighthouses, especially Point Lookout on the tip of North Stradbroke Island and Cape Moreton on Moreton Island, in great detail. What I didn't realize at the time was what a huge can of worms this research would open up, and how it would raise fundamental doubts about what Rippon actually saw the night his ship was sunk.

◇◇◇◇◇◇◇◇◇◇◇◇◇◇◇◇◇◇◇◇◇◇

I arrived in Brisbane at the start of July full of confidence in how the research was unfolding and in Gordon Rippon as a reliable source for the key position around which the search would be based. For a shipwreck hunter who relies on good-quality navigation information to do his job, Rippon was a dream source in terms of both his credibility and the evident care he took in preserving *Centaur*'s last-known position. When you consider his ordeal of being shipwrecked and then adrift on a raft for thirty-five hours, the fact that he was able to provide *Mugford*'s captain with such precise information immediately upon boarding the vessel and then remain on the bridge for a further four hours to assist with the search for survivors shows what a strong and professional man he was.

In addition to a progress meeting with Anthony Crack and the JSC, it was also planned that I would give a presentation at a public forum that was being held mainly for relatives and *Centaur* Association members. I would be speaking about the initial research I had conducted over the past few months in archives in the UK and Australia, and I also had the initial results of the reverse-drift analysis to share. All the information I'd been able to find in the archives seemed to back up Rippon's position, and I planned to discuss this in detail during my presentation. However, the source of the official position still remained unknown. One theory was that it came from the Japanese submarine, possibly by way of an intercepted communication, although I could find no information to verify this one way or the other. I was hoping that the additional research I planned to do in the National Archives of Australia in Brisbane and Canberra during this visit would shed more light on the mystery.

At my first presentation, to the JSC, I highlighted what I saw as the main operational risks for the search, which were both environmental in nature. My chief worry was the topography where the search would be conducted. It was our rotten luck that Rippon's position happened to coincide with the physical edge of the Australian continental margin, meaning that we'd be searching in extremely rugged terrain. In fact a recent CSIRO research cruise had produced an excellent bathymetric map for this area, revealing a major submarine canyon shaped like a fan, which the scientists had already provisionally named the Centaur Canyon. The range of water depths over the search area varied hugely, from about 500 to 4,000 metres, but my main concern was how steep some of the walls within the canyon were. Some had slopes greater than forty-five degrees, which would not only be impossible to search using a deep-tow sonar but were potentially dangerous. If the wreck of *Centaur* had actually sunk in this canyon, we were going to have a hell of a time finding it.

In addition to the rugged terrain, the search area was also the site of the largest and strongest oceanic surface current in all of Australia: the East Australian Current (EAC), a southerly flowing boundary current that was famously portrayed in the animated movie *Finding Nemo* as the oceanic superhighway that swept two of the central characters, Marlin and Dory, down from the Great Barrier Reef to Sydney. Speeds in the central core of the current average two knots, but can get as high as four under certain conditions. Together, the nasty seabed geology and the fast-flowing EAC was a nightmare scenario for any deep-tow search. I tried to put it into a realistic context for the JSC and said that it would be like conducting an aerial search for lost climbers in the Alps in the midst of a hurricane. If for whatever reason we happened to lose a sonar, I could imagine it being swept away like the two fish from *Finding Nemo*, never to be found again.

The other issue I had to contend with regarding the EAC was how survival island had managed to drift to the north-east from Rippon's position to the *Mugford* rescue position in the face of the southerly-flowing current. This was a conundrum that had stumped others and led people to doubt Rippon's navigation. Their argument was that it was impossible for the survivors to have drifted in any direction against the EAC, and for that reason Rippon's position had to be wrong as long as the *Mugford* position was correct, which everyone assumed it was. This was an overly simplistic view of the EAC, however, as boundary currents are complex, dynamic structures and do not always flow uniformly in one direction. In fact the EAC not only changes its position relative to the coast but can spin off eddy currents that appear to reverse the direction of flow.

Once again I turned to the oceanographer Dr David Griffin of CSIRO to help me understand the possible drift trajectories for survival island, in conjunction with the meteorological hind-casting by Len van Burgel. What I wanted to know was whether there were any oceanographic conditions, given the reconstructed wind field (ten knot winds from 131° true), that could result in a drift to the north-east. In order to come up with the answer, David had to essentially force his computer to search for occurrences throughout the fourteen-year BLUElink database in which a drifting object could have started close to Rippon's position and ended up close to the *Mugford* rescue position after thirty-five hours. While there was no way of proving that such conditions actually existed on 14–15 May 1941, I only wanted to know if it was possible, and if it was, the frequency of occurrence.

Fundamentally, in order for Rippon's position to be correct, there needed to be a cold-core eddy spinning clockwise in exactly the right place at the right time. Such eddy currents do exist, so we weren't creating a completely new phenomenon. But they are very rare. So rare in fact that of the 5,191 days David searched, he could find only 185 days, or just 3.5 per cent of the time, where these conditions existed. We had pushed the science as far as possible. Although the oceanography couldn't prove that Rippon was right, it certainly didn't prove that he was definitely wrong. The only other tantalizing piece of information I possessed to maintain my firm belief in Gordon Rippon was that in describing the conditions at the time of the rescue, one of *Mugford*'s crew actually noted that 'a north-easterly current was running'.

After meeting the JSC, I had some free time before the public forum, which I used to visit the Queensland Maritime Museum (QMM), who were helping me with the issue of visibility of the coastal lighthouses. Ian Jempson, ex-RAN, was CEO of the museum, while Jack Duvoisin was its resident lighthouse expert. In previous emails, Ian and Jack had questioned the range at which Gordon Rippon could have seen Point Lookout light. While Rippon's 3.30 a.m. position for *Centaur* was based on seeing the light from 23 nautical miles away, Ian and Jack had sent me technical information indicating that the maximum visual range was of the order of 16–17 nautical miles. Because the issue was highly technical and there were still a lot of unknowns, I decided to hold my judgement until we were able to meet face to face. But it was clear from the gist of the emails that both Ian and Jack had serious doubts about Rippon's navigation based on his ability to see this particular lighthouse.

Like the lighthouses he maintained, Jack Duvoisin spent his life looking out to sea rather than the other way around. Still, he considered himself to be an old salt and I was grateful to have the benefit of the encyclopedic knowledge he had gained from a career working as a lighthouse engineer for the Commonwealth Light Service and the Australian Maritime Safety Authority (AMSA). Although retired, he kept up his passion for lighthouses through his volunteer work at the QMM restoring navigation aids for display. Given the fundamental importance of Rippon's position being based upon a lighthouse sighting, I would eventually have a long correspondence with both Jack and Ian on the subject, and they were extraordinarily helpful with information right up to the day I left to start the search. I can safely say, however, that what they told me in that first meeting was truly worrying and left me with many sleepless nights over the next six months.

Based on the technical characteristics of the Point Lookout lighthouse alone, and setting aside any personal opinions about Gordon Rippon as a navigator or a man, Ian and Jack both believed that Rippon could not possibly have seen the light as he stated. There were a variety of factors involved, such as the meteorological visibility on the night of the attack, the extent to which background lighting both on land and on *Centaur* itself made seeing the light more difficult, and whether this light had a loom effect. But the central, overriding problem, which in fairness gave them good reason for their doubts, was that Point Lookout was a relatively low-power light that in theory didn't have the intensity to be seen from such a long distance away. We had to be certain we were looking at the characteristics of the light as it existed in 1943, but once all this information was retrieved from the archives, it only made their doubts stronger.

The intensity of these old lights is quoted in candlepower or candelas, terms I hadn't come across since my high-school physics class. The Point Lookout light, which had an elevation of 72.5 metres, used a four-burner acetylene lamp with a luminous intensity of 3,000 candelas. The distance the lights can be seen depends greatly on visibility: basically whether the air is clear, or foggy or full of sea mist. For example, with 10-nautical-mile visibility, Point Lookout can be seen from 11 nautical miles; but at 20-nautical-mile visibility the nominal range increases to 17 nautical miles. There was also the geographic range to consider, which is the calculated line-of-sight distance from the lighthouse to Rippon standing on top of *Centaur*'s monkey island, where he would have been when taking positions. This distance was 25.15 nautical miles, meaning that even with very good visibility of 20 nautical miles Point Lookout couldn't be seen from the geographic range.

Later on, Jack provided the diagrams and data to prove his point, which made it very hard for me not to begin doubting Rippon myself. He wasn't saying it was impossible for Rippon to have seen the light; just that the visibility and all other conditions had to be absolutely perfect on that night. This included a complete absence of background light from land, and being able to turn off *Centaur*'s floodlights when he was taking his sights. On balance, Jack thought it was far more likely that the light Rippon had sighted was not Point Lookout, but Cape Moreton, roughly 24 nautical miles to the north. Cape Moreton was a higher lighthouse, at 122 metres, and thus could be seen from further away; most importantly, it had a far more powerful electric lamp, with an intensity of 1,400,000 candelas. At 20-nautical-mile visibility this light could be seen from a distance of 46 nautical miles, well beyond Rippon's 23-nautical-mile range or even the calculated geographic range of 30.3 miles.

I got this final information from Jack the morning I was scheduled to speak at the public forum, and it made for extremely uncomfortable reading. I had previously considered Gordon Rippon a rock-solid source, and using his navigation as the basis for my search box was going to be a major part of the presentation I planned to give in just a couple of hours. I had no choice but to start considering the implications that Rippon was wrong, and what it meant in terms of the location and size of the search box, water depths, seabed geology, risk of failure, and the increased costs of having to search a larger area. In short, this unwelcome news threw all my preliminary planning up in the air and I had to quickly decide how to present it not only to the relatives and Anthony Crack, but also to the deputy premier of Queensland, Paul Lucas MP, whom I was to brief beforehand and who would be introducing me at the forum.

The last thing I wanted to do was to spook Paul Lucas with this bombshell about Rippon at our first meeting. Although he was standing in for Anna Bligh, he had taken a genuine personal interest in the project and asked good questions as I walked him through all the issues. I explained that a number of the survivors had reported seeing lighthouse lights while they were drifting, and that this would have to be a new avenue of research to see whether any of the sightings were credible and how they related to Rippon's position. If Rippon did mistake Cape Moreton for Point Lookout, it would also mean that survival island didn't drift northeast and I wouldn't have to invoke the rare clockwise eddy to explain this. On the contrary, a sinking position north of Cape Moreton was easier to explain because the drift direction would be south-east, more in keeping with normal EAC flow.

In the end, I got through the day by explaining the preliminary nature of my research and that more work needed to be done. The important thing was to not let this development throw the project off track. It certainly helped that Anthony Crack was unfazed and took it all in his stride. We were developing a very good working relationship and I was impressed by his calm acceptance of the situation: a necessary trait for someone so close to the heart of government.

Within weeks I had a revised research plan that I was working on to resolve this new uncertainty about Rippon's position. I hadn't lost faith in Rippon completely, as I had found nothing to change my original opinion of him: that he was an honest and well-regarded navigator who gave truthful answers when debriefed. I would just have to work harder to restore my confidence in his navigation.

⸻

Regardless of whether the search was going to be focused on Rippon's position or not, we still needed to organize a spread of deep-tow sonar and ROV equipment that was capable of working within the full extent of possible water depths from 200 to 4,000 metres. As with *Sydney*, I advised the government that it would be best to conduct the search in separate phases. Our aim was to start in early December 2009, which meant we had to work fast to select and hire a vessel operator, search contractor and ROV contractor. Because of the extreme depths involved, we expected that the search and ROV contractors would most likely be coming from America, but we hoped to find a suitable support vessel in Australia and ideally one based on the east coast.

Over the next few months while I assisted Anthony Crack's team with the procurements and provided technical advice to the JSC making the final contractor evaluations, I continued my research into the lighthouse question and began looking into claims made by people who believed they had witnessed the attack from land. Because the search was a hot topic in the local news, people began contacting Anthony's office claiming they had either seen the explosion or knew where the wreck was. I had the job of investigating these claims and handling the numerous offers from volunteers who wanted to join the search expedition itself. Based on the number of people willing to donate their time as electricians, deckhands, divers or even just to make the tea, the search had clearly caught people's imagination. As for those who were claiming to have witnessed *Centaur* exploding, the main thing I had to be wary about was exactly that – people's imagination.

I took each and every claim seriously, however improbable it seemed.

If it was someone saying they knew the position of the wreck, it invariably came from a trawlerman, or someone who knew a trawlerman. A typical example was the claim made by Dr Ross Evans, a Gold Coast GP, who told a reporter that 'the government will be looking in the wrong direction and waste their money' if the search was conducted near Rippon's position. Dr Evans believed the wreck was in the prawn grounds north-east of Cape Moreton. I had heard a similar story from others, telling how the local trawlermen knew of a spot on the sea floor where their nets got caught, which they called 'the beds' because they would sometimes pull up rusted iron hospital beds. One could only imagine how hospital beds came to be lying on the sea floor miles from the coast, but as Dr Evans refused to reveal the name of the trawlerman who gave him this information, there wasn't anything more I could do with it.

The second type of claim was more intriguing, as these were essentially oral testimony from people who were alive at the time and had carried their memories with them their entire adult lives. Most had no documentation or other witnesses to back up what they were telling me, so it came down to how closely their stories fitted the facts and whether I judged them to be reliable. Given how strongly I had rejected the oral testimonies that Glenys McDonald had relied upon for her mistaken belief about the *Kormoran–Sydney* battle location, I found it slightly ironic that I was now driving around Queensland to hear essentially the same type of accounts.

I needed to be cautious about such conflated memories, where people merged an unrelated experience in their life with a major news event simply because they had occurred at roughly the same time. It wasn't a question of whether they were being truthful, because in their minds they were absolutely 100 per cent certain of what they were telling me. One elderly man, who had been fifteen at the time, told me about the day he was crabbing with his father in Moreton Bay at 4 a.m. when they both saw a flash on the water to the north-east, between Bribie and Moreton Islands. The man said his memory of the event was 'as clear as a bell', but the location he gave me was completely impossible given where the survivors were rescued.

Invariably, the claims I had the most faith in were those in which a second independent witness was involved, or there was some contemporary documentation to back up the story. It also helped when the account had the ring of truth about it. Jean Elder, who lived at Bald Knob Farm, had been eight years old when *Centaur* was attacked. Although she was in bed at the time, her parents were both up, her father milking the cows and her mother, an early riser, doing the housework. The Elders were both Volunteer Air Observers, with a direct telephone line to report suspicious aircraft. Their farm was located on an elevated mountainside of the Blackall Range, which is known for its spectacular views of the coast, and because their house wasn't electrified until 1948, it would have been close to pitch dark with no moonlight at the time: an ideal setting to have seen a flash out at sea. This was definitely a claim worth checking, so when Erica Costigan, who had spent twenty-five years documenting every local story about the ship, brought it to me, I agreed to make the long drive up to Bald Knob to hear Jean out.

When we got to Jean's house, I could see there was real potential in her story being not only true, but also helpful to me in determining the likely location of the wreck. Although she herself had not seen the flash and was relating a family story that had been with her since childhood, she spoke with belief and a degree of detail that was far better than the other witnesses I had interviewed. Most importantly, she remembered quite clearly her mother reporting the flash by telephone straight away to an official and also referring to it in conversation with a neighbour a few days later. By then it was known that *Centaur* had been attacked, and Jean's mother had commented that the two things must have been connected. Standing where she had been when she had seen the flash, I could appreciate what a valuable clue this might be. The view of the sea from the Elders' house was through a narrow notch above the treeline that was only five degrees wide based on the compass measurements I took. The line of bearing (153° true) to the tip of Moreton Island where the flash was observed, if extended out to sea, was only three nautical miles north of Rippon's position! Was this the verification I was hoping to find?

This was potentially a very exciting clue because it was so specific in pointing to a relatively precise location at sea. Nevertheless, I only had Jean's word and needed to somehow corroborate the details of her story. When I looked into the name of the official her mother supposedly called, Eric Gilson Foxton, I did find army records that he was a captain living in nearby Maleny, as Jean had said. She'd thought he might be an intelligence officer, but also added that on the night of the flash he was staying with another family two kilometres away by the name of Kerr. It took me a while, but I eventually tracked down James Kerr in Sydney and went to interview him at his home.

James was just a few years older than Jean, but he confirmed important details of her story: namely, that Eric Foxton had been a great friend of his father's and was a regular guest at the house, and that the Elder and Kerr houses had been connected by a telephone exchange at Landsborough. James's mother was also a Volunteer Air Observer, so it made sense for her to regularly talk with Jean's mother when either of them spotted aircraft. This was as much corroboration as I would ever get, and although I'd hoped for more, it was enough for me to believe that Jean's story was more than likely to be true. For that reason I would factor it in to where I positioned the search box.

During my research in Australia, I kept a working copy of a chart I would update with various clues and positions, and sometimes use to help the people I was interviewing to visualize where everything was located. I still had the mysterious 'official position' marked on this chart, although no one knew for sure where it came from. Because this position (27° 17' south, 154° 05' east) was referenced on a number of later Japanese documents, the general belief was that it came from the attacking submarine *I-177*. I would have been happy if this was true, but eventually I was able to prove that it originated from an official based on shore in Brisbane three days after the attack. Since the position didn't come from either the *Centaur* or *I-177*, I decided it could be discounted. This wasn't the first time I'd found that an 'official position' was a fabrication, but it was surprising in this instance that it had somehow achieved this status in both Australia and Japan.

Meanwhile I was still struggling with the important question of whether Gordon Rippon was able to see the Point Lookout light from the distances needed for his fix at 3.30 a.m. Through the magic of the Internet I created an email group that included John Foley, Jack Duvoisin, Ian Jempson and Kevin Slade – a senior manager at the Australian Hydrographic Office who was also helping me – that allowed us to exchange information and ideas about this issue. In view of Jack's information that the maximum range for Point Lookout was in the order of seventeen nautical miles, we were stuck with the real possibility that Rippon never saw this light and mistakenly took his position from the more powerful Cape Moreton light instead. John thought we could get better information about the maximum range in 1943 from the AMSA archives, so he arranged for a friend to make a copy of their historical file on the Point Lookout light and send it to him. The file arrived while John was piloting a ship on the Torres Strait route, so I got first crack at reading through the two-inch thick stack of papers.

The AMSA file was essentially a record of every piece of official correspondence about Point Lookout dating back to the first requests for a light in 1927 because overseas shipmasters had complained about the absence of one along this stretch of coastline. It contained every piece of technical data about the tower and light, including the all-important Notices to Mariners detailing that the white light would flash three times every fifteen seconds. The visibility of the light was given as 22 miles. After the light was actually established on 9 February 1932, a lighthouse tender ship, the SS *Cape York*, was sent out to check its visibility the very next week. *Cape York*'s master, AS Stumbles, duly reported 'that the visibility of the Point Lookout light was checked and found to be 23.8 miles at 30 feet high water'. The file contained letters from two other masters who reported seeing the light from distances of 18–20 miles, with the latter commenting, 'there was a fresh NE blowing at the time but I had no difficulty in picking it up 20 miles off yet the weather was not too clear'. The fact that masters on ships smaller than *Centaur* could see the light from these distances proved that 17 nautical miles wasn't the maximum, and that Rippon's sighting from 23 miles was entirely possible.

With my confidence in Rippon restored, and the pessimism surrounding the visibility of the Point Lookout light lifted, I took a second look at the statements survivors had made about seeing lights on the two nights they were drifting on the rafts. Of those who were interviewed and asked about what lights they saw, fifteen replied that they did indeed see the lighthouse lights on at least one of the nights, and several saw the lights on both nights. Taking into account their low elevation, it was still geographically possible for the survivors to have seen the Point Lookout and Cape Moreton lights at 19 and 24 nautical miles respectively. Visibility would have had to be exceptional, although Rippon's sighting indicates it probably was. Of the survivors who saw the lights, two were merchant seamen whose statements I felt could be trusted.

Captain Richard Salt was a senior-ranked Torres Strait pilot (like John Foley), who had been coaxed out of retirement to take *Centaur* on one last trip. Despite burns to his hands and face, he got away from the sinking ship and wound up on survival island, where he optimistically predicted that they would be picked up by a passing ship. Having sailed up and down this coastline for nearly fifty years, he knew the lights, and the distances at which they could be seen, better than anyone. At 3 a.m. on the 15th, their second night adrift, Salt could see the glare of Cape Moreton light, but not the light itself. He explained in his interview that this meant they were over 27 miles from land, as he knew that Cape Moreton was a 27-mile light. He also added the very useful detail that 'I formed the opinion we were abreast of the light.'

At last I could see the disparate clues beginning to fall into place. Assuming that Rippon's position was approximately correct, and that a clockwise eddy did take the survivors on a looping course to where they were rescued by *Mugford*, then in the early hours of the 15th they would have been roughly abeam of Cape Moreton light as Captain Salt believed. Survivor William McIntosh's testimony added another significant piece to the puzzle, as he distinctly saw both lights on the night of the 14th after the ship sank: 'Taken fore and aft, I was sitting on the starboard side of the raft. I could see Cape Moreton 25-degrees north and Point Lookout about 25-degrees south.' I drew McIntosh's bearings to the two lights and was amazed to see that the intersection point was only three nautical miles north of Rippon's sinking position, and virtually in line with the Elders' sighting of the flash from Bald Knob Farm.

The three clues formed an incredibly tight cluster that was almost too good to be true given the extremes at which the observations were taken. It was the first time in all my shipwreck searches where every clue was based on the sighting of a distant light on the horizon. However, the fact that so many reliable witnesses reported seeing these lights did tend to indicate that the visibility was indeed exceptional. There was only one more sighting I needed to investigate, which was potentially the most promising of the lot because it was made by military personnel in an excellent position to have seen the aftermath of the attack. Only three men were involved, however, and my worry was whether any of them were still alive and healthy enough to speak to me.

The sighting had been made by army personnel attached to the 'O' Australian Heavy Battery, whose job was to prepare emplacements for establishing Rous Battery and a tented campsite on the ocean-facing coast of Moreton Island. Although the guns and searchlights hadn't yet arrived, they did have protective picquets posted at the battery position, which had a clear line of sight out to sea where *Centaur* was steaming up the coast. At the exact moment *I-177* was lining up to fire on *Centaur*, Lieutenant Russ Ward and Sergeant Dermot Reilly were doing the rounds of the picquets. They were chatting to the picquet, Sapper Rudi Glass, when *Centaur* exploded in a huge fireball that was seen by all three men. Lieutenant Ward ran up to the camp and woke his commanding officer, Major R. K. Fullford, who initially discounted the sighting but on Ward's insistence allowed him to make a telephone report of what he had seen. I was given an excellent written account of the story by Fullford's widow, but what I really wanted was to speak to one of the three men to verify the details, with the direction in which they saw the explosion most important.

I had no idea of the whereabouts of Ward, Reilly or Glass, but I did remember that shortly after I'd won the contract, Jan Thomas from the *Centaur* Association had forwarded an email to me from Fred Rubie, who was at Rous Battery in May 1943 and was willing to talk to me about what he knew. I met Fred at his home in Tocumwal, NSW, where he was able to confirm the basic details in Fullford's account. Unfortunately he wasn't able to add any new information about the incident, which was slightly disappointing, but just before leaving I asked if he had kept in touch with anyone he'd served with at Rous Battery. When he told me that he was still in contact with Russ Ward, I couldn't believe my luck.

I got straight on the phone to Ward and asked if we could meet. I needed to rearrange my travel schedule, but I was happy to fly back to Brisbane and make the four-hour drive down to his home in Yamba, Queensland, as long he could give me an hour or so of his time. I had grown so used to conducting my research in libraries and archives, it was exciting to be meeting people who were part of history as it was being made. Like so many others with a connection to *Centaur*, he readily agreed to tell me everything he remembered about 14 May 1943.

Three days later, I was sitting with eighty-nine-year old Russ and his wife Nancy in the lounge of their bungalow, listening to a story that was as vivid and clear in his mind as if it had happened the day before:

> I was the gun position officer and was recently promoted to lieutenant. I turned twenty-three two days before. I was in charge of two sergeants who were responsible for the individual guns. At 3.30 Reilly the sergeant and I went to inspect the sentries, which normally took us about an hour. We had finished with the gun positions and were heading over to the no. 1 searchlight emplacement where Rudi Glass was the sentry. Glass was facing seaward, and I was facing landward. All of a sudden Glass shouted, 'Shit, what's that.' I turned around and saw the explosion, which was a glare on the horizon, and then it died down. About two seconds later we heard this monstrous explosion. I knew something was drastically wrong so I ran up to the camp to wake Major Fullford who told me to phone in the report using the light field cable.

Russ's memory for detail was excellent. It was as if the experience had been permanently scared into his mind by the flash of that explosion sixty-six years ago. However, for him to be truly helpful to the search, I needed to know in fairly precise terms the direction in which it had happened. Russ didn't disappoint. He told me they were standing on a north-to-south path and the explosion was perpendicular to the path, which would make the direction approximately due east. When I pressed him whether he was sure, he added that it was possibly slightly north of east, but certainly not north of Cape Moreton. He estimated that the range in error was plus or minus 10°. When I plotted a line due east of Rous Battery, I was beyond delighted to see that it was barely one nautical mile away from Rippon's sinking position.

From the very beginning of my research, the Rous Battery sighting had had the potential to be one of the most important clues to the location of the wreck. These men were physically the closest of anyone to the incident, and what they had witnessed became part of the documented history of the *Centaur* story. There was always the small chance, however, that their story was the result of a conflated memory. Many otherwise honest people along this coastline were utterly convinced they themselves had seen something that in fact was physically impossible for them to have seen. Russ Ward was not one of these people. He was in the right place at the right time and immediately reported what he saw to a higher authority. I had needed to meet him to be 100 per cent certain of this. That was what I had been hired to do: to be absolutely meticulous in verifying every possible clue. But now that I had spoken to Russ, I had no doubts about the importance of his clue and where it told me we should begin searching for *Centaur*'s wreck.

To find a shipwreck on the first line of searching, you have to be either extraordinarily lucky or have the benefit of a fairly accurate sinking position.

Blue Water Recoveries once had an amazing spell where we found four deep-water wrecks on the trot, all on the first line. In those searches we obviously started with very good positions, but a fair amount of analysis and skill was also involved in picking the correct line out of the several we had to choose from. I doubt such a run of success with historic wrecks will ever be repeated. There was another time when I once found a wreck in twelve minutes. But as that was a small speedboat that had sunk in Baltimore's outer harbour in less than twenty-five feet of water, it can't be compared against *Centaur* and is only worth mentioning in the context of the absolute minimum time it can take to find a wreck by sonar when you know where the wreck is but can't see it.

Despite the odds against it, and the fact that I almost never contemplate finding a wreck so quickly, I actually thought with *Centaur* there was a real chance we could find it on the very first line given that all the clues indicated that Rippon's position was pretty much spot-on. While I kept such thoughts to myself so as not to raise expectations beyond what was reasonable, my plan for the search was designed to give us exactly this chance. In an overall search box of 19 by 21 nautical miles (399 square nautical miles) covering every worst-case scenario in the unlikely event that Rippon was wrong, I outlined a high-probability circle in red centred on a small dot that represented Rippon's position. In fact, I had plotted two dots. One was based on the position reported by the USS *Mugford* ('070° true from Pt. Lookout 24 miles'), which undoubtedly came from what Rippon had told *Mugford*'s captain, even though the position was probably worked out by the ship's navigator. I had calculated the second position (27° 16.6' south, 153° 58.3' east) as a dead reckoning from Rippon's 3.30 a.m. position based on *Centaur*'s speed (twelve knots) and course (due north) up until the time she was sunk. As the two dots were about a mile apart, I decided to run the first track-line right between them, knowing that both positions could be covered at the same time.

This was track-line no. 8 (the first of fifteen) in the centre of my search box, which I oriented 340°/160° to make towing the side-scan sonar over the rugged seabed as safe as possible. Williamson & Associates had again been chosen as the search contractor, and at 7.37 a.m. on 14 December 2009, their ageing but still very capable SM-30 sonar towfish entered the search box to officially start the Phase 1 wide-swath search for *Centaur*'s 94-metre hull amongst the mountainous terrain. It didn't take long to see just how technically difficult the search was going to be. Three steep submarine canyons, with walls as high as 600 metres, were carved into the section of continental margin we were searching. Within each of the fan canyons, narrower finger-like gullies descended down the slope faces. By the end of that first track-line I estimated that roughly 30 per cent of the seabed was virtually impossible to search, either because it was too steep or because acoustic shadows formed behind rocky ridges could not be insonified.

It was a little discouraging that our chances of finding the wreck were so seriously reduced because of the rough seabed. Nevertheless, I was grateful that the weather was ideal for towing operations, that the SM-30 sonar was working well, and that the other major operational risk I had feared would cause us problems failed to materialize. We owed our good fortune in this instance to one of those supposedly rare clockwise-spinning cold-core eddies that reduced the EAC from its customary two knots down to less than half a knot. Because David Griffin had advised that the eddy should persist for a further three days, I decided to keep searching the track-lines in order of priority before investigating any sonar targets.

The combination of depth and relatively short track-lines meant the pace of the search was pretty rapid. Depending on whether we were towing with the current or against it, the track-lines were taking between 5½ and 9½ hours to run, with most turns taking just a couple of hours. With so much geology to sift through, it was impossible for Williamson's sonar analyst to keep up the pace. I had my own image-processing system ready to monitor the images in real-time, but I was being constantly dragged away to deal with media issues. I wasn't supposed to be the main government spokesman for the search, but when the local *Courier Mail* newspaper began unfairly criticizing Anna Bligh for a website that was created to keep the public informed, I was asked to take on this role. So in addition to directing the search, I was writing a daily search diary for the website and doing interviews with a TV news reporter who came out by helicopter and hovered above us while a cameraman uplinked the video and my voice to the helicopter.

By day four of the search we had completed seven of the fifteen track-lines, and covered roughly 60 per cent of the search box at least once. As my aim was to cover all areas of the box a minimum of 200 per cent, we had effectively searched the entire high-probability area twice, even though we were less than halfway through the Phase 1 plan. By this time the influence of the cold-core eddy was diminishing and the EAC was rapidly increasing in strength, as David Griffin had predicted. In the space of less than twenty-four hours, the current speed had risen from two knots to four knots: the EAC was back with a vengeance. The only way the marine crew of our support vessel, the *Seahorse Spirit*, could stay on track in such conditions was to engage both her powerful bow and stern tunnel thrusters and 'crab' down the lines at an angle of forty-five degrees. I doubt anyone trying to sleep in the ship's lower decks that evening got any rest at all.

Mike Kelly, Williamson's sonar analyst, had carefully picked out six targets that matched the characteristics for *Centaur* I had given him. Using the CSIRO bathymetric data as a backdrop, he was able to overlay the side-scan sonar imagery on top, allowing the targets to be viewed in a three-dimensional perspective. This helped us assess whether they were geology or man-made. However, as none of the targets could be classified as a wreck at this point, I decided to stop the wide-swath search in order to collect higher-resolution imagery on each of the six. The spacing of the targets was such that we could image them all on one multi-leg track-line, but unfortunately we were at the wrong end of the search box to achieve this because of the strong southerly-flowing EAC. We needed to start back up at the northern end of the box, which meant running one more wide-swath track-line to get the SM-30 towfish in position. It was at the end of this track-line (no. 13), as the Williamson team were hauling in tow cable in preparation for the upcoming turn, that disaster struck.

Less than half an hour into the turn, with 2,000 metres of cable deployed, all signals and telemetry from the SM-30 towfish were suddenly lost. *Seahorse Spirit*'s speed was reduced to make some topside checks, but in the end it was decided to recover the depressor weight and towfish to the deck to determine what had caused the fault. As the last metres were being recovered, the lack of tension in the cable could mean only one thing: the depressor and towfish were both gone. The mechanical termination, whose sole function was to securely connect the cable to the depressor weight, had failed. We couldn't tell what had caused the failure, although a number of strands in the cable were snapped and the strong EAC current was undoubtedly a factor. All we did know was that the towfish was tethered to the depressor weight somewhere on the seabed in about 1,800 metres of water, and that they would play no further role in the search.

To their great credit, Williamson's team wasted no time in getting the spare AMS-60 sonar system ready to complete the search. Several attempts were made to acoustically interrogate an emergency system on the SM-30 towfish, but these plaintive pings were met with nothing but silence. It was a very sad end for a legendary sonar: a true workhorse that had located many important shipwrecks, including *Kormoran* and *Sydney*. There had been some talk amongst Williamson's team about the sonar being donated to a museum at the end of this project. Whether that would have happened will never be known. In one sense, the bottom of the ocean may have been the most appropriate place for the SM-30 to end her working life. For millennia, sailors have been traditionally buried at sea. Perhaps it is right that a deep-sea sonar should be afforded the same custom.

Despite the shocking loss, Williamson's team had the AMS-60 terminated and ready to be launched later that same day. Ironically, this sonar was better suited for what we were about to attempt. In order to investigate all six targets, a three-legged track-line was created that snaked its way through the search box from north to south. Of the six, only one was close to Rippon's position. It was a small target with no discernible debris field or acoustic shadow, and looked nothing like the dramatic images that had told me in an instant that we had found *Hood*, *Kormoran*, *Sydney* or numerous other shipwrecks. Even Mike had only picked it out during his second look through the imagery, designating it Contact 09. However, it did show some signs of having hard edges, which was a characteristic we normally associate with man-made objects and not geology.

The plan was to image all the targets with the sonar set to a 1.5-kilo-metre swath. We started at 5 p.m. on 18 December and expected the line to take us through to the following morning. There was no point in anyone trying to sleep anyhow, because both sets of thrusters would be going full pelt to counter the EAC, which had ramped up another notch to 4.2 knots. Mike Kelly was in his hut in front of his bank of monitors, ready to analyse each target on the fly, while I was shuttling from the processing system in my hut over to Mike's to compare notes. Mike liked to work in grey tones, while I preferred a colour scheme that I felt had more sensitivity and was certainly more appealing to the eye. One by one, each target was imaged, assessed and after much discussion discounted as geology, until only one remained: Contact 09.

This time the target looked more like a wreck, with a clear acoustic shadow and a small cluttering of debris nearby. It was definitely very promising, but it had a strange bend near its middle that required another look. I asked for the second line to be run at a different heading, more or less parallel to the long dimension of the target, that would allow us to accurately measure its length. This produced a better image, with very distinct angularity and shadow, but it still wasn't definitive.

I was pretty sure this was the *Centaur*, but the image wasn't good enough for me to be 100 per cent certain. The way the project was structured, I had the authority to order the transition to Phase 2, which was intended to confirm the identity of the shipwreck as *Centaur* by high-resolution side-scan sonar imaging alone. The government had also set in motion the shipping of a 6,000-metre-rated ROV system from America so that it was in Brisbane and ready to be used if the search was successful. A lot was riding on the quality of the image that would allow me to confirm that we had found the wreck, so I was determined to get absolutely the best one possible.

We had a problem, however. The target was sitting within a narrow gully, approximately 150 metres wide and 100 metres deep, which made obtaining that perfect image extremely difficult. As the *Centaur* itself was less than twenty metres high, not counting how much the hull had sunk into the muddy seabed, we couldn't get the image we wanted with the sonar being towed above the gully. We tried that twice with the AMS-60 set at the 1,000-metre swath, and neither image was an improvement on the previous one. In fact they were so poor they were useless. On one of the lines running uphill the towfish nearly collided with the seabed, missing it by only thirty metres. On the second line, the 4.2-knot surface current pushed the *Seahorse Spirit* so far south we missed the target by nearly 400 metres.

The only way to get the image I needed was to narrow the swath to 600 metres and tow the sonar downhill through the centre of the gully with the towfish below the top of the walls on either side. I likened the manoeuvre to that scene from the first *Star Wars* movie where Luke Skywalker has to fly down the equatorial trench, below the deck of the Death Star, in order to fire his proton torpedoes at it. The marine crew running the ship, and Williamson's team flying the sonar, would have to be inch-perfect to pull this off. On the first attempt the track was too good and the towfish passed directly over the wreck, at nadir, causing the target to be lost in the sonar's blind spot. At one point the altimeter on the towfish indicated that it got as low as fifteen metres off the bottom, which was much too close for comfort.

I still didn't have the image I needed, but if we pushed the envelope too far we risked a collision with either the wreck or the seabed. With the SM-30 sonar already lost, the last thing any of us wanted was to have the AMS-60 join its older cousin on the ocean floor. It was now early morning on 20 December and we had been running high-resolution lines for the past thirty-two hours. Each pass through the gully was a moment of nail-biting anxiety. Was the towfish on line? Was the altitude too low? Would the EAC surge, or drop off suddenly, causing the ship to lurch in one direction or the other? And the worst fear of all: would a critical piece of equipment fail at the crucial moment when the towfish was entering the gully? All we could do was learn from the previous lines we had run and make whatever small changes were possible to reduce the already high level of risk.

Because the towfish had got too close to the seabed on the previous run, it was briefly recovered to make sure there was no damage. The next line would be our sixth over Contact 09. We had learned that towing downslope produced better images and was safer than upslope, so the towfish was deployed on the western end of the track-line. As the trend of the gully was roughly west to east, the *Seahorse Spirit* would be crabbing down the line with the full force of the southerly EAC on its broadside. We factored in an especially long run-in to the start of the track-line to give the ship's drivers enough time to figure out the right heading and speed, as it was their job to effectively drive the towfish down the line. Once the towfish was on track, all that was left was for Williamson's team to control its altitude with the amount of cable they paid out from the winch. As the track-line was only 1,035 metres long, there was very little scope for corrections once the towfish had entered the gully.

The final adjustment we made was to shift the track-line from the previous run by twenty metres. If the towfish was where we wanted it, we could expect to see the target on the starboard channel illuminated against the southern wall of the gully. For all the good work that everyone connected with the project had done from the very beginning, it all rested on getting this image to confirm the target's true identity. By now we were all crammed into Williamson's hut, where the multitude of sonar and navigation displays would tell us the fate of the project. From the surrounding geology it seemed like the towfish was in the right place, but the target had yet to appear. As each second passed the question I asked under my breath was 'is the target coming, or did we miss it?'

And then it appeared. A black target about twenty metres to starboard in a zone where the AMS-60 produced its best imagery. Two seconds later, a stark white acoustic shadow appeared behind it. Because the towfish was so low, about seventy metres, the shadow was the dominant feature, which was what I wanted to see. It mimicked the angular contours of the target perfectly. This was no rock ridge: it was a shipwreck. And then came the detail that confirmed the wreck's identity: a break in the hull at the forward end where *I-177*'s torpedo had ripped through *Centaur*'s unarmoured hull. The damage from the resulting explosions was clearly catastrophic, but the bow was still tenuously connected to the rest of the ship. It took the AMS-60 sonar barely eighty-five seconds to show me what I so desperately wanted to tell the government and the relatives: the *Centaur* was found.

∞∞∞∞∞∞∞∞∞∞∞∞∞∞∞∞

Before calling Anthony Crack, I took a series of measurements of the wreck to be doubly sure it fitted the dimensions of *Centaur*. I had been so busy the last few days I had lost track of what day it was, but just as I was about to ring Anthony I noticed it was Sunday morning: a lovely time for him to receive such wonderful news. Anthony had worked very hard to make sure the search had every chance of success, so I was delighted he was the first one to hear. My phone call set in motion a series of formal announcements, the most important of which was an immediate joint statement from Acting Prime Minister Julia Gillard and Anna Bligh that the wreck of AHS *Centaur* had been located. The news spread very quickly, and within hours messages of congratulations began flooding in to the ship.

Anna Bligh rang me personally to express her thanks and appreciation to the entire team, which was a very nice touch as I was able to convey her message to everyone on board during our regular midday meeting. While the Williamson crew attempted (unsuccessfully) to locate the lost SM-30 towfish using the AMS-60 system, I was free to write up my last diary for the search. I was especially keen to provide additional information not covered in the official statement, including the position where the wreck was found, lying at a depth of 2,059 metres. It was approximately 1.75 nautical miles away from Gordon Rippon's dead-reckoning position, and in fact we had detected it on our first search track-line, although this wasn't immediately obvious. I wrote in my diary that this result was 'about the best testament you could make to a man whose life was spent navigating the high seas'. For all the doubts about whether he was able to see Point Lookout light, it was important to me that Rippon got the credit he deserved for our discovery, and I was delighted to see that a few newspapers made sure that happened.

The nice thing about finding the wreck when we did was that the

Centaur families didn't have to spend Christmas worrying about the out-come of the search. They could enjoy the holidays safe in the knowledge that the resting place of their relatives had been located. A message left on the government's website typified this feeling:

CONGRATULATIONS. I have just had the only Christmas gift I have ever wanted. I cannot believe this news. I'm just waiting for David's diary for today to come on the screen. How do I ever thank you amazing guys out there, this is just fantastic. Now I think I'm going to cry.

My most sincere thanks,

Sandra Bailey

Eldest granddaughter of LCPL Charles Richard LeBRUN. My hero.

The other benefit of finding the wreck before Christmas was that the crews who would be joining the ship for the Phase 3 video and photo-graphic documentation by ROV could return home to spend the holidays with their families. Although the actual search had been completed in less than a week, the work was nonetheless intensive and the break would do people good. I stayed in Brisbane to demobilize the *Seahorse Spirit* of the Williamson equipment and to ensure that the ROV system arriving from America cleared into Australia without any issues. Phoenix International, an underwater services company started by former colleagues of mine from Eastport and Oceaneering, had been awarded the ROV contract and had sent their 6,000-metre-rated *Remora* III system out to Australia by sea freight. We had taken a bit of a gamble in sending the ROV sys-tem out well in advance of the search, but in the end the timing worked out perfectly. Once the holidays were out of the way mobilization of the *Remora* III was scheduled to begin on 5 January 2010.

In the meantime, both the Commonwealth and Queensland govern-ments were able to celebrate our success and their decision to jointly fund the search. I was asked to speak at several press conferences, first with Anna Bligh as soon as the *Seahorse Spirit* arrived back in Brisbane, and then with Deputy Premier Paul Lucas at the start of the ROV mobilization. The only tricky moment came when I had to explain that confirmation of the wreck as *Centaur* was based on my interpretation of the sonar images and not on an actual picture. There was a slight rumbling amongst the reporters when one of them pressed me on this question, but my reply seemed to satisfy them as the issue never arose again. Although the imagery wasn't as clear-cut as it was with either *Kormoran* or *Sydney*, the measurements of the wreck and the evident damage perfectly matched what we knew about the *Centaur*. As no other ship of its size had sunk anywhere near this location, it was inconceivable that the wreck could be anything other than *Centaur*.

Press interest in the project was understandably high, especially from the *Courier Mail*, Brisbane's main newspaper, which had actively campaigned for the search to happen. Tuck Thompson, the *Mail*'s lead reporter on the story, was coming up with all sorts of novel angles to make sure that *Centaur* featured in every edition. He even organized for eighty-seven-year old Martin Pash to travel to Brisbane from his home in Melbourne so that he could congratulate me personally. Martin was one of only three living survivors from *Centaur*, and I'd made a point of visiting him during the research phase to hear his story. We'd spent hours talking about his life and the effect the sinking had had on him. Despite his humble beginnings as a steward in the merchant marine, he had gone on to do very well for himself in the haulage business and in property investment. He'd lived a good life, taking overseas holidays and buying a new car every two years, and he had the biggest television I had ever seen in a house: a monstrosity that was over a hundred inches wide, which had to be imported specially from Japan. However, when he talked about the *Centaur*, he would become emotional and cry. The passage of time had not relieved him of the terror he had felt then, and he still suffered from the occasional nightmare.

The other story the *Courier Mail* dredged up was whether illegal arms had been secretly loaded on board the *Centaur* in Sydney, making her a legal target for attack by the enemy. This mystery stemmed from a contemporaneous misunderstanding about the rifles and ammunition carried on board by the 2/12 Field Ambulance crew. The rifles were perfectly legal, but the resulting row led others to speculate that larger quantities of illegal arms were also stowed on board. Unfortunately, these rumours persisted and led others to dream up the idea that Japanese spies had somehow relayed the information to the submarine commander. If this unfounded speculation proved to be true, then it was Australia and not Japan who had violated the Hague Convention rules, an act that had condemned the ship's company to certain death. Of all the improbable stories about *Centaur*, this was the one that upset and haunted the relatives the most. Nevertheless, it wasn't in our brief to investigate every rumour, and either way we had no intention of penetrating *Centaur*'s hull, given its status as a war grave and protected wreck.

In fact we had a simple and straightforward objective for the ROV inspection: document the wreck, the damage it had suffered and any associated debris field. Two observers who had taken part in the search – Ed Slaughter, a maritime archaeologist representing the Queensland government; and Major Arthur Dugdale, representing the Australian army – would stay on for the final phase. I was also able to convince Anthony Crack that having John Foley on board would be a huge benefit given his superior knowledge of the ship. Tuck Thompson made a last-ditch effort to join us but I vetoed that on the basis that he didn't have the requisite safety training and we were running out of bunk space. I also didn't want any outside influences to distract from the job we had to do. Flying an ROV around a deep-water shipwreck can be tense at the best of times, so I didn't think it was fair on the pilots to add even more pressure to the mix. Tuck wasn't happy and he predictably criticized Anna Bligh for banning the media, but I had no doubt it was the right decision.

I wanted everyone's focus to be on getting the best possible pictures of the shipwreck for the sake of the relatives. Their hopes for the ROV expedition were at the forefront of my mind: in particular their wish for me to make sure a special memorial plaque they had created was laid on the wreck site. I had known about the plaque for months because it was an idea I had discussed with Jan Thomas and Richard Jones at our very first meeting. When I explained how the plaque for HMS *Hood* was designed to honour every person lost, they adopted the same idea and inscribed a CD with the names of everyone who served on *Centaur* as well as personal tributes from family members. The memorial plaque was unveiled to the media at the final press conference, held with Wayne Swan, the Treasurer of Australia, whose constituency of Lilley was just across the river from where we were mobilizing the ship. Wayne wanted to wish us well but he also asked me if I had any concerns for the next phase. I did. I was worried about the plaque being lost in the soft muddy seabed because of its weight. It was something I simply would not allow to happen and I needed a plan to make sure it didn't.

After a couple of false starts due to oil leaks, the *Remora* III ROV touched down on the seabed just after midnight on Sunday 10 January 2010, on the ridge above the gully where *Centaur*'s wreck was nestled. I had the pilots stop in this location about 800 metres from the wreck in order to conduct an important test. I wanted to see just how soft the seabed was, so I had the ROV carry a steel plate about the same weight and size as the *Centaur* Association's memorial plaque. When the ROV released the plate, I wasn't surprised to see more than half of its twenty-eight inches sink into the mud and total obscurity. I was determined not to let the association's plaque suffer the same fate.

The ROV was back on the move, descending into the gully towards the wreck. In my hut I had a duplicate of the navigation monitor and the ROV's sonar display, which allowed me to count down the metres to the nominal wreck position we had determined during the search. At 500 metres away I got a call from the bridge saying they had an oil leak in the starboard main engine, bringing everything to a halt. Fifteen minutes later, another call came in to say the leak was manageable and we could proceed. The *Remora* was now moving freely and quickly towards the target position: 400 metres, 300, 200, then 100 metres away. There were lots of individual contacts showing on the sonar display, but no obvious shipwreck. It was likely the search position was off, but by how much? The ROV was still too shallow, indicating that we needed to descend even deeper into the gully. At sixty metres away we came upon a huge boulder and my heart literally sank. For a few frightening seconds I had the worst feeling imaginable that I had got the search badly wrong. I shook my head and said to no one in particular: 'Don't tell me that's what we've found.'

After regaining my composure, I could see from the monitor that the seabed was falling off to our starboard side, so I asked the pilot to head in that direction. Thankfully it didn't take long and we came across no more boulders along the way. Eight minutes later, at 2.50 a.m., the hull of a shipwreck appeared out of the darkness and we were looking at an object that hadn't been seen for nearly sixty-seven years. As luck would have it, we came up directly upon a gaping great hole in the ship's port side. The twisted and torn shell plating was a wound that bore all the hallmarks of a torpedo strike. And in that instant we knew that we had found not only the *Centaur*, but also what had caused her death.

Because of the problem with the *Seahorse Spirit*'s starboard engine, we aborted the dive early after spending only an hour and forty-five minutes on the wreck. But in that brief time we filmed enough distinctive features to prove to everyone that it was indeed the *Centaur*. Despite the decades lying on the seabed, with all sorts of physical and chemical mechanisms trying to wash away its identity, the green band around the hull and the large red crosses on the bows showed the wreck was a hospital ship. It was a lot less apparent, but enough of the white '47' still peeked through the corrosion and rust-covered paintwork to identify it specifically as the *Centaur*.

True to the promise I'd made to him months earlier in return for all the assistance he'd given me, one of my first calls was to Chris Milligan in Canada. John Foley and I left the confines of our hut and stood out in the bright sunshine on the bridge wing of the *Seahorse Spirit* to get the best possible satellite reception. John asked if it was okay for him to break the good news to Chris, which was only right given that their partnership was the origin of the success all three of us were sharing. I am writing this now with the sad knowledge that John passed away in the summer of 2015, way too early for a man so full of energy and still in the productive years of his life. I remember him in many ways, but the moment I keep at the forefront of my memory is that conversation with Chris on the bridge wing as he regaled him with what we had seen of the shipwreck. I can hear Chris's excited voice flowing through the handset, while John's face beams with equal delight.

Over the next two and a half days we made three more dives to the wreck and spent a total of twenty-six hours conducting a full visual survey of the hull and the small debris field nearby. As I had surmised from the sonar images, the wreck was leaning over towards its port side at an angle of about fifty degrees, with the bow forward of the bridge nearly severed from the rest of the hull. The torpedo and oil tank explosions surely broke the back of the ship, and in view of the extensive damage to the hull, it was no surprise that *Centaur* sank so quickly. In fact the damage along the starboard side was considerably worse than the port side, suggesting that the secondary oil tank explosion was the main culprit in sending her to the bottom. The partially imploded foremast, squeezed and folded like a collapsed straw, was another indication of how rapidly she sank. The only real surprise to me was that the bow somehow remained attached to the hull and did not break away during the 2,000-metre death plunge to the seabed.

Judging by the crumpled keel, bent railings and collapsed decks, the ship hit the seabed at tremendous speed (certainly faster than her normal twelve-knot cruising speed), causing additional impact damage. Both the foremast and mainmast were down, as was the funnel, which made flying the ROV over the top of the wreck easier, as there were hardly any obstructions to concern the pilots. With the navigating bridge blown off by the explosions, and the wooden ceiling of the officer's house stripped away by years of biological and chemical degradation, it was possible for our cameras to peer down directly into the captain's quarters. Generally what you see in the interior of shipwrecks this age is an accumulation of grey-black detritus that makes identifying anything impossible. However, as *Remora* hovered above his private quarters, we caught glimpses of Captain George Murray's former life.

A white porcelain washbasin was the first object that caught my eye. I checked the detailed plans I had for the ship and could see that this was

Murray's bathroom. Very small movements of the ROV and the camera brought other objects into view, as *Remora*'s LED lights illuminated a scene that had been permanently dark for the previous sixty-six years. Next to the washbasin was Murray's bath, partially filled with debris and a brown sea anemone whose tentacles fanned out like the rays of the sun. The ROV moved a fraction further, allowing my camera to focus on two shiny chrome bath taps, and a large black valve that John Foley thought was used for the bathroom's hot-water supply. Opposite the bath was an open doorway that led to Murray's toilet and bedroom. Our lights reached far enough into the next room to see a single leather shoe lying on the floor.

Like most of those who were off duty at the time, Murray would have been asleep when the torpedo struck, and in rushing to the boat deck at the initial shock of the explosion would have had no time to put his shoes on. His priority would have been to get the lifeboats launched as quickly as possible, and knowing that every second counted, he evidently ran out barefoot. He was last seen with a group of men valiantly but unsuccessfully trying to launch lifeboat no. 2. I realised we were looking at the most private spaces of Murray's on-board life, but there was nothing ghoulish or disrespectful about the pictures we took. In my experience it is these types of pictures that best tell the human side of a tragedy and the toll in lives lost. In this case Murray's solitary shoe speaks to the urgency and panic that consumed the ship in the seconds after *Centaur* was attacked.

According to the ship's plans, the aft cabin on the port side of the officers' house was reserved for *Centaur*'s second officer: this was Gordon Rippon's cabin. We knew of course that Rippon had been one of those who had survived, but it was a compartment on the ship that John and I still wanted to find. Rippon featured prominently in Chris and John's book, and I felt I owed him a huge debt for leading me to the wreck. In fact, we all owed him for his superb navigation. So as the ROV traversed down the port side of the boat deck, I asked the pilots to stop at the third doorway. The door to Rippon's cabin, like the other two on this deck, was sprung wide open, clear evidence that it had been abandoned quickly, though the fifty-degree list of the wreck prevented us from having a close look inside. The only thing visible in the room, because it stood out against the dark background, was a white galatheid crab perched on the threshold. This crab, a common deep-sea dweller, would have been living off the wooden structure of the wreck it had made its home.

After documenting the damage to *Centaur*, our survey essentially consisted of flying around the entire ship while the video cameras recorded continuously and I took selective photographs with an independent still camera. I had a wide-screen television mounted on the wall of our hut, and had got carpenters to build a large wooden table where we laid out all the ship's plans and drawings. This was where John and I stood for hours on end, matching every feature we filmed against the plans I found in various archives. The ROV never entered the wreck but we did use the powerful high-definition cameras to zoom inside the converted cargo holds to see what they contained. There were plenty of hospital cots jumbled on top of each other, but no sign of any illegal arms.

Of all *Centaur*'s features, the one I most wanted to capture was the iconic red crosses that identified her as a hospital ship on a mission of mercy. I believed people would connect with the symbolism of the crosses if I could only photograph them the right way. The challenge was to capture their full seven-by-seven-foot size, while shining enough light on them to make out their colour, which was still vivid when the ROV was up close. We were working on the cross on the port bow, but getting the shot I wanted was proving to be impossible. With the ROV up close, there was enough light to bring out the colour, but at that distance we'd see less than half the cross. If we pulled back, we could frame the whole cross but our lights weren't powerful enough and the red would turn to black. We tried different light combinations at different distances, but getting the perfect balance was beyond the technology we had available. Photographers all around the world will have faced similar problems with light and distance, except that in the deep ocean there is zero ambient light, and as scuba divers well know, red is the first colour absorbed in seawater to the point where it disappears. Not willing to accept total defeat, I took a series of photos of different parts of the cross close up with the colour still visible, hoping that these could be stitched together at a later date into a single coherent image.

I was still absorbing my disappointment over the red crosses when Phoenix's pilot asked me where I wanted him to fly next. As we hadn't seen the mooring winches on the bow, I asked him to come up over the railing on the port side, which would bring the ROV above the bow forward of the breakwater. As the ROV gained altitude and the camera tilted down, an unmistakable shape appeared on the deck. For a split second I couldn't believe we could be so lucky, but the glint of bronze reflecting off our lights confirmed my first impression. It was the ship's bell, lodged upside down between two ventilation pipes. Actually, it was an amazing stroke of luck that it had fallen where it did, because if it had landed anywhere else, the fifty-degree list would have resulted in it rolling off into the mud, where it would have been lost forever.

The ROV was too high to make out any of the bell's details, so I asked the pilot to descend a bit in order to get a closer look. At the same time I zoomed in with my camera and began to focus on what appeared to be large block letters on the side of the bell facing our view. The letters were partly obscured by surface corrosion, and one was hidden behind the rust emanating from the base of a pipe, but we could clearly make out 'CENTAU' – all upside down, of course. Because the conversion into a hospital ship meant that the name of the ship was removed from where it normally appeared on the stern and bows, I had had no expectation of seeing it anywhere. Finding the bell this way was quite simply a miraculous double surprise. First that it had been caught between the pipes, and second that it had landed with the name facing up. The odds of that happening in the midst of the chaos and calamity to which the ship was subjected must be astronomical.

I had a second motive during this survey, which for the time being I kept to myself. It was to find a permanent and visually accessible place on the wreck to deploy the *Centaur* Association's plaque, a task they had entrusted to my care and which I took very seriously. Having seen the steel plate I used during the test disappear into the muddy seabed, there was no way I was going to allow that to happen to the association's precious plaque. With the wooden officers' house full of holes and the stern cluttered by the toppled mainmast and associated rigging, the best place appeared to be a clear area on the shelter deck starboard of the no. 1 cargo hold. Beneath the teak sheathing that was being slowly consumed by the galatheid crab and his friends, the shelter deck was made of steel, so it was permanent and clearly strong enough to support the weight of the plaque.

My chief problem, however, was that Ed Slaughter, the maritime archaeologist and historic shipwreck officer for the Queensland museum who was acting as the on-board observer for the government, was dead set against any physical disturbance of the wreck. I understood his responsibility and I personally made sure there was no accidental interference between the ROV and the wreck during the surveys we conducted. But in my mind he was missing the bigger picture of the importance and significance of the plaque to the association and the broader group of interested parties. In a very tangible way it would serve as a gravestone for the 268 people who had perished on the *Centaur*. And for all those families whose personal tributes were contained on the CD incorporated into the plaque, it physically connected them to the last resting place of their relatives. Besides, there was absolutely no conceivable way that a 29 lb bronze plaque laid on the deck would cause damage to a 3,066-ton shipwreck.

Once John and I were satisfied that we had full video and photographic coverage of the wreck, we moved out into the debris field to see what parts of the ship had landed there. Although the field was very small, it contained its own remarkable items, including hospital cots, leather boots, a teacup, a steamer trunk, and most incredibly, two army slouch hats. Of everything I photographed, it was a picture of one of these delicate felt hats lying on the seabed that prompted the biggest outpouring of emotion from the public. I'd thought the red crosses would be the image that summed up *Centaur*'s tragic story. Instead, it was that poignant picture of the iconic symbol of the Australian army that made people appreciate the fact that the loss of *Centaur*, above everything else, was a human tragedy.

In preparation for the final dive, I was quietly busy laying the groundwork to get official approval to place the association's plaque on the wreck, above the protestations of Ed Slaughter. Because this action would be legally seen as disturbing an historic shipwreck, which was the brand-new status of *Centaur* due to our discovery of its location, I had to submit an application to the Department of the Environment, Water, Heritage and the Arts (DEWHA). Jan Thomas and Richard Jones had already canvassed the association membership and there were no objections to my proposal, while Anthony Crack was also in my corner, lining up Paul Lucas to ring Peter Garrett, the DEWHA minister, to explain the situation. At one point I was so consumed with completing the application that I actually had to suspend the ROV operations for an hour to give it my full attention.

Ed Slaughter wasn't too pleased with the approach I was taking. I can only assume it was because he thought I was questioning his authority, whereas I just felt the matter was so important it needed to be decided by someone higher up the chain. The breakdown in my relationship with Ed was unfortunate, but I was convinced I was doing the right thing for everyone involved, especially the *Centaur* Association who entrusted me with this solemn responsibility. A mere two hours after my application was submitted, I got the answer I wanted in the form of DEWHA Permit No. 14, authorizing me to place the memorial plaque on the shipwreck. If any more proof was needed that the government agreed enthusiastically with my proposal, it came from Garrett himself, who issued a statement the same day declaring that his department had approved the application in record time so as not to delay the project.

The following morning, at 5 a.m., the *Remora* III ROV was on the seabed, moving into position towards the wreck with the memorial plaque secured to the bottom frame of the vehicle. A rope lanyard tied to two corners of the plaque was held in the jaws of *Remora*'s twin manipulator arms. We had talked through the plan several times, so now it was just a matter of executing it. I'd even taken the extra step of lining the back of the plaque with rubber to prevent any galvanic corrosion if the bronze made contact with steel parts of the wreck.

The delicate business of laying the plaque on the sloping foredeck was in the hands of Paul Nelson, the Phoenix supervisor, who was controlling the manipulator arms and had to release the lanyard from the jaws at exactly the right moment. When the ROV got to the spot I'd picked out, I gave the order for Paul to release and he did so expertly. Despite the steep angle of the bow, the plaque remained firmly in place and the lanyard floated up as designed. A great deal of thought had gone into creating this plaque and I knew that people were going to be very happy to see that it was now permanently in place exactly where it belonged. It was in a perfect position, with nothing nearby to detract from its simple beauty and the words and emblems carefully chosen by the *Centaur* Association.

IN MEMORY OF SHIPMATES,
RELATIVES, COLLEAGUES AND FRIENDS
WHO PAID THE SUPREME SACRIFICE ON A
MISSION OF MERCY, 14 MAY 1943
2/3 AHS *CENTAUR* ASSOCIATION
2010
LEST WE FORGET

Although the loss of AHS *Centaur* didn't carry the same national significance as that of HMAS *Sydney*, the discovery of *Centaur*'s wreck was a momentous event for Brisbane and for the relatives spread throughout Australia. For me personally, it was another successful discovery of an important shipwreck, and a wonderful opportunity to use my skills again to help people find their relatives and obtain some peace and comfort from answers to questions they never thought would be resolved.

On 2 March 2010, a national service of thanksgiving and remembrance was held at St John's Cathedral in Brisbane in memory of those who had served on *Centaur*. The service was an opportunity for the families to gather and remember their relatives, whether they perished that day in May 1943 or lived to carry the physical and emotional scars for the rest of their days. The cathedral was jam-packed with over 700 people. An indication of the importance the government and the military attached to the service was evident in the long list of dignitaries who attended, including Prime Minister Kevin Rudd, Queensland premier Anna Bligh, Governor General Quentin Bryce, Queensland governor Penelope Wensley, Chief of the Army Ken Gillespie, Chief of the Navy Russell Crane, and a whole raft of lesser VIPs and organization representatives.

Martin Pash spoke of his narrow escape, and was given a standing ovation for his courage in recounting such a terrible ordeal. Kevin Rudd delivered a moving speech in which he told the relatives:

> The *Centaur* was a vessel of mercy and it was sunk without mercy by a vessel of war, and its wreck will forever be a sepulchre for those 268 souls who perished, a memorial to the 75 merchant navy personnel, 193 Australian service men and women, doctors, nurses, orderlies, cooks and stewards. Designated now as a war grave. Now protected ever more from intrusion. Forever now a sacred place. Forever now a reminder that the preservation of your freedom was purchased with the blood, the sweat and the tears of those who came before us. You now know the final resting place of your loved ones after so many years.

Jan Thomas, whose face was etched with the pain of losing her father at such a young age, said of the service that 'Having the nation stop to remember is a big step forward for us.' She wrote a longer note to me after the project was finished. I suspect her words come closest to articulating the range of complex feelings many relatives feel when the period of time between loss and resolution is so long.

> It is a time of very mixed emotions for us. Excitement that the detective work has solved the mystery, and trepidation at facing old and unresolved emotions. Also relief that *Centaur* can now be properly protected – not a vague order over an empty bit of ocean – and there will be no further false claims to the distress of still grieving people.

I don't believe in closure. Trauma of this sort doesn't go away, it has shaped our lives and become part of who we are, but finding *Centaur* will be a huge help in managing our loss, and healing the wounds.

Knowing that we had been able to mark the grave with a plaque on the deck was of enormous significance, and we thank you for your guidance and encouragement in creating that plaque.

Not only knowing where they are, but knowing that they will be protected. We can now say, at least figuratively, that we are bringing them home.

Later that year I received a letter that caught me completely by surprise. It was from the Australian High Commissioner in London explaining that I was to be awarded an Honorary Order of Australia Medal (OAM) for service to Australia for locating HMAS *Sydney* II and AHS *Centaur*. Having lived in England for so many years, I knew all about the honours system, and after the last *Centaur* press conference Anthony Crack had told me that reporters had asked him whether the country should give me an award in recognition for what I had achieved. But in my heart I was still a kid from New Jersey, where we didn't have medals, so it did take me some time to fully appreciate the importance of the award, especially as it was an honorary one and they are infrequently given.

For various reasons my investiture ceremony didn't take place until the summer of 2012, when Governor General Bryce was visiting London to attend the Queen's Diamond Jubilee. The governor general had previously been governor of Queensland, so she was well informed about the *Centaur* and we were able to have a proper chat about the shipwreck discovery at the reception afterwards. Seeing the investiture also helped my three young children to understand why their daddy was away from home for such long periods of time. I am very proud of the award and it holds great significance for me, particularly as it is connected with two historic events in Australia and two defining moments in my own career. I can only speak for myself, but I would like to believe that Australia has benefited from our relationship as much as I have.

A final surprise, which I value and cherish as much as the OAM, was given to me by the *Centaur* Association, who had made a limited number of miniature replicas of the memorial plaque we laid on the bow of the wreck. They give these to organizations that hold regular memorial services or in other ways continue to tell the story of the *Centaur*, to display as land-based public focuses for the grave site that no one will ever be able to visit. I was incredibly honoured, therefore, when the Association made an exception and offered one to me in appreciation for locating the *Centaur* and encouraging them to have the plaque made. Their letter, which was signed by the association's president and secretary, Richard Jones and Jan Thomas, ended with the words:

We trust you will accept this replica as a token of our gratitude for all you have done for us, without which *Centaur*'s final resting place and the final resting place of our family members would remain a mystery.

VII

Esmeralda

VASCO DA GAMA'S SECOND
ARMADA TO INDIA

Esmeralda

SUNK 30 APRIL 1503

One question I am often asked when people find out that I work as a shipwreck hunter is: 'Have you ever found any treasure?' Although the treasure I personally associate with most of the shipwrecks I have worked on is of the historical type, I do know what people mean when they ask this: coins and trinkets made of gold, mounds of silver bars piled up like loaves of bread, or precious jewels the size of cat's eyes.

You will be hard pressed to find anyone, if they are being honest, who can say they have never dreamt what it would be like to find a fabulous treasure. It is a perfectly natural desire born from childhood games digging for imaginary hoards buried in the garden. To some extent we all possess an inner Indiana Jones that would love to experience the adventure and excitement of uncovering extraordinary riches in an exotic faraway land. And there have been a few high-profile recoveries of shipwreck treasure in the past that have given people hope that they can make their dreams come true. Mel Fisher, whose divers found the mother lode of silver, gold and emeralds from the Spanish galleon *Nuestra Señora de Atocha* in just fifty-five feet of water in the Florida Keys, springs to mind.

329

So the question about treasure is understandable. The reality, however, is very different. Treasure salvage of shipwrecks is a highly speculative and complicated business, fraught with risks. I'd venture to say that for every successful project where a profit or break-even was achieved, there have been ten to twenty failures where all the money invested was lost. And of the projects that have been successful, a good number of those wind up in courtroom litigation, with the hard-earned profits frittered away in legal fees. The case of the SS *Central America*, in which not a single penny has been returned to the investors despite the recovery of gold coins and bars worth some $150 million, is a prime example of how even a successful project can go badly wrong.

The salvage of treasure from shipwrecks is also an ethical minefield centred squarely on the issue of whether it is right for shipwrecks to be commercially exploited through the selling of artefacts for profit. Archaeologists, who have been pitted for more than two decades against treasure hunters in an increasingly bitter and divisive battle, argue that historic shipwrecks are fragile and rare time capsules that ought to be preserved *in situ* and only disturbed under certain conditions. In their view, for-profit ventures are incompatible with good archaeology and should be outlawed. When the ethical considerations of one profession demand the criminalization of another, the fighting is bound to get bloody. Some have even likened the archaeological community's crusade against treasure hunting as a holy war. In such a caustic environment suspicion, scepticism and prejudices abound.

It was in this environment that I struggled for many years to mount one of the most challenging and ambitious projects of my career. The difficulty wasn't the depth of the wrecks. At Blue Water Recoveries we already held the Guinness World Record for the deepest shipwreck ever found, at 5,762 metres, and had located several others deeper than 5,000 metres. The wrecks I was keen to find this time were so shallow you could reach them free-diving. Nor would scores of relatives be counting on me to find their loved ones, as with *Derbyshire*, *Hood*, *Sydney* and *Centaur*. No, the challenge with this project was that I was stepping way out of my comfort zone, and into a realm of expertise in which I had no formal training or practical experience. I would be treading on the patch of academic archaeologists, which they guarded jealously and with deep suspicions about anyone with a commercial background like mine. It made no difference that this was a project about passion, not profit, and that nothing would be sold or commercially exploited. I would still wind up in the cross hairs of numerous archaeologists keen to throw cold water on our findings despite the considerable scientific proof we eventually published.

Finally, I didn't pick just any ordinary historic shipwreck on which to cut my archaeological teeth. These wrecks would be the oldest pre-colonial vessels ever found and archaeologically excavated. They were from Europe's Golden Age of Discovery and had a direct connection to one of the world's most famous explorers, national hero of a country that would view me as an unqualified outsider messing around with its most important cultural heritage. They were archaeological treasure of the highest order. If I was ever to unleash my own inner Indiana Jones, this project was it.

Vasco da Gama ranks alongside Christopher Columbus and Ferdinand Magellan as one of the most famous and accomplished explorers from the European Age of Discovery, a period when extensive overseas exploration by a handful of competing seaborne powers led to the rise of global trade in concert with the building of colonial empires. Da Gama was already a noble, a cavalier of the Order of Santiago and a *fidalgo* of the royal household, when in 1497 he was chosen by the Portuguese King Dom Manuel I to be captain-major of Portugal's most important exploration east to find the direct sea route to India. Since their discovery of Madeira Island in 1418 and the Azores archipelago in 1427, Portuguese ships commanded by their most skilled explorers and navigators had ventured south into the Atlantic Ocean, reaching successively distant locations down the west coast of Africa, in a quest to conquer new lands and to find Christian allies who could help in their costly conflict against the Moors. In 1488, a small Portuguese fleet commanded by Bartolomeu Dias finally rounded the Cape of Good Hope at the southern tip of

Africa, but it ventured no further because his crew became frightened and refused to go on. The Portuguese knew that India was a land rich in valuable spices, especially pepper, and it was the ultimate objective in their plans for economic and territorial expansion. However, it wasn't until the ambition of Dom Manuel was matched by the bravery of Vasco da Gama that they reached their goal.

Da Gama's discovery of the sea route to India in 1498, with an armada of just three ships, has been described by historians as one of the most important events recorded in the history of mankind. This daring and courageous exploit of seamanship and navigation gave Portugal virtually total control of the rich spice trade with India. By the middle of the sixteenth century they dominated the world's trade, with a hundred-year monopoly in the Indian Ocean over their Dutch and English rivals. Along with the discovery of Brazil in 1500 by Pedro Álvares Cabral, it was one of the crowning achievements in Dom Manuel's reign, helping to cement his reputation as 'Manuel the Fortunate' and one of Portugal's most important rulers ever. For da Gama, it meant more riches and titles, including being named the first Admiral of the Indies.

Three years later, Dom Manuel called on da Gama once again, this time to sort out a serious problem with his newly established *feitoria* (factory) in Calicut. The factory had been established by Cabral, who had commanded the second armada to India and had been tasked by Dom Manuel to negotiate a treaty with the Zamorin of Calicut allowing the Portuguese to openly trade for spices. Although the treaty was signed and some trading did take place, Cabral's policy of conducting hostile trade, and anger amongst the Arab merchants over the heavy-handed practices of the Portuguese, led to a violent conflict in which fifty-four Portuguese from the factory were slaughtered. In revenge, Cabral bombarded Calicut with his heavy guns, killing as many as 500 people. Cabral's mission failed on numerous counts. Of his thirteen ships, only five returned with cargo while five others were lost at sea; he failed to establish good trading relations with the Zamorin; the factory at Calicut was massacred; and it was discovered that the Zamorin were Hindus and not primitive Christians as the Portuguese originally believed.

A fourth armada was subsequently organized (the third armada had already left Lisbon, unaware of Cabral's news), with the primary aim of retaliation against the Zamorin of Calicut. This was to be a military mission and the largest armada to date: twenty ships brimming with heavy iron and bronze cannons that would give the Portuguese a powerful advantage over whatever enemy they faced on land or at sea. Dom Manuel appointed da Gama captain-major and gave a special assignment to another important nobleman, Vicente Sodré, da Gama's elder maternal uncle. Vicente and his brother Brás Sodré would lead a smaller sub-squadron of five ships (three naus and two caravels) with independent instructions from the king to 'guard the mouth of the Strait of the Red Sea', effectively blockading Arab trading ships from sailing onwards and delivering their cargoes for trade with Europe. There was no doubt about Vicente Sodré's position in the hierarchy of the twenty captains below da Gama, as he was named to assume the role of captain-major if anything happened to his nephew and the other captains were given the directive to follow Sodré's orders as if da Gama had demanded them himself.

Da Gama's armada reached the Malabar coast of India in late August of 1502 and wasted no time in taking revenge against a large merchant ship carrying Muslims returning from a pilgrimage to Mecca. Despite the offer of a considerable ransom and pleas to spare the lives of the passengers, da Gama had the ship bombarded and watched it burn with every man, woman and child still on board. Likewise, he turned down all offers of peace and settlement with the Zamorin, and when they rejected his ultimatum for the expulsion of the entire Muslim settlement from Calicut, he had his ships bombard the city for two days. He also ordered each ship in his fleet to hang captured Moors from their masts and parade them along the coast so people watching from the seashore could see. To inflict even more terror, he had the captives' hands and feet cut off and sent in a boat to shore while he let the bodies wash in on the tide. A letter he sent with the boat to the King demanded payment not only for the loss of the Portuguese goods in the factory, but also for the gunpowder and cannon balls his ships expended during the bombardment.

Despite these appalling acts of brutality and violence, the Zamorin would not be cowed into a subservient trading relationship with the Portuguese. The rulers of Cochin, Cannanore and Qulion were more amenable, however, and da Gama was able to finalise commercial treaties with each and establish permanent factories in the former two ports. Vicente Sodré's squadron primarily served a military function during this period and found themselves in the thick of all the actions against the Zamorin. When da Gama left for Cochin, for example, Sodré remained behind to blockade Calicut and prevent ships and supplies from entering the port. Da Gama had a similar role in mind for his uncle when he set sail for Lisbon at the end of the year with all his trading ships full of cargo.

It had always been intended for Sodré's squadron to remain behind as a permanent coastal patrol in the Indian Ocean, guarding the entrance to the Red Sea. His nephew, however, had a more pressing concern. Worried that the Zamorin would attack his newly established factories in Cochin and Cannanore, da Gama ordered Sodré to base his patrol in this area, where he could guard the ports and still be able to capture Arab ships trading between Kerala and the Red Sea in keeping with Dom Manuel's wishes. Based on his later actions, it would appear that Sodré was more interested in enriching himself and his brother than in complying with his nephew's orders. Whatever his reasons for leaving the Malabar coast and sailing up to the Gulf of Aden, it set in motion a train of events that would ultimately lead to the loss of two valuable ships along with many lives.

If Vicente Sodré's decision to abandon his guard duties was based on a desire to capture rich trading ships as prizes, he picked a good location in the Gulf of Aden. Abetted by his brother Brás, he plundered five Muslim-owned vessels and, continuing their barbarous streak, killed all the crews and burnt the ships. The ruthless and murderous behaviour of the Sodré brothers was matched only by their greed in keeping the best of the stolen goods for themselves, which angered the other captains in the squadron and fermented a poisonous atmosphere amongst the crews.

With their holds full of pillaged cargo, Vicente took his squadron to the Khuriya Muriya Islands off the south-eastern coast of Oman to shelter from the south-west monsoon and to repair the hull of one of the smaller ships, a caravel. They anchored in a bay off the largest of the five islands (known today as Al Hallaniyah), which was inhabited by a small indigenous population of Arabs who existed largely on the fish and turtles they caught. Despite their avowed hatred for Moors, the Portuguese actually enjoyed friendly relations with the islanders. They stayed for weeks, bartered for food and provisions, and some of the sailors even indulged in the affections of the married women while their men were away fishing.

In May, the fishermen knew from the behaviour of the fish that a violent storm from the north was about to hit the island. They warned the Portuguese that they should move their ships to the leeward side or risk them being lost in the heavy seas that would come. Vicente took advice from the other captains and his Arab pilots, but in the end decided to ignore the warnings, suspecting that the fishermen just wanted to get rid of the Portuguese as they were unhappy with their women entertaining the sailors. The Portuguese were confident that their iron anchors would hold the three large naus in place, but they took precautions by putting out additional moorings at their bows and by moving the smaller caravels to a safe location on the other side of the island.

When the winds came, as the Arab fishermen had accurately predicted, they were sudden and furious and were accompanied by a powerful swell that tore the Sodré brothers' ships from their moorings and drove them hard against the rocky shoreline, smashing their wooden hulls and breaking their masts. Brás Sodré's ship, the *São Pedro*, was driven sideways onto the shore, allowing many of the men not killed by the falling rigging to scramble to safety over the collapsed masts. Vicente's ship, the *Esmeralda*, however, sank with its sails to seaward, causing everyone to be lost including Vicente himself. Although Brás initially survived the wrecking of his ship, he later died of unknown causes, though not before he had his two Arab pilots killed – including one who was the best pilot in all of India, left to him by his nephew da Gama – in misplaced revenge for the death of his brother.

After burying their dead on the island, the surviving Portuguese salvaged everything they could from the wrecks – even the nails – before setting fire to the hulls. Pêro de Ataíde, the captain of the third nau, took command of the three remaining ships and sailed them back to India, where he met Francisco de Albuquerque, a more senior captain from the fifth armada, and handed over seventeen cannons and two barrels of gunpowder that they had salvaged. Albuquerque took command of the two patrol caravels and sailed to Cochin, which, without the protection of Sodré's squadron, was under attack from a large army from Calicut. Ataíde was allowed to return to Lisbon in his nau, but succumbed to illness and died in early 1504 after his ship was wrecked near Mozambique. Shortly before he died, he wrote a five-page letter to Dom Manuel describing the loss of the two naus and the irresponsible behaviour of the Sodré brothers.

The carelessness and greed of Vicente and Brás Sodré came at a considerable cost to the Portuguese Crown and the reputation of Vasco da Gama. Two valuable ships full of cargo were lost and the Zamorin siege of Cochin disrupted the nascent Portuguese spice trade along the Malabar coast. Da Gama's relationship with Dom Manuel suffered, and he was shunted aside in favour of other advisers when it came to Indian affairs. In fact he was no longer a welcome figure at court and had to wait sixteen years before Don Manuel conveyed upon him the feudal title of Count of Vidigueira. Even that honour only came because of his threats to defect to the Spanish court, which Dom Manuel could not allow under any circumstances.

As for the Sodré brothers, their actions received general condemnation when news of what had happened finally reached Portugal. Pêro de Ataíde's original letter, which survived the great earthquake and fire of 1755 that destroyed much of Lisbon and the country's historical documents, ensured that at least one first-hand account could be relied upon for the basic facts. Ataíde placed the blame for the wrecking of the *São Pedro* and the *Esmeralda* squarely at the feet of Vicente Sodré for his arrogance in refusing to heed the warnings about the impending storm. Because of the connections to Vasco da Gama, the unhappy account of this tragedy also featured in the history of Portugal's maritime explorations by several sixteenth-century chroniclers. The story has since been told and retold multiple times over the centuries by historians writing in other languages.

∞∞∞∞∞∞∞∞∞∞∞∞

With such a famous and well-documented story about two very important shipwrecks from one of the earliest Portuguese armadas to India, you might expect that somewhere along the line a group would have tried to locate the wreck site. Maritime archaeologists in Portugal take great pride in documenting their underwater cultural heritage, and for the 1998 Expo held in Lisbon commemorating the five hundredth anniversary of da Gama's arrival in India, the government even financed the excavation of an early-seventeenth-century shipwreck (the 'Pepper Wreck' of 1606) from the Tagus River. Even though Blue Water Recoveries had no practical archaeological experience and no archaeologists on our staff, we realized that such early shipwrecks, if found, could contain rare artefacts of a type never before seen, and fill a gap in the knowledge about how the Portuguese conducted maritime trade and warfare at the turn of the sixteenth century. In late 1997, when we decided to switch our focus away from twentieth-century shipwrecks and to work in Oman, where we knew we could obtain the necessary permissions, finding the *São Pedro* and the *Esmeralda* was at the top of our list.

Our research understandably began in Lisbon, at the Biblioteca Nacional and the Arquivo Nacional Torre do Tombo, but over the space of five months we discovered important related documents in England, Venice and Rome. Including Pêro de Ataíde's letter and the sixteenth-century accounts written by the Portuguese chroniclers Gaspar Corrêa, João de Barros, Fernam Lopez de Castanheda and Damião de Góis, we were able to compile a rich collection of more than thirty accounts on which to base a decision about where to search for the wrecks. The main analytic work involved a careful deconstruction, and line-by-line comparison of the details provided in Ataíde's letter against the four chroniclers' accounts. As with all our wreck searches, our aim was to extract the key pieces of information – basically geographical clues – that pointed to the precise location in which the two ships sank.

Of my small team, Peter Cope, an ex Royal Navy submariner and keen sailor, was given the primary responsibility for this research project. Peter and I had a similar approach to the way we incorporated the rigorous analysis of information and sources into the decision-making process of where to search. For example, with the four Portuguese chroniclers, who all wrote their versions of the story years after it took place, we wanted to understand as much as possible about their relationships to the royal court and other key figures like Francisco de Albuquerque, and most importantly, the access they might have had to survivors from the Sodré squadron and/or to original documents that have since been destroyed.

Once we were certain the correct location of the wreck site was in the Khuriya Muriya Islands of Oman, and not the Yemen, where a different group was apparently prepared to search, it boiled down to what part of Al Hallaniyah – the only one of the five islands ever to be populated – best matched the various geographical clues we had extracted from the documents. Collectively, the clues described a very specific geographic location that included a large cove sheltered from the west winds; good holding for anchors; a beach suitable for careening a caravel; a steeply banked shoreline; a well for drinking water to support the indigenous population; and an anchorage exposed to storm winds from the north. We were able to narrow the possible location even further after finding a UK Admiralty chart produced in 1837 that showed two anchorages and nearby wells situated on opposite ends of the northern coast of Al Hallaniyah.

We began making plans for a reconnaissance search of the island, which unfortunately included finding a replacement for me, as I had a rather important event to attend in May of 1998. Because the south-west monsoon starts in June, we decided to conduct the search in May, which happened to be the same month I was to be married to my fiancée Sarah. As much as I wanted to direct the search as I had directed the research phase of the project, I wasn't about to put off our wedding date, and accepted that someone other than me would be enjoying this adventure with Peter. That person turned out to be Alex Double. Alex was working as a supply teacher in the UK when he first came to be interviewed at BWR, but what really interested us about his background was his long experience as a skipper and dive instructor on live-aboard diving vessels in the Red Sea and Indian Ocean. I doubt we could have found anyone more perfect for the role, so he was quickly signed up and arrangements were made for the two of them to fly out to Oman in mid May.

The basic plan for the two-week expedition was for the team to conduct visual and metal-detector searches at the first location, an anchorage on the north-west coast of the island where the current village was based, before moving over to the second anchorage in Ghubbat ar Rahib Bay if further searching was necessary. We had excellent contacts in Oman, in part because the main investor in BWR had a close connection with Sultan Qaboos bin Said al Said, which ensured that whatever permissions and logistical support the team needed were arranged straight away. It was fairly evident that without such government support you would be unable to operate in the remote location of the islands, which probably explains why no one had ever mounted a similar search before us.

The next time I heard from Peter it was via a fax forwarded from our office. Sarah and I were on our honeymoon on the Amalfi Coast of Italy, but I had left instructions with the office to contact me if something urgent came up. BWR was a close-knit company and my secretary Carole, who was also part of our research team, knew whether something was important enough to merit disturbing me. This news certainly was. After finding nothing but modern rubbish at the first anchorage, Peter and Alex had hiked over to the other side of the island and, standing on a hilltop above Ghubbat ar Rahib Bay, immediately saw that the anchorage there matched the geographical clues to a T. Although the sea was almost flat calm, they watched as a slight swell was funnelled into one location and broke as a three-foot wave, and figured that would be the most likely spot for a ship to be driven on to the rocky shoreline in the midst of a raging storm from the north.

After convincing the *wali* of the island to let them borrow one of his fishing boats, they came back to the bay the next day with their diving gear, and chose to swim along a compass course that brought them to the suspect location in about five metres of water. They had been diving for less than twenty minutes when Alex spotted something he knew was not a natural feature. It was a perfectly smooth, round rock poking just a few centimetres above the sand. The two of them dug around the rock to extricate it and saw it was a large stone cannon ball with the letters VS neatly chiselled into the surface. They checked the letters again, several times, by tracing their fingers over the carving. Peter had zero archaeological training, so calling him an amateur would be being generous. But he had spent the past seven months reading everything he could find about Vicente Sodré, so he wrote on his slate the obvious thing that popped into his mind to show Alex: 'VS – Vicente Sodré.'

Peter told me afterwards that they did a little celebratory dance underwater once they realized what they had found. I knew exactly how excited they must have felt, especially after finding the wreck site so quickly where their intuition told them it would be. As further confirmation that this actually was a wreck site, and not an isolated artefact, they went on to find sixteen more of the large cannon balls and numerous smaller ones lying on the surface or partly buried in the sand. Before leaving the site, they also found an iron ship's nail on the beach, which fitted with the story of the Portuguese salvaging everything they could before setting fire to the hulls.

As soon as Alex and Peter returned to the UK with a handful of the cannon balls for analysis, we started planning a larger expedition to map the site, conduct trial excavations and search for possible wreckage in deeper water. The expedition took place in October and November of 1998 with the team expanded to five, including a Portuguese maritime archaeologist named António Camãrao who had been helping us in Lisbon with the historical research. All our divers received basic training in archaeological survey methods and we shipped out our own rigid inflatable boat to serve as the survey and diving vessel. For the second time I missed out, even though I was chiefly responsible for the survey, as we had another project in America that was in crisis and my co-directors decided I needed to be there to solve that problem.

Peter came back from the second expedition with hundreds of photos of the island and wreck site, which I appreciated from a technical point of view but which also made me sick with envy. The location was stunning. From the deep blue expanse of the L-shaped bay to the bare limestone and granitic cliffs eroded by wind and weather, it appealed to me on so many different levels as a marine scientist. Add to that the adventure of working in an incredibly remote and pristine environment, and the excitement of being the first to uncover and hold 500-year-old artefacts from an important sixteenth-century shipwreck, and I felt I'd been cheated out of the experience of a lifetime. My deep personal disappointment at not being part of the 1998 expeditions was tempered by the fact that they were so successful. More artefacts were recovered from the trial excavations (mainly lead-covered iron shot), indicating that much more cultural material was buried there and that everything was of the right age to support our growing belief that we had found the place where the Sodré ships had been wrecked.

It was clear the wreck site had enormous potential and that we had only scratched the surface in terms of what other fascinating artefacts could be buried there. However, we had also reached a point where to proceed with a full excavation we would need a bigger team of qualified and trained archaeologists, a bigger ship to support the operations, and therefore a bigger budget. The driving factor was that the majority of artefacts were bound within a very large concretion that needed to be carefully broken down to extract potentially delicate objects without causing them any harm, something for which our team didn't have the necessary skills. We had done a very good job with our research and in finding the correct site, but we would risk all that if we handled the full excavation badly. Part of our rationale in taking on the project was that we wanted to see whether it was viable for a commercially orientated company like BWR to work on historic shipwrecks within the ethical boundaries of the archaeological community. We always expected that at some point we would need partners in the archaeological community, and that time had come.

Perhaps I was too close to the project and overly enamoured with its importance and potential, but I fully expected that in reaching out to the archaeological community, especially in Portugal, we would get a favourable reaction from interested partners. Boy, was I mistaken. There was the odd lukewarm response, but half our letters went unanswered and the rest turned us down outright. I could accept that funding in Portugal was very tight, but we even struggled to get people from academic circles in the UK to show any interest. Had we overestimated the historical and archaeological importance of these shipwrecks? I didn't believe so, but the lack of interest in our pleas for help with the project was telling.

I was disappointed for our company and for everyone who had worked to make the two expeditions a success. It seemed such a waste of good work that what we had started couldn't be finished. On a personal level I was gutted that I'd never get to visit the island or dive on the wreck site. I would never close my mind to a future opportunity for the project to be resurrected, but for the time being I had to let the Sodré shipwrecks and Ghubbat ar Rahib Bay slip away from my imagination.

After the termination of the Oman project, I had plenty of things to occupy my time. For four years BWR ran an offshore vessel used to survey undersea routes for telecommunication cables, which saw me travel around the world winning contracts that resulted in the vessel working in the Gulf of Mexico, the Bahamas and the Caribbean, Atlantic, Pacific and Indian Oceans. I worked as a consultant managing a number of the shipwreck searches covered elsewhere in this book. And in our personal lives Sarah and I had what can only be described as a head-spinning start to family life, going from no children to three in less than a year. At the end of 2008, there was also a major change in the ownership of BWR. The original owner had died suddenly the previous year, resulting in the company being transferred into my name. As a director/employee I had had a limited say in the type of projects we pursued. Now the pressure was squarely on my shoulders to run the business as I saw fit.

It took me a few years to build up a cash reserve, but as soon as I could devote some free time to it, I began thinking about how I could resurrect the Sodré shipwrecks project. People needed to see my passion and the potential for the project themselves, so I decided to rely on personal visits as opposed to letters and phone calls, which hadn't worked the first time. This involved multiple flights to Oman, Washington DC and California throughout 2012, but by January the following year I had an agreement with Oman's Ministry of Heritage and Culture (MHC) to co-manage the project with them on a collaborative basis; $30,000 in grant funding from National Geographic's expeditions council and the Waitt Foundation to help pay for expedition costs; a fully equipped live-aboard dive vessel that I'd hired for a two-week reconnaissance survey; and a crack team of professional archaeologists and surveyors lined up to join the project.

Because of the long gap in time since the 1998 expeditions, I decided to start with a repeat reconnaissance survey to map the shallow-water wreck site with greater precision, and a high-resolution search of the anchorage in Ghubbat ar Rahib Bay using the best quality side-scan sonar, magnetometer and GPS positioning equipment operated by industry professionals. I was fortunate to recruit Dave Parham, an associate professor from Bournemouth University, to be the archaeological director of the project, and Dr Bruno Frohlich, a forensic anthropologist from the Smithsonian Institute, to lead the investigation of possible burial sites on the island. I was also delighted that Alex Double was available to join my ten-man team as a guide and photographer and to provide diving instruction to the two MHC staff from their newly established underwater archaeology programme who would be working alongside us.

We boarded the *Saman Explorer* in the small coastal town of Mirbat, southern Oman, before making the overnight transit across Khuriya Muriya Bay to Al Hallaniyah Island. I had hired the 34-metre dive vessel on an exclusive basis for the last two weeks in May, along with a second ten-metre boat to be used for the geophysical survey, which was towed behind the larger vessel. Our small convoy, including two Zodiacs, arrived in Ghubbat ar Rahib Bay in the early hours of the morning and promptly anchored up to the wreck of the *City of Winchester*, a British cargo steamer that had been captured by a German raider and scuttled in the bay on 12 August 1914, making it the first vessel to be lost in World War I. *Winchester's* position placed us squarely in the centre of the bay, facing the rocky shoreline where the Sodré ships were wrecked and with the impressive Ra's al Hallaniyah, a 500-metre-high wind-eroded limestone bluff, off to our starboard side.

Normally when I go to sea it is to some distant point in a vast and endless ocean with no landmarks to reference the eye. The only way to know where you are in the world is by the lines of latitude and longitude that appear on a navigation display: a type of digital reality. However, when I rose at daybreak that morning I knew exactly where I was, as I'd been dreaming about this location for the past fifteen years. I don't normally get emotional about places. Al Hallaniyah was a ruggedly beautiful place: the way the pristine blue waters contrasted with the buff-coloured cliffs carved and striated by various mineralogies was especially striking. I'd seen many other beautiful places in the world. The difference was that I had never wanted to come to a place more than here. Part of it was because of wanting to complete unfinished business and prove doubters wrong. But mainly I wanted to see and feel it for myself. That was what drove me to make sure that my experience of the island was more than just a virtual one based on viewing someone else's photographs.

By 8 a.m., Dave and Bruno's teams were already working in the water and on the island, and the side-scan sonar was being deployed to begin the geophysical survey of the bay searching for possible wreckage in shallow-water gullies seaward of the wreck site. At the time, we were still working on the premise that the second of the two Sodré shipwrecks might have sunk in the deeper waters of the bay. If this was the case, we hoped to detect buried iron objects using a sophisticated caesium magnetometer, or with a side-scan sonar in the unexpected event that wreckage was exposed on the surface. By mid-morning, the first important question was answered. The stone cannon balls that littered the gullies were still there, remarkably in almost exactly the same positions they had been in 1998.

Our concern about the cannon balls was because Al Hallaniyah was experiencing a mini-transformation in the form of a $100 million infrastructure project intended to turn it from one of the most remote inhabited islands in the world to a tourist destination. The project included construction of a new harbour on the western end of the island and a modern tarmacked road through the mountainous interior to the beach located a few hundred metres from the wreck site. The construction was well under way when we arrived, and we could hear and feel the explosions of dynamite blasting a path for the road while we dived. The plans also included a car park at the beach (we later used it for football matches against the ship's crew), which would be completed the following year. This meant that Ghubbat ar Rahib Bay, once accessible only by an arduous boat trip from the mainland, could now be reached by car.

This modernization completely changed the risk profile for the wreck site. Anyone living on the island, or visiting it, could now drive down and then make the five-minute swim from the beach to where the artefacts were lying in a mere two to six metres of water. Scuba gear wasn't even necessary, as the cannon balls could be seen from the surface and easily reached by someone free-diving with a snorkel. For the past few years boats like the *Saman Explorer* had brought tourist divers to the bay, mainly to dive on the *Winchester*. How long would it take before they found the wreck site and began disturbing the artefacts, or worse, removing them as souvenirs? Although recovering artefacts wasn't part of our plan for this phase of the project, that position had to be re-evaluated now that we could see a clear risk to the site.

As project director, with overall responsibility for the expedition, I had the luxury of shuttling between the three different teams to oversee their operations and the discoveries each was making. I sped around the bay in a Zodiac, helping Bruno map and excavate potential burial sites in the morning, when the temperatures were still bearable, and diving with team member Peter Holt in the afternoon to investigate sonar and magnetometer targets detected during the survey. So little was known about the bay that our work had a tangible feeling of exploration about it. We mapped the seabed, discovered an unknown steamboat associated with the *Winchester*, and found two ancient stone anchors that harked back to an earlier era of foreign visitors to the island. The work was physical, exhilarating and enormous fun. Even though most of us had volunteered our time to be part of the expedition, we all felt as though it was an extraordinary adventure holiday we would gladly have paid to be part of.

As a reconnaissance survey our objective was to collect data and information needed to plan for a future full excavation of the site if it was warranted and approved by the MHC. The increased risk profile meant that a full excavation was now more than likely. I had brought a hydraulically powered water dredge, which Dave and his team used to make some trial excavations to get a feel for the quantity and type of artefacts that were buried in the gullies and the amount of rock and sand covering them. Other than the stone cannon balls, ceramic sherds and more lead-covered iron shot like we had found in 1998, I wasn't expecting that anything else remarkable would be uncovered given the superficial nature of these excavations. I was completely unprepared therefore when Dave came up to me at lunch on only our second day on site to tell me that Jessica Berry, one of our keenest archaeological divers, had found a ship's bell lodged under a large boulder in little more than four metres of water.

To say I was gobsmacked would be an understatement. The fact that a bell had survived in the turbulent waters of the gullies for more than 500 years (one of the gullies was so rough we named it Washing Machine Gully) was truly incredible. According to Jess, it appeared to be in reasonable condition given its age and where it was found. She had taken a couple of pictures, which showed the intact bell mouth facing outward from its position under the boulder. Had it really been there since the day the ship sank? Why had the Portuguese been unable to recover such a valuable object when they had clearly been intent on salvaging anything of value from the wrecks? And if the ship's bell had been missed, what else had been left behind? Finally, would the bell have any writing on it to help us identity the shipwrecks?

I had to see it for myself. We ate lunch quickly, zoomed over to the wreck site and followed Jess to where she had made her miraculous discovery. The sight of the bell was even more unbelievable than Jess's close-up photo had indicated, because it was so openly exposed you didn't have to look hard to see it. I wanted to check its condition so I reached my hand into a gap under the boulder to feel how much of the bell was still there. It was fractured across near the crown, but other than that both sides were present and basically intact. Whatever else we uncovered during the rest of our time on site, we had an absolute trophy of an artefact that justified resurrecting the project. It was also more than fitting that Jess had made the discovery because she was one of the project sponsors, having paid for travel costs through her not-for-profit Maritime Archaeology Sea Trust (MAST).

While the bell was the highlight, every aspect of the expedition greatly exceeded my expectations. Ayyoub Al-Busaidi, MHC's supervisor of underwater archaeology, was equally excited to have a significant project upon which to develop his programme. The MHC agreed with our recommendation that all artefacts lying at or close to the surface should be recovered, which we did at the end of the second week after buying every single cleaning bucket we could find in Mirbat to hold them. Most of the sixty-nine recovered artefacts were cannon balls, including twenty-six large ones all carved with the letters VS. Another important artefact was a matched pair of heavy bronze sheave wheels. Everything we saw and recovered was entirely consistent with a European ship from the early sixteenth century.

We could not say for certain whether the wreckage was from the *São Pedro* or the *Esmeralda*, or from either for that matter, as such a definitive determination required extensive scientific and archaeological analysis. However, the indications were very positive. For one thing, we could find no historical evidence of any other Portuguese ship from the early sixteenth century sinking in this location. And in Dave Parham, I now had a respected academic archaeologist – arguably the most active in the UK over the past decade – who shared my confidence. Finally, we had Jess's bell, which a careful examination on land showed to have a raised inscription possibly containing some characters or numbers. A crust of corrosion made it impossible to determine exactly what was written there, but I could see enough to suggest that it might contain an important clue.

A full excavation of the interconnected gullies where wreck material was believed to exist was going to be a very big job. Our estimate was that a minimum of 950 cubic metres of sand and countless boulders – some weighing as much as a small car – would have to be shifted off site to reach the artefacts that had worked their way down to the level of bedrock as wave after wave of pulsating water rushed over the site. There was also the large concretion to break down, which because of its size (40m³) could not be removed as a coherent whole. We'd need a more powerful excavation system than the small water dredge we'd used for the trial excavations and a bigger team of trained archaeological divers, including some professional divers to safely shift the huge volumes of rock out of the gullies. The expedition would also need to be longer: twenty-two days on site compared with ten the previous time. All of this dictated that for the next expedition, in the immortal words of Roy Scheider's Captain Brody in *Jaws*, we were going to need a bigger boat.

The MHC's first suggestion of a suitable boat was an interesting one, and it got everyone on my team a bit excited. It wasn't a boat actually: it was a ship, and a very big one at that. In fact it was a super yacht. The *Fulk Al Salamah* was one of two super yachts used by Sultan Qaboos and the royal family. At the time I inspected her tied up alongside in the harbour in Muscat's Old Town, she was being used as a support vessel for the sultan's newest yacht, *Al Said*. The *Fulk* was an older yacht, built in 1987, and was soon to be replaced by a new support vessel of the same name. She was plenty big at 136 metres and 10,800 tons, and she was equipped with all the small boats, diving equipment and cranes we would need. We could have made her work for the expedition and she would have served as a very comfortable floating hotel. In truth, however, she wasn't ideal, so I wasn't too disappointed when the MHC reported that she wasn't available because of a scheduling conflict, but that they had another ship lined up.

The RNO *Al Munassir* was nothing like the *Fulk*: there were no carpeted staterooms, no fine china and crystal glasses, and no luxurious lounges. She was a basic 64-metre utilitarian landing craft, with one small mess for eating and one bunkroom where all eighteen of us would sleep in triple bunk beds. But in every other way she was absolutely perfect for the job. She had a shallow draft so could get in close to the wreck site, and virtually the entire length of the ship was one long working deck. This was going to be a dirty, messy job with up to thirty dives being made each day, so every inch of deck space would be needed for the staging of diving equipment and the cleaning and recording of recovered artefacts.

After months of planning, the expedition was scheduled to start in mid April 2014 with mobilisation of the *Al Munassir* at the Said Bin Sultan naval base north-west of Muscat. National Geographic had agreed to sponsor me for a second year, and the MHC significantly increased their involvement, with more divers participating and extra funding to buy equipment they could use on future projects. This capacity building, in the form of training the MHC divers and equipping them to conduct underwater archaeology projects around the country, was a secondary but important part of our mission for the MHC. We also made sure that the entire project was being conducted in full compliance with the UNESCO Convention for the Protection of Underwater Cultural Heritage. Although Oman was not a signatory to the 2001 convention, which established specific rules such as prohibiting the trade in artefacts, we mutually agreed that adherence to it would be a cornerstone of our project.

Our plans were shaping up very well, and the only minor problem we faced was that a Portuguese archaeologist seemed to have made it his mission in life to aggressively criticize our project and make baseless claims about my role as its director. Alexandre Monteiro, a doctoral student at Nova University in Lisbon, made his first attack on a widely used public forum for underwater archaeology. He described our 2013 expedition as an example of 'bad archaeology' and made false claims about our discoveries, stating that we had found 'almost zilch, some stone shots, a fragment of a ship bell' and that we had 'no datable objects, just smoke and mirrors'. It was an extraordinarily prejudicial attack that would have been easier to accept if it had been made in ignorance. But Monteiro had somehow got hold of a copy of one of our reports, so he should have known that the ship's bell was virtually whole, not just a fragment, and that the thirty-seven stone shot we had recovered was actually a large number given that these were from the surface only. He also seemed to miss, either by design or pure carelessness, the significance of the bronze sheaves, pointing to a European ship of an early age.

Monteiro liked to present himself as being in the know, but what he didn't know was that we were making progress deciphering the inscription on the ship's bell, and that it would make a mockery of his claim that we had no datable objects. I had taken the bell to the Warwick Measurement Group at Warwick University to see if their industrial CT scanner could reveal what was written behind the layers of corrosion. Dr Jay Warnett performed the analysis, which took several three-hour runs in the CT scanner with different orientations of the bell to get the clearest image of what was written. After the first run, he emailed me to say he could clearly see the number 8. The next day, using a different scanning strategy, he revealed a 9 to the left of the 8. He tried one more scan and thought he could make out a 4 to the left of the 9, although it was less clear than the first two numbers.

One of the most common things written on ship's bells are dates: generally the date when the ship was made or launched. We only had a partial date, but the final X-ray image indicated that it was '498', which if confirmed would suggest a date for the ship as 1498, which fitted with the Sodré squadron leaving for India in 1502. It was a stunning use of state-of-the-art technology. The images would also be of immense value to the conservator to guide her hand in carefully lifting away the centuries of corrosion that covered the inscription. Ironically, one of Monteiro's central criticisms against me was that I was scientifically unqualified to lead this project. This was just the first of many opportunities to prove him wrong.

<hr />

The three-day transit from the naval base to Al Hallaniyah gave us plenty of time to run through the excavation plan with the whole team, including *Al Munassir*'s crew. In contrast with the gentle pace of the 2013 reconnaissance survey, I had to establish a demanding schedule to make the most of our time on site and to best utilise the large team of divers I had available. Initially this came as a shock to some of my Omani colleagues, and tensions resulted. Omanis are beautiful, friendly people. Anyone who has spent time in their fascinating country will tell you the same thing. Our work cultures were very different, however, and I needed them to understand why I had to insist that every minute we had in the field was spent productively collecting data. Many months had gone into the planning of every last logistical detail, and we only had one chance to get it right. Once we arrived at the island we were on our own, with no cell phone coverage, no medical assistance and no means outside of what was available on the ship to correct problems.

My worst fear was an incident with either of the two rigid inflatable boats (RIBs) we'd be using to shuttle divers from the *Al Munassir* over to the wreck site and other crew to different parts of the bay. We were totally reliant on these workhorse boats for all our transportation. Lose one and we could kiss goodbye to half of what we wanted to achieve. Lose both and the project was over. I didn't have to wait long for that fear to be realized. On the very first trip over to the wreck site, whilst carrying three of our most experienced divers and a full load of scaffolding pipe and dive gear, a nervous crewman judged the entrance to the gullies all wrong and ran the boat hard aground on the exact same rocks where the *São Pedro* had wrecked. Fortunately the divers saw what was about to happen and bailed from the boat before it hit the rocks. With the engine killed, the boat swung sideways to the onrushing waves and was promptly swamped.

It was frankly a quite disastrous start, which could have been worse but for the quick thinking of the divers. By the time I made it to the island in the second RIB, the team had already unloaded all the scaffolding pipes and were bailing out the boat. Its propeller was destroyed, but thankfully the hull and floats were fine other than some gouges in the fibreglass. In the end the incident cost us a day of lost time, as the only way to get the heavy scaffolding pipes out to the wreck site was to float them using a system of jury-rigged water bottles that Peter Holt devised. In the overall scheme of things we were lucky nobody was injured or that the boat wasn't permanently damaged. But we needed to use the accident to impress upon people the inherent dangers and to make some changes in how the boats would be operated thereafter.

While the RIB and the scaffolding pipes were being tended to, the next big crucial moment for the project was taking place in the sky above the deck of *Al Munassir*. The MHC had hired a large diesel-powered air compressor, which would be used to drive the multiple airlifts needed to conduct the excavation operations. Compared with the small water dredge we used in 2013, the switch to the large airlift system was almost an industrial scale step-up in our capability to excavate the gullies. To make the system work, however, the one-ton compressor had to be placed as close as possible to the wreck site, and the only thing that could carry it to this location was a heavy-lift helicopter provided by the Royal Omani Air Force. The entire excavation plan relied on having the compressor in exactly the right spot on the beach, where I was waiting to receive it, and from where I could see that the helicopter was struggling to maintain its position over the anchored ship. As the helicopter was waved off from the lift and headed back to its base in Salalah, I wondered whether I had just witnessed the end of the project before it even started.

When I got back to the ship, I learned from the captain that strong crosswinds had caused the lift to be aborted and that the helicopter had been running low on fuel. A second attempt would be made two days hence, when the helicopter was again available, but we'd have to hope for lighter winds that day. The first thing I checked that morning when I woke was the winds. They had decreased to 5–10 knots: the lift was on. At exactly the agreed time, the helicopter came into view and settled into position above the foredeck of the *Al Munassir*, where the crew were waiting to secure the lift sling to the cargo net holding the compressor. The *Al Munassir* was unanchored this time, moving slowly ahead, which was a better arrangement for the pilots. I had made a landing zone for the compressor out of seaweed just a few metres from the water's edge; the pilot came in low and hit it perfectly.

There is a moment in every big project like this when everything is finally in place and the work can begin in earnest. Until that happens, you really can't be sure what, if anything, will be accomplished. When the helicopter released the sling holding the compressor, that was the moment for me. Eight months of dedicated planning and preparation had gone into getting all the people, equipment and vessels together for this moment. It was way too early to celebrate; the next three weeks would determine whether the expedition would be a success or not. But as I stood on the beach with the sun beating down and watched the helicopter disappear, I could begin thinking about the artefacts the Portuguese had left behind five centuries ago.

As soon as the archaeologists began excavating the site, we started finding more stone cannon balls. They ranged in size from small ones you could hold in the palm of your hand to large ones bigger than a bowling bowl, and were made of two different rock types. They were the first firm evidence that we were dealing with the remains of a warship well equipped for conflict. Sodré's squadron had boasted the latest in European naval technology and ordnance, which the people of Calicut had been simply unable to defend against. When the Portuguese bombardment began, and these large stone shot started to rain down on them, crashing through the roofs and walls of their homes, they would have been shocked by the brutal effectiveness of this new form of warfare. The full extent of the weaponry became evident when the excavators reached the large concretion.

The concretion was situated at the intersection of three gullies leading into Washing Machine Gully. When the upper layer of loose sand was removed, five bronze powder chambers were revealed welded into the rock-hard surface. These bronze chambers were what held the gunpowder used to fire the breech-loading guns on the ship. After a gun was fired, the used chamber would be taken out and a fresh one inserted with another shot to allow relatively rapid firing. They would have been valuable pieces of equipment, so it was surprising that the Portuguese had left them behind when you consider that, according to Pêro d'Ataíde's letter to the king, they had salvaged at least some of the guns.

Dave Parham and his team of archaeological divers took control of excavating the concretion, which was a laborious and tedious but ultimately satisfying job considering the number of exciting artefacts it produced. Before anything was touched, a photographic mosaic of the upper surface was made, and then the divers got to work with chisels, airlifts and, in the case of the most heavily concreted areas, pneumatic hammers. After eight days of intensive work, the 40m³ concretion was reduced down to the level of bedrock, but not before it had yielded in excess of 1,000 individual artefacts. Hundreds of lead, iron and stone shot had to be physically chiselled out, which caused the site to sound like a stonemason's yard when all the hammers were going.

As each team returned to the *Al Munassir*, the team standing by would help them unload their diving kit and crowd around to see what new objects had been recovered. An amazing array was retrieved from the concretion, including belt buckles, a beautiful set of eight spoons, a door lock, a ring made out of stone, trade beads, even peppercorns and cloves that had survived encased within the protective cocoon. There was rigging and other parts of the ship, but no timbers and only fragments of wood, which was unsurprising given the poor conditions for wood preservation. Ultimately the number of bronze breech chambers swelled from the five that were first seen to a remarkable nineteen, which surely represents the largest single collection of this type from a wreck site.

We had become so used to seeing the divers return with common artefacts like ordnance or ceramic sherds that the first gold coin caught us completely by surprise. It was found by Jess and Ahmed Al-Siyabi, MHC's most hard-working diver, as reward for excavating an area of Washing Machine Gully, and was the first of twelve gold *cruzados* we recovered, several of which had to be fished out of crevices in the bedrock after metres of sand and rock was removed. The coins were as shiny and crisp as the day they were struck. It is this remarkable characteristic of gold, that however long it is immersed in seawater it will re-emerge looking exactly like it did the day it was lost, that helps explain its universal allure and wonder. We did not treat the coins as treasure, although in archaeological terms they were exactly that given their value in helping to accurately date the shipwreck. However, with no numismatic experts on board I had to wait until returning home before learning about their age and significance, although they were clearly from the reign of Dom Manuel or his predecessor João II.

After the area of the wreck site that produced the first seven coins was completely excavated, I thought no more would be found. I was personally on hand to film one coin being recovered, and detected a second with my metal detector to show the team where to dig, but I hadn't recovered one myself until I got very, very lucky working an area of spoil adjacent to where the concretion once was. While the rest of the team got on with the heavy excavation work, I gave myself the job of scouting new areas or scanning spoil heaps for objects that might have been missed. The spoil heaps accumulated at the point where sand and rock was ejected out of the airlifts, and they generally contained small pieces of lead: nothing to get too excited about. That was exactly what I thought I had found at the start of one dive when my metal detector started ringing in my ear. I could tell by the strength of the signal that the object was close to the surface, and it only took a few waves of my hand to expose what at first looked like a clump of coal. I picked it up between my fingers and spun it around until the edge of a gold coin caught my eye. There wasn't just one edge though: there were five!

I could see enough of the coins to tell that what I had found was five *cruzados* like the other seven that we had already recovered. I knew immediately that this clump had been sucked up the airlift and that I was incredibly fortunate to have gone back over this particular spoil heap. In fact missing it altogether would have been just as easy as finding it. The black clusters attached to both sides of the stack of *cruzados* were curious. As I studied them, I could see that they too had the shape of coins and realized they must be made of silver. The silver had completely oxidized, hence the black colour, but it hadn't obliterated the face of the largest coin on one end. Turning the coin one way, then the other, I could just about make out a large central cross that looked like the same cross I had seen on Manueline buildings throughout Portugal.

With each recovery we made, the physical evidence confirming the wreck site as one of the Sodré ships was mounting. The *cruzados* were the most definitive proof pointing to a Portuguese origin, but an object

I found two days before the expedition's end topped everything. Once again I was scouting a new area with my trusty metal detector to find the most promising place for the excavators to work next. I was close to where Jess had found the bell when the probe I was using detected something and began to ring. This object was buried deeper so I had to pass the probe over the spot several times to see exactly where the signal was strongest. Fortunately the sand was very clean and loose, so it was easy to fan it away. I had been fanning for several minutes, with the pit getting deeper and wider, but it seemed I was no closer to whatever was down there. I had been finding a lot of modern fishing weights made of lead on the site, which were always deeply buried, and thought this was yet another one when a disc started to appear. It was lying flat and was quite heavy, so I was able to gently slide it out from the sand without any trouble.

I immediately knew this was an extremely important object, and that if it was able to speak to me it would be shouting 'I am Portuguese.' On either side of a central reinforced hole were two of the most important national symbols in the reign of Dom Manuel, which still grace the Portuguese flag to this day. At the top of the disc, close to what I recognized as the remnants of a suspension ring, was the Portuguese royal coat of arms. At its bottom was the legendary *esfera armilar* (armillary sphere), a personal emblem chosen by Dom Manuel that became an iconic symbol of the Portuguese maritime empire during the Age of Discovery. Dom Manuel disseminated the *esfera armilar* by integrating it in royal orders, religious architecture and important objects of a permanent nature. Whatever the identity and function of this rare disc made of copper alloy, the inclusion of the *esfera armilar* marked it as an object of high status that was probably handled by only select figures on the ship, such as the commander, captain and master pilots.

<hr>

Although the coins and the copper disc were the most sensational discoveries made during the 2014 expedition, it was the archaeological and scientific analysis of more than 2,800 artefacts, including those recovered in 2013 (and in 2015, when the final large-scale excavation was conducted), that allowed us to confirm that they came from one of the Sodré shipwrecks: in all probability Vicente's *Esmeralda*. We decided to announce our discovery to the world by way of an academic article in one of the most respected peer-review journals in the field of nautical archaeology: the *International Journal of Nautical Archaeology* (*IJNA*). I hoped the project would get good coverage by the media, but I wanted that coverage to be off the back of an academic paper subjected to independent peer review. Had we announced the discovery earlier, for example after the bell was discovered, I have no doubt we would have been roundly criticized for making empty claims without having them independently verified.

As a non-archaeologist directing a project of this magnitude, I knew that some academics would be lining up to slam me if I put a foot wrong. I had recently seen what had happened to Barry Clifford, an American shipwreck hunter famous for finding the pirate ship *Whydah*, and took this as a lesson about what not to do. In the space of one year Clifford made high-profile announcements that he had found Christopher Columbus's long-lost *Santa Maria* off the coast of Haiti and silver treasure from the pirate William Kidd's *Adventure Galley* off Madagascar, only to have both discoveries debunked as false by a UNESCO team of archaeological experts called in to investigate his claims. It is hard to tell why Clifford decided to go public on the strength, or weakness, of such flimsy findings. Whatever his reasons, he was swiftly condemned by UNESCO and paid a serious price in terms of damage to his credibility and legacy. It was a mistake I was not about to repeat.

In advance of the *IJNA* paper, I had decided to build a scientific case for the origin and age of the shipwreck through a number of analytical studies on the artefacts. Some objects were easy to assess: with the bell, it was a matter of just having it professionally cleaned and conserved to reveal the date hidden behind the layers of corrosion. As foretold by the CT scans, the partial date on the bell was shown to be 498, and a letter M was also revealed. Although the M was in the right position to have possibly spelled a version of *Esmeralda*, the other letters were completely worn away and no amount of technology could bring them back. Still, we had the date and it confirmed that the shipwreck was chronologically correct with Sodré's squadron leaving Lisbon in early 1502.

The next thing I looked at was whether it was possible to determine the origin of some of the raw materials used in making the artefacts, on the presumption that in preparing their ships for the voyage, the Portuguese would have mainly relied on materials sourced locally, as opposed to trying to obtain them en route to India. We had lots of objects made of lead, for example, a material that can be traced back to the ore deposit where it was mined by comparing isotopic ratios that are specific to each ore deposit. Pioneers in the field of lead provenancing, like Dr Zofia Stos-Gale, formerly of Oxford University, had been building a large database of isotopic ratios specific to more than 6,000 lead and copper ore deposits around the world, and using that information to help archaeologists study the ancient use and trade in lead.

Luckily for me, Zofia happened to live about ten miles from my office, and after meeting me and hearing about our project happily agreed to help me with the study. I had arranged for pure lead samples to be extracted from fifteen of the artefacts we had undergoing conservation in the UK and sent them off to a top laboratory at Durham University for the analysis. Zofia was careful to explain that her interpretation of the data could not be used to say with absolute certainty that the lead came from a specific mine, in part because not every possible ore deposit in the world was included in the existing database. What she could say, however, was that the lead used in making the fifteen objects I had sampled was consistent, to a very high degree of analytical error, with ancient ore deposits in Spain, Portugal and England. Knowing the source of the lead was another piece of corroborating evidence for the origin of the ship, but I still wanted something more specific and turned my attention to the stone cannon balls.

Based on the considerable number we recovered (157 of all sizes) and the range in size, the mere presence of the cannon balls proved that the wreck site was that of a warship. However, could the stone used in making the balls tell me more about the origin of the ship? Having a basic background in geology, I felt on more solid ground investigating the source of the stone, but the final analysis would still rely on the advice of a number of geological experts I was able to recruit in Portugal. Once again samples were chosen and sent off for analysis at various labs to test for age, geochemical and trace element compositions. Most of the balls we recovered were made from a meta-igneous rock that upon first examination could have come from a number of locations in central and southern Portugal. At one point I even visited some quarries with a local geologist to see if we could pin down the source. The real breakthrough came when I had one of the cannon ball rocks dated and the age came back as 550 million years, which geologically narrowed it down to an area of central Portugal called Abrantes.

In addition to being a precise match for the rock, Abrantes ticked a lot of boxes from a practical point of view. For one, the Tagus River running right through the centre of the region would have been the ideal way to transport the cannon balls 140 kilometres downriver to the waiting ships in Lisbon. Abrantes was also a well-populated area at the time, where stonemasons and other associated workers would have been in plentiful supply. Even Dom Manuel based his court in Abrantes for a considerable period, with two of his children being born there.

I was beginning to feel I had enough archaeological and scientifically sound evidence to prove my case, what with the date of the bell, the disc marked with the *esfera armilar* and the Portuguese/European origin of the raw materials used in equipping the ship. Even the fact that we found no cannons on the site fitted with what was known about the Sodré wrecks: that all the guns were eventually recovered in two documented salvages. I still had the coins, though, and was curious what story they would tell, so I contacted a Portuguese numismatist and sent him photographs of the gold *cruzados* for his expert opinion. The first thing I learned was that in the reign of Dom Manuel, silver coins were actually a much better tool for dating than gold ones because more information existed about when new specimens were created. As the only silver coin in the clump that could be made out was the large one with the cross, I sent him the best photo I had. His reply was like something from the script of a Hollywood movie:

Dear Mr Mearns,

You are about to make an outstanding discovery, one that will put your name in the annals of Portuguese numismatics: the lost silver coin of King Emanuel I, called 'INDIO' (the Indian), struck in 1499 after the return of da Gama's first voyage to India.

I was sitting at my kitchen table when this email arrived and was so struck by its contents that I told my wife, 'You need to read what this says because emails this good don't arrive every day.' In fact I can't remember ever getting one with a more memorable opening line.

As it turned out, this identification of Dom Manuel's fabled silver coin, the índio, was indeed correct, although more than a year would pass before it could be confirmed by another expert, João Vieira, sent by the Bank of Portugal to the UK to help me identify all the silver coins. In that time I asked the same conservator who had worked on the bell to separate the five gold *cruzados* from the silver clumps. This resulted in three coin clumps and better exposed one side of the índio as well as one side of an older coin made in 1475 known as the *real grosso*. With the coins separated, Jay Warnett was able to get high-quality X-ray images of all three with his CT scanner, and these were the images that João Vieira helped me analyse using Jay's powerful computer software.

Other than the two coin faces that were now visible, no other coins could be seen. In fact, the other two clumps looked like amorphous pieces of molten metal: there was no way of knowing the tremendous history hidden deep inside them. Only the power of Jay's computer allowed us to peer into them and reveal every delicate face. Starting at one end of a clump and moving methodically through to the other, we were able to examine each slice of the X-ray image, including taking extremely precise measurements. It was intensive and painstaking work, but an amazing privilege to unpick history this way using twenty-first-century technology. In the end we were able to confidently identify all twenty-four coins in the three clumps, which collectively provided the strongest evidence of all that the artefacts were from a ship that sank in 1503.

As for the silver índio, the more I learned about the coin the more

I understood why finding it on the site was a discovery of outstanding historical importance. For one thing, it was extremely rare: so rare in fact that only one other known specimen exists in the world, held in the collection of the National Historical Museum of Brazil. It is why the índio has in the past been called the 'lost' or 'ghost' coin of Dom Manuel. There were even rumours circling amongst numismatic scholars that perhaps the Brazil índio was fake. Happily, the coin I found ended that speculation once and for all.

What really made the índio special, however, was that it was especially conceived and commissioned by Dom Manuel for the new trade he saw happening with India after da Gama's discovery of the sea route. Two coins were created: a gold one called the *português* and the silver índio. The reverse side carried a large cross of the Order of Christ – the first time it was ever used on a Portuguese coin – and this was the cross I saw when I first found the clump. Encircling the cross was the Latin inscription: IN HOC SIGNO VINCES ('With this sign, you shall conquer'). While the *português* went on to be a very successful and influential coin, with other countries copying its weight and design, the índio didn't find the same favour, and was replaced in 1504 by a heavier silver coin called the *tostão*. In keeping with the 1502–03 date of Sodré's squadron, no *tostãos* were amongst the twenty-four coins we identified.

The very short life and limited circulation of the índio obviously explains why it is so rare. Only the armadas that sailed to India between 1500 and 1503 could have carried the coin. It makes the índio a very precise and powerful dating tool. And in view of the fact that none of the other armadas lost a ship in the Khuriya Muriya Islands, its presence on the wreck site is virtually a fingerprint of one of the Sodré ships. Short of finding the name Sodré stamped on an object, which was a totally unreasonable expectation, the índio was my smoking gun.

The response from the public and the media to the MHC's announcement of the discovery – which was made in mid March of 2016, on the same day the *IJNA* paper came out – was overwhelming. Our story was covered in more than 500 news articles and appeared in 183 websites about archaeology and history, while videos produced by National Geographic were viewed more than eight million times. The publishers of *IJNA* keep track of media mentions for academic articles using a numerical score: our article scored the highest ever for an *IJNA* article (more than six times the nearest article) and the second highest recorded for the whole subject of archaeology. The pay-off from this huge interest was that it drove scores of people to download and read the article, which was a key objective of our media strategy. Unsurprisingly, we ended 2016 as the most downloaded article to appear in *IJNA*, nearly three times the total of downloads for the next nine articles combined.

As I always wanted the project to be judged on the basis of its academic output, the success of the *IJNA* article was very satisfying. Since the announcement, several more peer-review articles have either appeared in the press or been accepted for publication, and a number of papers have been presented at maritime archaeology conferences. However, our commendable publishing record and the positive media attention that was generated didn't stop the project from being criticized in certain academic circles, either publicly or behind the scenes. Ironically, some of the criticism was that the media attention wasn't warranted and that the connection of the Sodré wrecks to Vasco da Gama had been overplayed. However, no one – including our main critic Alexandre Monteiro, who had adopted a biased stance against me and the project from the very beginning – could argue against the archaeological evidence and scientific analysis we presented in the *IJNA* paper. Monteiro childishly called the array of technological methods we employed 'mambo jambo', which said more about his poor comprehension of the paper than anything else.

Monteiro aside, the academic feedback we got was generally very positive. Although people wanted to see more interpretation of the site and how the assemblage of artefacts would help in the understanding of Portuguese expansion in the Indian Ocean at this critical time, they appreciated that this first paper was an interim report focused on identification of the wreck site. The team of independent archaeologists, historians and other scientists I've been able to attract to the project is growing, so it is very much a case of watch this space as more studies are completed. For example, we are still seeking opinions about the rare copper alloy disc marked with the *esfera armilar*.

I initially thought the disc was a component of an early navigation instrument called an astrolabe. Astrolabes are one of the most prized artefacts found on shipwrecks, with every new one that is discovered meriting academic discussion. The size and shape of the disc I found, combined with the central reinforced hole and the suspension ring, are all characteristics suggestive of an astrolabe. On the other hand, the thin disc is completely unlike other marine astrolabes, which are almost always heavy weighted instruments made with cut-outs designed to keep them still in the wind while an astronomical sight is being taken. Whatever the true identity and function of this disc, it is undoubtedly very rare, and quite possibly unique, simply owing to the early age of the ship it was on. And in the end, that is precisely why I persevered with the Sodré shipwrecks: to discover something so special it could rewrite history.

VIII

USS *Indianapolis* and *Endurance*

WAITING TO BE FOUND

ENDURANCE

USS *Indianapolis* and *Endurance*

I've never really had any difficulty answering the inevitable question 'What would you like to find next?', as the list of fascinating and worthwhile discoveries still to be made is long enough for me to spend several more lifetimes hunting shipwrecks. In fact the list has grown over the years rather than got shorter. So it is time and money, as is the case in most professions, that generally controls what I do next.

One reason the list keeps getting longer is because of requests I routinely receive from people wanting me to find a shipwreck that is personally significant to them. Quite a few have been for nationalistic reasons. For example, the Spanish journalist and historian Dr Santiago Mata has asked me to help their navy find the *Baleares*, a heavy cruiser that was sunk by destroyers of the republican navy in March 1938 during the Spanish Civil War. *Baleares* was hit by at least one torpedo that caused her forward magazine to explode, with the loss of 765 men when the ship sank off Alicante at a depth of 2,500 metres. The author of numerous books about World War II, Santiago wants to investigate the wreck as he feels there are holes in the official account. In particular he wants to find out why so many lives were lost when the ship remained afloat for three hours and two British destroyers were on hand to rescue survivors.

More commonly, the requests are from family members who want to know more about how their relative died or simply where their remains are buried. At first I was surprised that many of these requests came from people who had never met the relative they wanted me to find, yet generally they were the most determined of all to have such questions answered. I saw this with my friend Garry Baverstock, whose uncle Ernie Baverstock died a hero in X turret of HMAS *Sydney*, fighting *Kormoran* to the very end. Although Garry was born long after the ship went down, Ernie still loomed large in the Baverstock family and in Garry's life (his brother was even christened Sydney). In the run-up to the search for *Sydney*, Garry became heavily involved as one of the most proactive relatives, advocating tirelessly that the full truth of the action with *Kormoran* had to be established, whatever the outcome. He saw this as his personal responsibility in memory of his uncle and as the custodian of Ernie's story on behalf of his family.

I have seen the same determination, multiplied by a factor of ten, from a truly remarkable woman who first contacted me six years ago wanting to locate the wreck of the US destroyer *Strong* (DD-467). In terms of the wider history of the US naval war in the Pacific, the sinking of the *Strong* on 5 July 1943 after being hit by a Japanese long-lance torpedo fired from a record distance of twenty kilometres hardly merits a footnote. She wasn't a major warship; she wasn't involved in a significant battle with the enemy that merited its own name; and the loss of life (46 of *Strong*'s crew of 329 were killed), while unquestionably tragic, was at the lower end of the scale for US naval casualties. Although history appears to have forgotten the *Strong*, Tammi Hedrick Johnson, an anthropologist and Kentucky mother of two, aims to put that slight right in the name of her great-uncle Ensign William 'Billy' Hedrick.

After seeing his name on a family headstone at the tender age of ten, Tammi was inspired to learn everything she could about the uncle she would never meet. Her quest to find his resting place has become a lifelong mission that is now well into its fifth decade, and that has led her beyond the personal story of Billy's life to exhaustive research into the *Strong* itself, the ship's company and the attack that led to her loss in Kula Gulf of the Solomon Islands. She feels it is her destiny to find the wreck to make sure that Billy's sacrifice and the history of the ship are not forgotten. Having been won over by her passion, commitment and seriousness of purpose, I have helped Tammi where I can and have no doubt that one day she will succeed. We have been close to finding *Strong* before and I would dearly love to see that happen for her sake and that of all her supporters.

A number of submarines also feature on my list, although generally they could prove to be the hardest to find because of a lack of precise information about where they sank. The story of the French submarine *Surcouf* demonstrates the problem. The *Surcouf* was a cruiser and the largest submarine ever built, so locating her should be fairly straight-forward even in the deep waters of the western Caribbean where she was lost with all hands in early 1942. The problem, however, is that the only clue to where she sank rests on the hazy account of an American freighter, the *Thompson Lykes*, which collided with something submerged in the water in the general vicinity in which *Surcouf* was operating but didn't stop because they feared it was a German submarine. Given the inherent doubt in the freighter's story, no one has been confident enough to mount a search for the wreck

A similar situation exists for the Polish submarine ORP *Orzel* (*Eagle*), which many Poles want to see found so the crew can be recognized for their incredible bravery in their fight against Germany in World War II. After Germany's invasion of Poland, *Orzel* was on patrol in the Baltic when the captain decided to bring her in to Tallinn, Estonia, because of a faulty air compressor and because he was unwell. Although Estonia was neutral at the time, Germany pressured the Estonians to seize the submarine by force and to disarm it and confiscate all navigational charts. Jan Grudzinski, the lieutenant commander, assumed command of the boat and hatched an escape plan with the rest of the crew before all their torpedoes were offloaded. After overpowering their Estonian guards, the crew made a midnight escape whilst under machine-gun and artillery fire from other ships and from shore. Initially they patrolled the Baltic, but with their supply of fresh water running low they made an intricate and treacherous passage across the North Sea without the aid of navigation charts, diving by day and proceeding at night. Despite the strains and stresses of the forty-day voyage, they arrived in Scotland in excellent health and spirits. The Royal Navy was so impressed by *Orzel*'s escape that they awarded Grudzinski an honorary DSO for his leadership.

Following a refit, *Orzel* was assigned to the Royal Navy's 2nd Submarine Flotilla and was employed on patrol duties in the North Sea. The crew received high praise for their fighting spirit, which included sinking an important troop transport carrying hundreds of German troops for the secret invasion of Norway. It was during her seventh patrol, in early June of 1940, that *Orzel* was lost without trace or any clue about what might have happened to her. The boat's last orders were to proceed to the Skagerrak, and some people suspect she must have hit a mine during that transit. But as no signals were received from the crew after they left Rosyth, there is a very large corridor with almost endless possibilities where the wreck may lie. Although previous attempts to locate *Orzel* have failed, others are determined to continue searching. Until those efforts are successful, the Polish navy and family members of the crew have to be satisfied with the tribute paid by Admiral Sir Max Horton, who at the time of the boat's loss was chief of the Submarine Service: 'The passage of the *Orzel* from the Baltic to the United Kingdom is in itself an epic of determination, seamanship and endurance, which will long be remembered in the annals of the British Submarine Service.'

The final submarine on my list is *AE1*, Australia's very first submarine, which was lost in the islands of Papua New Guinea in all probability after colliding with a submerged reef, earning her the unfortunate distinction of being Australia's first major naval loss in World War I. There were no survivors from her mixed crew of thirty-five RN and RAN submariners. *AE1*'s operational history was brief, with her sole battle credit stemming from the surrender of Rabaul the day before she was lost. She was last seen on the afternoon of 14 September 1914 by HMAS *Parramatta* just west of Duke of York Island, which has provided searchers with a starting point to look for the wreck. The area has proved difficult to search, however, due to the extreme topography and magnetism associated with the volcanic nature of the islands. As a result, the fourteen in-water searches that have been conducted over the years from 1976 to 2015, while ruling out a few potential locations, have all come away empty-handed.

The best chance of finding *AE1* rests with a group of volunteers following a model similar to the one HMA3S used in their search for the *Sydney*. Their company, named Find *AE1* Ltd, is led by Peter Briggs, a retired RAN rear admiral, and Dr Michael White, a retired barrister who also served in the RAN before becoming a lawyer. They are a well-organized group and the recent search they conducted was a far better effort than most of the earlier attempts. But they are faced with a seriously tough problem that is crying out for the type of government support and professional approach we had with *Sydney* and the *Centaur*. I certainly believe that the wreck of *AE1* is findable with the right technical approach and search team, but it will depend on Peter and his colleagues convincing sponsors, or the government, to fully back their plans.

It is probably by choice more than anything else that the wrecks on my list pretty much all fall into the category of 'hard to find'. That could be because all the easy-to-find wrecks have already been found, or because my natural tendency is to gravitate towards difficult search problems like *Sydney*. As technology has improved, so has our ability to find shipwrecks, especially ones that have previously been elusive either because of their small size or due to environmental complications. For example, the best chance anyone will have of finding *AE1* given the large search area and difficult seabed conditions is if they use a synthetic aperture sonar system with co-registered bathymetry and imagery deployed from an autonomous underwater vehicle (AUV). This is one of a variety of technologies that will transform wreck finding in the future.

Over the past five years, a number of very famous shipwrecks that were on not just my list, but everyone else's too, have been found. These include the Italian battleship *Roma*, the fabulously valuable Spanish galleon *San José*, the ships of Sir John Franklin's lost expedition HMS *Erebus* and HMS *Terror*, and the Japanese super battleship *Musashi*, which I had a hand in locating. These discoveries reinforce my belief that we are living through a golden age of shipwreck hunting, in which the development of advanced underwater technology has coincided with an increased public interest in the oceans and our shared maritime history. It is interesting that each of these shipwrecks had been the target of previously unsuccessful searches before they were ultimately found – which leads me to one of the very top shipwrecks on my list: the USS *Indianapolis*.

USS *Indianapolis*

THE FORGOTTEN SHIP

SUNK 30 JULY 1945
880 died
316 survived

The story of the USS *Indianapolis* is tragic on so many levels it is hard to know where to begin. On the evening the '*Indy*' was sunk, 30 July 1945, the crew would have had no idea of the absolutely crucial role their ship had played in possibly the most momentous event of the twentieth century. Of the 1,196 men on board, only *Indy*'s commander, Captain Charles B. McVay III, was aware of the top-secret cargo they had carried at record speed from San Francisco to Hawaii, and then on to the Pacific island of Tinian, where it was assembled and loaded onto the B-29 bomber *Enola Gay*. The cargo was so important that McVay was told that in an emergency it was to be saved at all costs, even above the life of the ship itself.

The consignment entrusted to McVay and the men of the *Indianapolis* was the heart of the atomic bomb that the US would go on to drop on Hiroshima. Having successfully delivered it to Tinian, the *Indy*'s secret mission was over; the movement of the ship was no longer time-critical or cloaked in secrecy. McVay's next set of orders was to refuel at Guam and then proceed to Leyte in the Philippines for an extended period of training. However, fatal flaws in how the US Navy handled the routing of combatant vessels meant that *Indianapolis* had become, to all intents and purposes, a forgotten ship. So when the Japanese submarine *I-58* slammed two torpedoes into the cruiser's starboard side just after midnight, causing her to sink in twelve minutes, those who were not killed outright by the explosions and fire or who were unable to escape the steel coffin of *Indianapolis*'s hull were forced to suffer an agonizing experience so horrifically cruel many wished they had perished with their shipmates.

Of the nearly 1,200 men on board *Indianapolis*, perhaps as many as 900 survived the attack and sinking. A relatively small number of those who went into the water alive would not have made it through that first night. These were the most seriously injured men, with extensive burns and other trauma. For them the end was mercifully quick. By daybreak, the survivors were scattered across the water, clinging to whatever means of support they were able to find among the flotsam that *Indianapolis* had left behind. The ship had sunk so fast, there was no organized abandoning; no ship's boats were launched and not every man was able to grab a life jacket or belt before they either jumped overboard or were swept away as the sea closed over the rapidly sinking hull. Some would have been carrying minor injuries – burns, broken arms and twisted ankles – while nearly all would have been covered with a thick layer of oil that burned their skin and made them violently sick.

Survival now depended on how quickly rescue ships could reach their location, which was roughly at the midway point of their 1,171-nautical-mile voyage from Guam to Leyte. If the radio operators had been able to transmit a distress message giving their most recent position, the chances of survival for the large majority of the men would have been very good. Whether a distress message was actually received or acted upon is a highly contentious issue that haunts the navy to this very day. What we do know, however, is that no ship was sent to rescue *Indianapolis*'s men. Worse than that – and this is the most controversial failing and the crux of the tragedy – none of the navy operational commands who had been informed of *Indianapolis*'s routing was responsible for noting whether the ship had actually arrived in Leyte as scheduled. So when the *Indy* didn't arrive she wasn't listed as overdue and no action was taken to investigate her whereabouts. In fact, the marker plotting her assumed progress towards Leyte was removed from the plotting board at 11.00 local time on 31 July (the ETA) even though she clearly hadn't arrived.

The bottom line of this bureaucratic bungling between the various commands and fleet commanders was that *Indianapolis* wasn't just forgotten, she was effectively erased from the face of the earth. The scene had been inadvertently set for the worst sea disaster in American naval history. In total, 880 of *Indianapolis*'s men lost their lives, with as many as 580 suffering the most horrific and slowly painful deaths due to exposure, dehydration, drowning and shark attacks. The 316 who did survive owed their lives to a bomber pilot, Lieutenant Wilbur Gwinn, who on a routine patrol searching for Japanese ships and submarines spotted an oil slick containing some thirty men splashing wildly to attract his attention. Gwinn's report at 11.20 on 2 August, eighty-three hours after the *Indy* had sunk, set in motion one of the largest air and sea rescue missions conducted by US forces during the war. It was given top priority by the navy's most senior commanders, who, upon learning that the survivors were from the USS *Indianapolis*, immediately realized what a black stain this would be on their records.

Captain McVay, who was luckier than most, having crawled into a life raft he had found on the first night, was one of the last to be rescued, just before midday on 3 August, nearly four and a half days after his ship had sunk. He was as confused as anyone as to why it had taken so long before the rescuers came looking for him. McVay was taken to a naval hospital on the island of Peleliu, arriving there on the 4th. A press conference was organized, with reporters flown in on the 5th. By now McVay would have been informed about the huge number of men killed under his command. He would have known there was nothing he could have done differently to avoid the torpedo attack or save the ship. In his comments to the press, his frustration was clearly directed at others:

We were due at our anchorage at 1100 hours. I should think by noon or 1300 they would have started to worry. A ship that size practically runs on train schedule. I should think by noon they would have started to call by radio to find out where we were, or if something was wrong. So far as I know, nothing was started until Thursday [2 August]. This is something I want to ask somebody myself. Why didn't this get out sooner?

On Monday 6 August, *Enola Gay* dropped the atomic bomb on Hiroshima, leading to the surrender of Japan and the end of the war. McVay and his men had done their duty, and despite their horrific ordeal could take pride in the role they had played in shortening the war. On 15 August, the day that all of America was celebrating Japan's formal surrender, the country also learned about the loss of the *Indianapolis*. The navy had decided to embargo the story until after President Truman's historic announcement that the war was over. For anyone who noticed the story of the lost cruiser amidst all the celebrations, they may have accepted it as a necessary cost to end the war.

Despite the navy's decision to use the end of the war as a good day to bury the bad news about *Indianapolis*, they inexplicably decided to prolong the scrutiny of this sad event by court-martialling Captain McVay on a charge of 'hazarding his ship by failing to zigzag'. It was an extraordinary decision that would hoist upon McVay the dubious dishonour of being the first American commanding officer ever to be court-martialled for losing his ship as a result of enemy action. To add to his ignominy, the navy even flew *I-58*'s commander, Mochitsura Hashimoto, from Japan to give evidence in court against him. McVay's orders had clearly stated that the decision to zigzag was at his discretion, dependent on weather conditions. Hashimoto, however, testified that zigzagging would have made no difference in making his attack.

McVay was cleared of the charge of failing to order his men to abandon ship in a timely manner, but he was found guilty on the charge of hazarding his ship by not zigzagging. The press and many other people saw that he was being made a scapegoat for the errors of others. Even Congress criticised the navy for conducting a hasty trial before a detailed investigative report was completed. The navy had backed itself into a corner and struggled to find a way out of the quagmire they had created. In the end they remitted McVay's sentence, but not his conviction. His naval career would continue, albeit with the black mark of the court-martial on his record and the loss of so many men on his conscience. It was the loss of the men that deeply troubled him, and he was never able to rid himself of the guilt he felt, no matter how unjust the accusation. He also suffered in his personal life, with the loss of his second wife to cancer and a grandson he adored to a brain tumour. On 6 November 1968, at the age of seventy, McVay took his own life using his service revolver, and thus became the final casualty of the *Indianapolis*. In his hand he held the tiny toy sailor he carried for good luck.

Nothing further could possibly be learned about the loss of the *Indianapolis*, or the conduct of Captain McVay, from investigating the shipwreck. When found, it will be very badly damaged, testament to the devastation that two torpedoes can cause and how the fate of the ship was instantly sealed the moment they struck. So why search for it if there are no questions to answer or mysteries to solve?

In 2000, a search for the wreck did take place with the approval of the USS *Indianapolis* Survivors Organization. Four crew members from the *Indy* participated in that search, which unfortunately was plagued by equipment failures. I played a small role assisting the leader of the search, Craig Newport, who was a friend and former workmate at Eastport International. The intention then, as it will surely be in the future, was to use the wreck as a powerful visual tool in telling the story of the *Indianapolis* and the price her men paid in completing their secret mission to deliver the bomb that helped end World War II. Back in 1945, the deaths of 880 men were deliberately overshadowed by the announcement that the war was over. It is about time that wrong was righted.

The men of *Indianapolis* haven't been completely forgotten with the passage of time. A number of books, films and television productions have come out that tell the full story. The plight of the crew reached its highest point in public awareness in the 1975 film *Jaws*, when Captain Quint, the crusty shark hunter played by Robert Shaw, delivers what is now considered one of the most famous monologues in movie history, describing how he was on the *Indianapolis* and how the survivors were attacked by swarms of sharks. More recently, a documentary film ten years in the making (*USS Indianapolis: The Legacy*) tells the story almost entirely through the first-hand accounts of the largest number of survivors ever interviewed. The survivors have been having reunions since 1960 and still assemble every year to pay tribute to their shipmates, even though the youngest amongst them is now ninety and there are only about twenty of them still alive.

Despite the failure of the 2000 search, I have no doubt that the wreck of *Indianapolis* will be found before too long. It will be a challenging search because of the extreme depth and the rugged seabed geology in that part of the world. However, as each year passes, more ultra-deepwater search systems are being built, and various groups have set their sights on finding the *Indy*. I would love to be involved if the call ever comes, and I have no doubt that images of the wreck will be able to convey the story of the men's sacrifice better than anything else. The wreck is, after all, the grave site of the majority of men who died in the attack, and although their physical remains won't be seen, their presence will be felt everywhere.

Shackleton's *Endurance*

THE GREATEST SEARCH OF ALL

SUNK 21 NOVEMBER 1915
All survived

For a shipwreck hunter who likes challenges, there can be no greater challenge than the *Endurance*. The shipwreck of Sir Ernest Shackleton's polar yacht, sunk in the heart of the Weddell Sea during his International Trans-Antarctica Expedition (ITAE) is, without question, the hardest there will ever be to find. I have studied every single aspect of searching for *Endurance* for more than fifteen years and I consider it to be the ultimate shipwreck search challenge. In mountaineering terms it is the K2 of shipwrecks.

Finding *Endurance* is the ultimate challenge for one reason alone. It is not because of the depth: even though 3,000 metres is quite deep, I have found many wrecks that are deeper. It is not because *Endurance* was made of wood: thankfully the two main types of mollusc that bore their way into wood structures cannot survive in the cold, deep waters of the Weddell Sea. And it is not because we don't know where *Endurance* lies: I have analysed every single piece of navigation information recorded by Shackleton and his crew, in particular the logbooks of Captain Frank Worsley, and I judge the data regarding the sinking position to be of excellent quality. The reason is unlike any other difficulty I have faced searching for wrecks. It is the same reason why *Endurance* sank in the first place, and it is also what makes the challenge very special. The challenge is the ice.

The Weddell Sea of Antarctica is almost permanently covered by a constantly moving pack of ice that can be as thick as two metres. The pack ice froze *Endurance* in its icy grip, then crushed the stern, allowing water to flood in. Shackleton and his twenty-seven men had no option but to abandon the ship and set up a temporary camp nearby with tents, three lifeboats and all their provisions. As the ice floe drifted north, the sun began to heat the ship, causing the ice holding it to melt and the hull to sink deeper. Finally, on 21 November 1915, the Weddell's frozen grip broke and *Endurance* plummeted to the bottom, 3,000 metres below. Shackleton was the hardest of men, but even he was gutted by the loss of his ship. The few lines he wrote in his diary reveal the depth of his anguish: 'At 5 p.m. she went down by the head. The stern, the cause of all the trouble, was the last to go under water. I cannot write about it.'

The pack ice remains the overriding danger to any ship – even the world's most powerful and modern icebreakers – that enters this part of the Weddell Sea. The first lesson all ship captains learn about navigating in ice is to avoid the ice: skirt around it, or penetrate it through leads that open up between large ice floes. However, to find *Endurance* we will not have the luxury of avoiding the ice. On the contrary, we will have to ram our way through as much as 250 nautical miles of it before reaching the search area. Once we get there, the truly hard part begins, because to conduct the search and then film the wreck we will need to remain stationary above it while thousands of tons of ice move inexorably in the opposite direction. The success of the entire expedition will hinge on whether the icebreaker we use can resist that overwhelming force.

At this point you might ask: is *Endurance* really that special that it warrants attempting something so ridiculously hard? I am clearly biased, but I think that without a doubt it is. In a nutshell, it is one of the world's most famous and iconic ships, whose loss set the stage for arguably the most epic survival and rescue story of all time, led by one of the greatest explorers in history. Even people who don't follow exploration will have surely marvelled at the exceptional and timeless photos of *Endurance* taken by the famed photographer Frank Hurley. And they will undoubtedly recognize the name Shackleton as the legendary leader who by his dogged and relentless efforts saved all twenty-seven of his men in the most dire and hopeless of circumstances.

While we plan to conduct meaningful science at the wreck site to make sure a legacy is created from this unique expedition, I believe it will be the photographs we take of *Endurance* that excite and inspire people around the world. Using the best sonars, cameras and lighting we'll be able to create a virtual three-dimensional reconstruction of the shipwreck that will fascinate and amaze people. In the same way that Hurley's photographs have, for a century, inspired people to gaze in awe at the beauty of Antarctica, my hope is that the images we bring back of the wreck site and the bottom of the Weddell Sea will have the same lasting effect, standing alongside Hurley's photographs to tell the story of *Endurance* and Shackleton's ITAE expedition to new generations.

People say I have a dream job. Well, if I do, this is the project I dream about. It is the one I most want to complete and the one I have worked hardest to make happen, with the help of scores of supporters. In addition to a large sum of money, the main thing we need is the right type of heavy icebreaking ship that can also double as an expedition vessel. There are very few such ships located in the southern hemisphere, which has led to me travelling the world looking at various options. During this time, I have had the Shackleton family, led by Sir Ernest's granddaughter Alexandra 'Zaz' Shackleton, firmly in my corner. Zaz is now a lifelong friend. Of all the information I uncovered during my research, the discovery that pleased me most was that the family are still the legal owners of the shipwreck.

I am not the hero-worshipping type, but Shackleton has definitely become a hero to me, and I have used his example in the past when I have found myself in tough spots. I have learned from him the virtue of waiting, of having patience. It is not a natural characteristic for me, but it was the only way I got through all the problems and setbacks we had during the *Sydney* search without rushing into a bad decision. I just had to remind myself of Shackleton's genius in naming the spot where he and his twenty-seven men lived on an ice floe for over three months as they drifted north: he called it Patience Camp. The effort to obtain the funds and ships we'll need to search for *Endurance* has certainly taken a lot of patience, but I can wait. I learned that from Shackleton.

---

* The wrecks of *AE1* and the USS *Indianapolis* were successfully located in 2017.

Afterword

Whenever I am asked about the influences on my career as a shipwreck hunter I credit the 1985 discovery of *Titanic* by Bob Ballard and WHOI for ushering in a new age of deep ocean exploration that paved the way for other famous shipwrecks to be found. In the eyes of most people, including me, *Titanic* still occupies the top spot as the most famous of all shipwrecks. However, long before *Titanic* was ever found a group of scientists and engineers from the US Naval Research Laboratory (NRL) quietly set about conducting a mission of national importance that truly opened the doors to the deep ocean for the first time. Sadly, it was another catastrophic loss that inspired their groundbreaking work.

On 10 April 1963, the US nuclear submarine *Thresher* (SSN 593) sank in the North Atlantic while conducting deep diving tests. *Thresher* was the first, and worst, instance of a nuclear submarine loss at sea with all 129 of the boat's crew killed. The magnitude of the loss was such a shock to the Navy that a plan was literally developed overnight to find the *Thresher*. Before then the deepest shipwreck found was less than 450 metres and the method of searching comparatively primitive, essentially dangling a man within a sealed observation chamber fitted with viewports. *Thresher*'s depth of 2,560 metres necessitated an entirely new towed, multi-sensor approach to searching and from this tragedy NRL's (and the United States') deep ocean search capability was born. Because most of NRL's missions using the search ship *Mizar* were classified, the world knows véry little about what they achieved and the record depths they reached.

NRL's pioneering development of deep tow technology, photographic techniques and search methodology in the 1960s and early 70s provided the foundation for Ballard's and WHOI's success in the 1980s. Ballard continued to rely on the same type of photographic search systems pioneered by NRL, and the experience he gained from his own surveys of the *Thresher* and *Scorpion* (another sunken US nuclear submarine found by NRL in 1968) wreck sites helped him refine the search for *Titanic*. Although Ballard's method of finding *Titanic* was not very different from how NRL located the *Thresher* twenty-two years earlier, the worldwide interest generated by *Titanic*'s name and the famous story of her sinking created a game-changing moment in history. Suddenly, the entire world knew the deep ocean was accessible to man without ever realizing that the greatest ocean depths had been conquered more than two decades earlier.

The discovery of *Titanic* was a watershed moment without which my own discoveries of *Lucona*, *Derbyshire* and *Hood* might not have happened. The companies I worked with to find those wrecks employed more modern and more efficient sonar-based search systems, but the breakthrough from NRL's early age of deep ocean shipwreck hunting was more about a realization of what was possible. After *Titanic* the deep ocean was no longer the hidden and unreachable realm the majority of people presumed it was. It didn't matter that NRL had found and filmed a wreck at 5,010 metres in 1970 (the *LeBaron Russell Briggs*) or that a top secret CIA project recovered parts of a Soviet ballistic missile submarine (*K-129*) from a depth of 5,030 metres in 1974. Because these record-setting achievements were shrouded in government secrecy, it was as if they had never happened and thus their impact beyond the world of the US Navy was virtually non-existent.

In the same way that NRL's groundbreaking work helped shape Ballard's Navy-funded programme at WHOI, my career as a shipwreck hunter was also indirectly boosted by Ballard's discovery of *Titanic* and *Bismarck*. Even as late as 2004, when I first briefed the RAN Chief about the chances of finding the *Sydney*, I referenced the successful location of *Titanic* during my presentation. As anyone who has ever tried to do something particularly challenging has probably learned, it's a darned sight easier to get others, especially financial backers, to believe in your plan when someone before you has proved that what you're about to attempt is actually achievable.

I've been incredibly fortunate that the arc of my career has coincided with a period in history where technology has caught up with mankind's ambition to explore the deep ocean with greater precision and capability. Having cut my teeth on one of the earliest commercially available side-scan sonar systems when image interpretation was considered more an art than a science, I am grateful to have endured long enough to see my industry mature. New technologies that were at the very early prototype stage when I started out, like autonomous underwater vehicles (AUV) and multi-beam echo-sounding sonars (MBES), are already proving their potential to completely revolutionize the gathering of data at sea.

There is however, a much broader need for these technologies beyond shipwreck hunting that, with further advancement and expansion, will have an impact on everyone's lives. It is often said (because it is true) that we know more about the surface of the moon than we do about the surface of our own planet. That's because 71% of the earth is covered by the oceans, and the seabed that is hidden below this watery curtain has barely been surveyed or explored. Current estimates are that only 10% to 15% of the ocean floor is mapped to a resolution of 100 metres. Compare that to 98% of Venus, 100% of the moon at higher resolution and 100% of Mars with much of it at 20-metre resolution, and you start to realize how much more we know about outer space than we do about our own inner world. In a future where the world's population will be increasingly dependent on its oceans, we can no longer afford to live in ignorance about the largest part of our planet.

There are many aspects of the oceans that need to be measured and understood on a global scale, but experts agree that mapping the ocean floor should be made a priority. In the summer of 2016 a remarkable meeting took place in Monaco where 150 scientists, scholars and business associates endorsed the objective of comprehensively mapping the ocean floor by the year 2030. It is a bold but achievable objective (called Seafloor 2030), and a necessity if we are to continue to exploit the oceans without damaging them anymore than we already have. GEBCO, the world's only international organization with a mandate to map the ocean floor, will be responsible for compiling the data into a global bathymetric database with the ultimate objective that the database is shared in order to 'make the seafloor public'. AUVs and multi-beam sonars will undoubtedly be at the forefront of that data collection.

At that same meeting in Monaco another bold challenge was laid down. This one was in the form of a cash prize – a $7,000,000 XPRIZE to be exact. The prize, sponsored primarily by Shell with a $1,000,000 bonus prize offered by the US National Oceanographic and Atmospheric Administration (NOAA), challenges competing teams to develop new technologies to autonomously map the deep ocean floor in high resolution at rates faster than currently possible. The competition runs for two years (up to twenty-five teams will compete in two rounds of at-sea testing) and it is hoped the end result will accelerate the innovation needed to bring forward the goal set by Seafloor 2030. As a marine scientist and explorer I am honoured to have been selected by the XPRIZE scientific advisory board to serve as one of the judges for a competition that is close to my heart.

Some people choose their careers, but I like to think that my career chose me. It happened the moment I switched on my university's ancient EG&G 259-4 side-scan sonar recorder for the first time and watched the geology of the West Florida shelf reveal itself as if the ocean had been drained of water. Thirty-five years later I still have that same heightened sense of expectation whenever I switch on a side-scan sonar, as my scientific training and curiosity draws me forward to the screen in the hope that I'm about to see some long-lost object or an interesting geological feature. Creating the global bathymetric database – the base map for 71% of our planet – is the closest we will all come to that experience of draining the oceans of water and revealing what lies below. Having 'discovered' the wreck of *Athenia* from the safety and comfort of an office using an on-line database like that contemplated by GEBCO and experiencing the same feeling of excitement with that discovery as if I was on a ship over the top of the wreck, I am a strong believer in their goal to democratize the seabed.

As a kid growing up in the streets of New Jersey, in the shadow of Manhattan, I know just how lucky I've been to have such a passion for my life's work. When the history of the time period covered within this book is written, I believe there is a good chance it will be called the 'Golden Age of Shipwreck Hunting'. The evidence for that is seen in the number of shipwreck discoveries made by professional and amateur groups all around the world appearing in the news on a regular basis. I am proud to have contributed to that rich history with the wrecks I've found and the stories I've told. The adage that with improved technology anything lost in the oceans can be found regardless of the depth is certainly truer today than when I started out. One thing that hasn't changed, however, is that you still need to be looking in the right place.

Bibliography

I MV *Lucona*: MURDER AND FRAUD ON THE HIGH SEAS

Wolfgang Blum, 'Wie wurde die Lucona versenkt?', *Die Zeit*, 22 September 1995

'Eastport International Solves the M/V *Lucona* Sinking Mystery', *Waves*, September–October 1991

Peter R. Limburg, *Deep-Sea Detectives, Maritime Mysteries and Forensic Science* (ASJA Press/iUniverse, 2005)

R. C. Longworth, 'The Lucona Affair', *Chicago Tribune*, 15 March 1989

Elizabeth Nash and Geraldine Norman, 'Ship explosion ignites political time-bomb', *The Independent*, 26 November 1988

Hans Pretterebner, *Der Fall Lucona* (Hans Pretterebner Verlagsgesellschaft m.b.H. & Co. KG, Vienna, 1987)

Michael Z. Wise, 'True Confections: Baker is Guilty', *The Washington Post*, 13 March 1991

II MV *Derbyshire*: LOST WITHOUT TRACE

Anon, 'MV *Derbyshire*, Report of Court No. 8075, Formal Investigation' (HMSO, London, 1989)

'A report by the Rt Hon. Lord Donaldson of Lymington to the Secretary of State for Transport to assess what further work should be undertaken to identify the cause of the sinking of the MV *Derbyshire*' (Department for Transport, 1995)

'A report into the circumstances attending the loss of MV *Derbyshire* which foundered on or about 9 September 1980 in position approximately 25° 30' North, 130° 30' East with the loss of 44 lives' (Department for Transport, 1986)

R. E. D. Bishop, W. G. Price and P. Temarel, 'A Theory on the loss of the MV *Derbyshire*', *Transactions Royal Institute of Naval Architects*, 133 (1991), 1–27

M. Dickinson, 'Report of the ITF *Derbyshire* Mission' (International Transport Workers' Federation, London, 1994)

D. Mearns, 'Search for the Bulk Carrier *Derbyshire*: Unlocking the Mystery of Bulk Carrier Shipping Disasters', International Conference *Man-Made Objects on the Seafloor*, Society for Underwater Technology, London, 1 February 1995

D. Ramwell and T. Madge, *A Ship Too Far – The Mystery of the Derbyshire* (Hodder & Stoughton, London, 1992)

The Honourable Mr Justice Colman, 'Report of the Re-Opened Formal Investigation into the Loss of the MV DERBYSHIRE' (The High Court of Justice [Admiralty Court], 2000)

UK/EC Assessors' Report, 'MV Derbyshire Surveys' (Department for the Environment, Transport and the Regions, 1998)

III   HMS *Hood* and KTB *Bismarck*:
SEARCH FOR AN EPIC BATTLE

Iain Ballantyne, *Killing the Bismarck* (Pen & Sword Maritime, Barnsley, 2010)

David Mearns and Rob White, *Hood and Bismarck* (Channel 4 Books, London, 2001)

Burkard von Müllenheim-Rechberg, *Battleship Bismarck* (Arms and Armour Press, London, 1991)

Norman Polmar and Michael White, *Project Azorian* (Naval Institute Press, Annapolis, MD, 2010)

Mike Rossiter, *Ark Royal* (Bantam Press, London, 2006)

Bruce Taylor, *The Battlecruiser HMS HOOD* (Seaforth Publishing, Barnsley, 2004)

IV TSS *Athenia*: THE FIRST CASUALTY OF WORLD WAR II

Francis Carroll, *Athenia Torpedoed, the U-Boat Attack that Ignited the Battle of the Atlantic* (Pen & Sword Maritime, Barnsley, 2012)

Cay Rademacher, *Drei Tage im September: Die Letzte Fahrt der Athenia 1939* (Mare, Hamburg, 2009)

Melanie Wiggins, *U-Boat Adventures: Firsthand Accounts from World War II* (Naval Institute Press, Annapolis, 2013)

Andrew Williams, *The Battle of the Atlantic* (Bantam Press, London, 2006)

V HMAS *Sydney* (II) and HSK *Kormoran*:
SOLVING AUSTRALIA'S GREATEST MARITIME MYSTERY

Commissioner T. R. H. Cole, *The loss of HMAS Sydney II* (Department of Defence, Canberra, Australia, 2009)

T. Detmers and J. Brenneke, *Hilfskreuzer Kormoran* (Koehlers Verlagsgesellschaft, Munich, Germany, 1959)

T. Frame, *HMAS Sydney – Loss & Controversy* (Hodder & Stoughton, Rydalmere, NSW, Australia, 1983)

G. H. Gill, *Australia in the War of 1939–1945, Royal Australian Navy 1939–1942*, Vol. 1 (Australian War Memorial, Canberra, Australia, 1957)

P. Hore and D. L. Mearns, 'HMAS Sydney – An End to the Controversy', *Navy Historical Review*, Vol. 24, No. 4, December 2003

M. McCarthy, *HMAS Sydney (II)* (Western Australian Museum, Welshpool, WA, Australia, 2010)

G. McDonald, *Seeking the Sydney: A Quest for Truth* (University of Western Australia Press, Nedlands, WA, Australia, 2005)

D. L. Mearns, *The Search for the Sydney* (Harper Collins, Sydney, Australia, 2009)

M. Montgomery, *Who Sank the Sydney?* (Leo Cooper and Secker & Warburg, London, England, 1983)

W. Olson, *Bitter Victory – The Death of HMAS Sydney* (University of Western Australia Press, Nedlands, WA, Australia, 2000)

W. Olson, *HMAS Sydney (II) – In Peace and War* (Wesley John Olson, Hilton, WA, Australia, 2016)

Royal Australian Navy, Sea Power Centre, *HMAS Sydney II, The Cruiser and the Controversy in the Archives of the United Kingdom,* ed. Captain Peter Hore (Royal Navy, Defence Publishing Service, Canberra, Australia, 2001)

P. Schmalenbach, *German Raiders – A History of Auxiliary Cruisers of the German Navy 1895–1945* (Patrick Stephens, Cambridge, England, 1979)

B. Winter, *HMAS Sydney – Fact, Fantasy and Fraud* (Boolarong Publications, Brisbane, Australia, 1984)

VI AHS *Centaur*: SUNK ON A MISSION OF MERCY

R. Goodman, *Hospital Ships* (Boolarong, Brisbane, Australia, 1992)

D. Jenkins, *Battle Surface – Japan's Submarine War Against Australia 1942–1943* (Random House Australia, Milsons Point, NSW, 1992)

David L. Mearns, 'A Quest for Australia's Wartime Wrecks', *The Explorers Journal*, Vol. 88, No. 2, p. 22–27 (2010)

C. S. Milligan, *Australian Hospital Ship Centaur* (McGill University, Montreal, Canada, 1981)

C. S. Milligan and J. C. H. Foley, *Australian Hospital Ship Centaur, The Myth of Immunity* (Nairana Publications, Queensland, Australia, 1993)

A. E. Smith, *Three Minutes of Time: The Torpedoing of the Australian Hospital Ship Centaur* (The Lower Tweed River Historical Society, Queensland, Australia, 1991)

VII *Esmeralda*:
VASCO DA GAMA'S SECOND ARMADA TO INDIA

Pêro d'Ataíde, *Carta de Pero de Atayde a El-rei D. Manuel, Fevereiro 20, 1504,* ANTT, Corpo Cronológico, Parte I, Maço 4, No. 57

J. Barros, *Ásia de João de Barros: Dos feitos, que os Portuguezes fizeram no descubrimento, e conquista, dos mares, e terras do Oriente,* Decada I, Livro VII, Capitulo II (1552)

F. L. Castanheda, *História do descobrimento & conquista da Índia pelos portugueses,* Livro I, Capitulo LIV (trans. 1582; 1st edn 1551–60)

G. Corrêa, *Lendas da Índia,* Livro I, Capitulo VI (Lisbon, Academia Real de Sciencias, 1858; written *c.*1550s)

Bibliography

Bailey W. Diffie and George D. Winius, *Foundations of the Portuguese Empire 1415–1580* (University of Minnesota Press, Minneapolis, 1978).

D. Góis, *Chrónica do Felicíssimo Rei Dom Emanuel*, Parte 1, Capitulo LXXIV; Parte 4, Capitulo LXXXVI (Lisbon, 1567)

A. Gomes and A. M. Trigueiros, *Portuguese Coins in the Age of Discovery 1385–1580* (Lisbon, 1992)

Livro das Armadas da Índia, 1497–1640 (manuscript in the Arquivo Nacional Torre de Tombo)

Livro das Armadas da Índia, c.1568, Memória das Armadas que de Portugal passaram à Índia esta primeira é a com que Vasco da Gama partiu ao descobrimento dela por mandado de El-Rei Dom Manuel no segundo ano de seu reinado e no do nascimento de Cristo de 1497 (manuscript in the Academia de Ciências de Lisboa)

David L. Mearns, Dave Parham and Bruno Frohlich, 'A Portuguese East Indiaman from the 1502–1503 Fleet of Vasco da Gama off Al Hallaniyah Island, Oman: an interim report', *International Journal of Nautical Archaeology*, 45.2: 331–351 (2016)

S. Subrahmanyam, *The career and legend of Vasco da Gama* (Cambridge, 1997)

A. M. Trigueiros, *Apareceu O 'Índio' de D. Manuel I*, Moeda, 21.2, 55–9, 1996

VIII   *Indianapolis* and *Endurance*: WAITING TO BE FOUND

Stephen Harding, *The Castaway's War: One Man's Battle Against Imperial Japan* (Da Capo Press, Boston, MA, 2016)

Dan Kurzman, *Fatal Voyage: The Sinking of the USS* Indianapolis (Atheneum, New York, 1990)

Margot Morrell and Stephanie Capparell, *Shackleton's Way* (Nicholas Brealey, London, 2003)

Pete Nelson, *Left for Dead: A Young Man's Search for Justice for the USS* Indianapolis (Delacorte Press, New York, 2002)

Richard F. Newcomb, *Abandon Ship: The Saga of the USS* Indianapolis, *The Navy's Greatest Sea Disaster* (HarperTorch, New York, 2001)

Ernest Shackleton, *South* (William Heinemann, London, 1919)

Frank A. Worsley, *Endurance* (Jonathan Cape and Harrison Smith, New York, 1931)

Afterword

Robert D. Ballard, *The Discovery of Titanic* (Hodder & Stoughton, London, 1987).

Sherry Sontag, Christopher Drew and Annette L. Drew, *Blind Man's Bluff: The Untold Story of Cold War Submarine Espionage* (Hutchinson, London, 1999)

Acknowledgments

The original idea for this book came from Tom Gilliatt, Publishing Director for Allen & Unwin in Australia, some years ago after I had located the wreck of *Centaur*. During those initial discussions Tom made a pretty convincing case that people would be interested in a broader book about how someone becomes a 'shipwreck hunter' and what a career in this unusual profession would entail. I hope he was right, but I take full responsibility for whether I've managed to live up to his expectations and have been able to convey what a truly exciting thing it is to make such remarkable shipwreck discoveries. To look back thirty years and more of one's professional life takes time, however, so I am most grateful to Tom for his patience in waiting until I could manage the break needed to complete this book.

I would also like to thank Clare Drysdale, the Editorial Director of Allen & Unwin, for her enthusiasm and encouragement throughout the project. It was important to me that this book was published in both Australia and the UK. I thank my agents, Heather Holden-Brown and Margaret Gee, for finding the right publisher to make that happen. I would also like to thank Alison Cathie who introduced me to Heather, and was extremely kind and generous with her time in taking an interest in my writing career.

Writing this book allowed me to appreciate the importance of my teachers and mentors in helping me develop as a scientist, a professional and as a person. I owe an enormous amount to Gregg Rice,* formerly of Fairleigh

Dickinson University, Al Hine and Peter Betzer from the University of South Florida College of Marine Science, and to Don Dean who gave me my chance at Eastport International and who showed by his example how to lead a team.

I was extremely fortunate to have joined two pioneering companies at the precise moment they were developing ground-breaking technologies to reach the greatest depths of the ocean. However, what really made Eastport International and Blue Water Recoveries truly special were the talented people who worked there. Both companies were small enough that your co-workers were also friends. While their names might not appear in the book the following people played big parts in several of the stories I've told. At Eastport International: Mark Wilson, Bill Lawson, Craig Bagley, Larry Ledet, Carl Overby, Ron Schmidt,* Paul Nelson, Greg Gibson, Terry Carroll, John Finke, Jerry Marenburg, Don Dean, John Kreider, Larry Mocniak and Roy Truman. At Blue Water Recoveries: Bob Hudson, Mark Cliff, Carole Menzies, Jim Mercer, Julian Cope, Peter Cope and Lyle Craigie-Halkett.

There are numerous other colleagues and friends who have made important contributions to the projects covered in this book. Although I haven't the space to name them all, I would like to acknowledge and thank the following people in particular.

Lucona: Irv Bjorkheim and Larry Robinson; *Derbyshire*: Mark Dickinson, Rob White and Rory McLean; HMS *Hood*: Ted Briggs,* Julian Ware, Sarah Marris and Lindsay Taylor; HMAS *Sydney*: John Perryman, Wes Olson, Barbara Poniewierski, Peter Hore, Garry Baverstock, Glenys McDonald, Keith Rowe, Don Pridmore, Patrick Flynn, Mack McCarthy and the Finding Sydney Foundation Directors; AHS *Centaur*: Chris Milligan, John Foley,* Anthony Crack and Arthur Dugdale; *Esmeralda*: Alex Double, Dave Parham, Bruno Frohlich, Hassan Al Lawati, Ayyoub Al Busaidi and Ahmed Al Siyabi; *Endurance*: Terry Garcia, Kristin Rechberger and Alexandra Shackleton.

Finally, to my wife Sarah and our children Sam, Alexandra and Isabella, my eternal thanks and love.

(*deceased)

Index

395

Index